Advances in Global Change Research

Volume 70

This book series is indexed in SCOPUS.

Advances in Global Change Research

Aims and Scopes

This book series is aimed at addressing a range of environmental issues through state-of-the-art and/or interdisciplinary approaches. The books in the series can either be monographs or edited volumes based, for example, on the outcomes of conferences and workshops, or by invitation of experts. The topics that the series can consider publishing include, but are not limited to:

Physical and biological elements of earth system science, in particular

- Climate change
- Biodiversity
- Sea-level rise
- Paleo-climates and paleo-environments

Social aspects of global change

- Environmentally-triggered migrations
- Environmental change and health
- Food security and water availability
- Access to essential resources in a changing world

Economic and policy aspects of global change

- Economic impacts
- Cost-benefit analyses
- Environmental governance
- Energy transition

Methodologies for addressing environmental issues

- Planetary data analysis
- Earth observations from space
- Proxy data analyses
- Numerical modeling
- Statistical analyses

Solutions to global environmental problems

- Sustainability
- Ecosystem services
- Climate services
- Technological/engineering solutions

Books in the series should be at least 200 pages in length, and include a table of contents.

Images supplied in colour will be reproduced in both the print and electronic versions of the book at no cost to the author/editor.

Manuscripts should be provided as a Word document and will be professionally typeset at no cost to the authors/editors.

All contributions to an edited volume should undergo standard peer review to ensure high scientific quality, while monographs should also be reviewed by at least two experts in the field.

Manuscripts that have undergone successful review should then be prepared according to the Publisher's guidelines manuscripts: https://www.springer.com/gp/authors-editors/book-authors-editors/book-manuscript-guidelines

Filip Aggestam
Editor

Governing Forests Through Transformative Change

Emerging Policy Pathways and Perspectives

Springer

Editor
Filip Aggestam
ORBIS Green Policy and Governance
Forest Policy Research
Network—European Forest Institute
(FPRN-EFI)
Vienna, Austria

ISSN 1574-0919 ISSN 2215-1621 (electronic)
Advances in Global Change Research
ISBN 978-3-032-19071-0 ISBN 978-3-032-19072-7 (eBook)
https://doi.org/10.1007/978-3-032-19072-7

This Springer imprint is published by the registered company Springer Nature Switzerland AG
The registered company address is: Gewerbestrasse 11, 6330 Cham, Switzerland

Acknowledgments

This book is the result of the hard work of the chapter authors and reviewers, whose expertise and insightful contributions made this project possible. Thank you! I would also like to express my gratitude to Lorena, Frida and Felix for their support, and for occasionally reminding me not to take things too seriously.

Live long and prosper!

Contents

Contributors

Filip Aggestam ORBIS Green Policy and Governance, Forest Policy Research Network—European Forest Institute (FPRN-EFI), Vienna, Austria

Tanya Baycheva-Merger ICLEI European Secretariat GmbH, Freiburg Im Breisgau, Germany

Aron Buzogány BOKU University, Vienna, Austria

Alexandru Giurca European Forest Institute, Governance Programme, Bonn, Germany

Daniela Kleinschmit Albert-Ludwigs-Universität Freiburg, Freiburg, Germany

Michael Köhl University of Hamburg, Institute for Wood Science—World Forestry, Hamburg, Germany

Sandra Liebal Chair of Forest Policy and Forest Resource Economics, TU Dresden, Tharandt, Germany

Markus Lier Natural Resources Institute Finland (Luke), Joensuu, Finland

Stefanie Linser Department of Economics and Social Sciences, BOKU University, Institute of Forest, Environmental and Natural Resource Policy, Vienna, Austria;
Centre for Bioeconomy, BOKU University, Vienna, Austria;
EFI Forest Policy Research Network, Vienna, Austria

Philipp Mack Chair of Forest and Environmental Policy, University of Freiburg, Freiburg, Germany

Kit Prins Independent Expert, Geneva, Switzerland

Helga Pülzl BOKU University, Vienna, Austria;
European Forest Institute (EFI), Joensuu, Finland

Lena Riemann Chair of Forest and Environmental Policy, University of Freiburg, Freiburg, Germany

Andrzej Talarczyk Forest and Natural Resources Research Centre, Warsaw, Poland

Johanna Trummer School of Agricultural, Forest and Food Sciences, Forest Science, Bern University of Applied Sciences, Zollikofen, Switzerland

Rina Tsubaki European Forest Institute, Governance Programme, Bonn, Germany;
Forest and Nature Conservation Policy Group, Wageningen University and Research, Wageningen, The Netherlands

Gerhard Weiss University of Natural Resources and Life Sciences, Vienna, Austria

Jerylee Wilkes-Allemann School of Agricultural, Forest and Food Sciences, Forest Science, Bern University of Applied Sciences, Zollikofen, Switzerland

Ivana Zivojinovic University of Natural Resources and Life Sciences, Vienna, Austria

Abbreviations

ACF	Advocacy Coalition Framework
AI	Artificial Intelligence
CAP	Common Agricultural Policy
CBD	Convention on Biological Diversity
CEI-BOIS	European Confederation of Woodworking Industries
CEPF	Confederation of European Forest Owners
CEPI	Confederation of European Paper Industries
CICES	Common International Classification of Ecosystem Services
C&I	Criteria and Indicator
CO_2	Carbon Dioxide
COFFI	Committee on Forests and the Forest Industry
COGECA	General Confederation of Agricultural Cooperatives
COPA	Committee of Professional Agricultural Organisations
COST	European Cooperation in Science and Technology
CPA	Critical Policy Analysis
CSD	Commission on Sustainable Development
CFRQ	Collaborative Forest Resources Questionnaire
DG	Directorate-General
EAFRD	European Agricultural Fund for Rural Development
EAGF	European Agricultural Guarantee Fund
EC	European Commission
ECE	United Nations Economic Commission for Europe
EEA	European Environment Agency
EFC	European Forestry Commission
EFFIS	European Forest Fire Information System
EFI	European Forest Institute
ELO	European Landowners' Organization
EMF	Ellen MacArthur Foundation
EOS	European Organisation of the Sawmill Industry
EP	European Parliament
EuropaBON	European Biodiversity Observation Network

ERCC	Emergency Response Coordination Centre
ETS	Emissions Trading System
ESG	Environmental, Social, and Governance
EU	European Union
EUDR	EU Deforestation Regulation
EUROSTAT	Statistical Office of the European Union
EUSTAFOR	European State Forest Association
EUTR	EU Timber Regulation
FAO	United Nations Food and Agriculture Organization
FECOF	Federation of European Communal Forest Owners
FES	Forest Ecosystem Services
FISE	Forest Information System for Europe
FRA	Forest Resource Assessment
FLEGT	Forest Law Enforcement, Governance and Trade
FSC	Forest Stewardship Council
GDP	Gross Domestic Product
GHG	Greenhouse Gas
GIS	Geographic Information System
GPS	Global Positioning System
ha	Hectares
HWP	Harvested Wood Products
IAF	International Arrangement on Forests
ICPDR	International Commission for the Protection of the Danube River
ICTs	Information and Communication Technologies
IEEP	Institute for European Environmental Policy
IFF	Intergovernmental Forum on Forests
IIASA	International Institute for Applied Systems Analysis
ILUC	Indirect Land Use Change
IMF	International Monetary Fund
INFORMAR	Integrated Forest Management Learning Architecture
IPBES	Intergovernmental Science-Policy Platform on Biodiversity and Ecosystem Services
IoT	Internet of Things
IPCC	Intergovernmental Panel on Climate Change
IPF	Intergovernmental Panel on Forests
ITTA	International Tropical Timber Agreement
ITTO	International Tropical Timber Organization
IUCN	International Union for Conservation of Nature
IUFRO	International Union of Forest Research Organizations
JRC	Joint Research Centre
LBA	Legally Binding Agreement
LIFE	Financial Instrument for the Environment
LLM	Large Language Model
LULUCF	Land Use, Land Use Change, and Forestry
MAES	Mapping and Assessment of Ecosystems and their Services

MCPFE	Ministerial Conference on the Protection of Forests in Europe
MRV	Monitoring, Reporting, and Verification
MS	European Member States
NACE	Nomenclature of Economic Activities
NbS	Nature-based Solutions
NBSAP	National Biodiversity Strategies and Action Plans
NDC	Nationally Determined Contributions
NFI	National Forest Inventory
NWFP	Non-Wood Forest Products
OECD	Organisation for Economic Co-operation and Development
PEBLDS	Pan-European Biological and Landscape Diversity Strategy
PEFC	Program for the Endorsement of Forest Certification
PES	Payment for Ecosystem Services
PESTEL	Political, Economic, Social, Technological, Environmental, and Legal (analysis)
REACH	Registration, Evaluation, Authorisation and Restriction of Chemicals
RED I/II/III	Renewable Energy Directive I/II/III
REDD+	Reducing Emissions from Deforestation and Forest Degradation
SDG	Sustainable Development Goals
SEEA	System of Environmental-Economic Accounting
SEMAFOR	System for the Evaluation of the Management of Forests
SFC	Standing Forestry Committee
SFM	Sustainable Forest Management
SLU	Swedish University of Agricultural Sciences
SoEF	State of Europe's Forests
SQLite	Structured Query Language Lite (database)
SWD	Staff Working Document
TEEB	The Economics of Ecosystems and Biodiversity
TFEU	Treaty on the Functioning of the European Union
UEF	Union of European Foresters
UN	United Nations
UNCCD	United Nations Convention to Combat Desertification
UNCED	United Nations Conference on Environment and Development
UNDP	United Nations Development Programme
UNEP	United Nations Environment Programme
UNFCCC	United Nations Framework Convention on Climate Change
UNFF	United Nations Forum on Forests
UNSPF	UN Strategic Plan for Forests
USSE	Union of Forestry Societies in Southern Europe
VCA	Value Chain Analysis
WB	World Bank
WRI	World Resources Institute
WPF	Working Party on Forestry
WWF	World Wildlife Fund

Part I
Introduction and Conceptual Framing

Chapter 1
Forests, Policy and Governance: Framing the Transformation Challenge

Filip Aggestam

Abstract As multifunctional ecosystems, forests reside at the intersection between different policy domains, including climate, biodiversity, agriculture, energy and trade. Forests are, for this reason, central to addressing some of the interconnected challenges of the triple planetary crisis: climate change, biodiversity loss and pollution. This chapter defines key terms used throughout the book, including "forests", "forest policy", and "governance". It distinguishes between forest-focused and forest-related policies as well as between strategic (non-binding) and specific (binding) policy instruments. It also examines how diversity in policy approaches across sectors and levels contributes to the fragmentation of European forest governance. Transformative change is conceptualised as a systemic process that reconfigures policy capacities, steering mechanisms, and value structures. This framework is used, alongside complementary analytical approaches to analyse how policies evolve, interact, and influence forest policy outcomes.

1.1 What is a Forest?

A forest is typically characterised as an ecosystem dominated by trees and other woody plants. They provide multiple services, such as being a carbon sink and mitigating climate change, generating food, purifying water, and supporting people who rely on them for their livelihood, subsistence and culture. Forests are also home to over 80 per cent of all terrestrial biodiversity, making them important in relation to biodiversity and nature conservation (FAO, 2020, 2022). Forests are not homogeneous, they vary significantly in terms of their composition, structure, and ecological function. These variations are due to differences in climate, geography, soil types and varied societal demands and forest use. This results in a wide range of forest types, including tropical rainforests, boreal forests, temperate woodlands and peatlands. Each forest type plays a distinct role in sustaining biodiversity, regulating the

F. Aggestam (✉)
ORBIS Green Policy and Governance, Forest Policy Research Network—European Forest Institute (FPRN-EFI), Vienna, Austria
e-mail: aggestam@orbis-gpg.com

F. Aggestam (ed.), *Governing Forests Through Transformative Change*, Advances in Global Change Research 70, https://doi.org/10.1007/978-3-032-19072-7_1

climate and supporting other ecosystem services. Additionally, diverse forest types respond differently to environmental pressures such as deforestation, pollution, and land use (see Box 1). In other words, the ecological and functional variations associated with different forest types require a nuanced understanding of what constitutes a "forest", particularly when discussing and analysing forest-related policies. This includes the varied impacts across forest governance frameworks and levels.

One challenge in forest policy research is the lack of a universally agreed definition of what constitutes a "forest", including the diverse range of perspectives on what we mean when talking about forests (Aggestam, 2024b; Aggestam & Giurca, 2021; Aggestam & Pülzl, 2020). This implies that definitions of forests vary significantly depending on the perspective being considered, such as forests being studied from ecological, legal, or socio-economic perspectives. For example, from an ecological standpoint, a forest can be defined by the presence of trees and vegetation and the interactions between flora and fauna within the natural environment. This definition may include factors such as canopy cover, tree height, and species diversity (FAO, 2020, 2022). However, forest ecosystems can also include a wide range of non-tree vegetation, soils, and soil organisms, as well as different types of wildlife, which contribute to the complex and functional understanding of forest ecosystems.

Box 1 Different types of forests

- **Tropical forests** are biodiverse ecosystems that contain more than 50 per cent of the world's species. However, they are highly vulnerable to deforestation and the conversion of land for agricultural use. The rapid loss of tropical forests significantly contributes to global carbon emissions and biodiversity loss (Lewis et al., 2015). International policy efforts, such as the Reducing Emissions from Deforestation and Forest Degradation (REDD +) initiative, target tropical regions to incentivise forest conservation while supporting socio-economic development (Pauly & Tosteson, 2022).
- **Boreal forests** are essential for carbon storage due to their soil organic matter reserves, however, they are particularly sensitive to climate change. Increased temperatures lead to more frequent forest fires and permafrost melting, releasing stored carbon into the atmosphere (Bradshaw et al., 2009; Lecina-Diaz et al., 2018). Policies for boreal forests tend to focus on climate adaptation, fire management, and protection.
- **Temperate forests** tend to face different pressures, such as intensive logging, agricultural expansion and urbanisation. They are important for timber production, but also play a significant role in providing ecosystem services, such as water regulation and recreation. Sustainable Forest Management (SFM) practices in temperate zones commonly focus on balancing timber production with conservation and recreational uses (Luyssaert et al., 2008; Millar & Stephenson, 2015).
- **Peatlands** are technically wetlands but are sometimes categorised as forest ecosystems due to their vegetation cover. They are critical carbon sinks, storing large amounts of carbon in their soils. However, when drained or degraded, they can become significant sources of carbon emissions (Loisel et al., 2021; Turetsky et al.,

2015). This makes peatland conservation a high priority in forest policy, particularly in countries with extensive peatland areas, such as Indonesia.

From a legal or institutional perspective, the definition of a forest may focus on land use, property rights, institutions, and the regulations governing forest management. In this context, a "forest" may be legally classified based solely on tree cover, regardless of whether the ecosystem functions as a natural or managed forest (Chazdon et al., 2016; Lund, 2002). Some legal definitions[1] include minimum thresholds for tree density, canopy cover, and land area that are used to classify an area as a forest.[2] Such legal definitions can play an important role in governance as they determine how forests are reported on and which forest areas fall under national, regional, or international forest management regimes and policies. Moreover, socio-economic and cultural perspectives also shape how forests are understood. For example, forests may represent more than just an ecosystem, e.g., refering to cultural practices and spiritual values (ECE, 2020; Sikor & Lund, 2009). This highlights the human dimension of forest use and underscores the importance of these perspectives in forest policy and governance. Understanding these varied perspectives is therefore an important starting point when discussing forests, conducting forest policy research, or developing policy instruments and governance systems that concern forests.

1.1.1 Defining What We Mean by Forest Policy and Governance

Forest policy refers to the legislative, regulatory, and institutional frameworks that shape the management and use of forests. Forest policy can cover a variety of interconnected objectives and areas of work, including timber production, biodiversity protection, climate change mitigation and the provision of ecosystem services (Aggestam, 2024b). Historically, forest policies have been designed with national borders in mind and have focused primarily on ensuring a sustainable timber supply for industries such as mining and shipping. However, the way forests are used has changed over time, and there is now greater recognition that forest management is not only about timber. Consequently, forest policy has shifted towards more integrative frameworks that attempt to balance multiple objectives and recognise the environmental, socio-economic, and climate-related importance of forests (Arts & Buizer, 2009; Edwards & Kleinschmit, 2013).

[1] Variations in definitions can affect international forest reporting and policy-making. For example, variations can make it difficult to compare countries regarding the use of statistics and international policy-making due to the lack of a common language.

[2] The Food and Agriculture Organisation (FAO), for instance, defines a forest as "*land spanning more than 0.5 hectares with trees over 5 m tall and canopy cover of at least 10 per cent, or trees capable of reaching these thresholds* in situ". The definition excludes land that is predominantly used for agriculture or urban development (FAO, 2020).

By contrast, **forest governance** covers the broader system of processes, actors (both public and private) and institutions (both formal and informal) that influence forest management and policy implementation. These can, for example, include governmental agencies, non-governmental organisations (NGOs), civil society, forest-based industries and local communities. Forest governance is inherently multi-level, stretching from the local to the international. It also covers the norms and values associated with forest use (Arts et al., 2014; Chaffin et al., 2014; Pülzl et al., 2013). Furthermore, as with forest policy, traditional top-down governance models are increasingly being challenged by adaptive and integrative models that can account for the complex range of factors that affect forest use and management (Agrawal et al., 2008, McDermott et al., 2010).

The definitions of forest policy and governance vary significantly depending on the context in which these terms are used. In research, for instance, a common distinction is made between **forest-focused policy**, which refers to policies that explicitly target forests, and **forest-related policy**, which refers to policies that affect forests indirectly, for example, from other sectors such as energy or agriculture (Aggestam & Pülzl, 2018; Aggestam, 2024b; Winkel et al., 2013). This distinction reflects the multifunctional use of forest products and services. This distinction is important when evaluating policy impacts (e.g., the metrics used to assess effectiveness). Along these lines, researchers have emphasised the need to recognise these differences in order to improve policy coherence (Rayner et al., 2010). Furthermore, variations in the definition of forest policy, whether forest-focused or forest-related, can affect the research process by shaping its scope, methodologies, and conclusions drawn from empirical material. In turn, interpretations of policy impacts are also influenced by the definitions used in research.

These examples emphasise the significance of definitional distinctions in policy research and the need for fostering a shared understanding of basic terms such as "forest policy" and "governance", while also distinguishing between forest-focused and forest-related policies. This book categorises policy frameworks into two broad types: strategic and specific policy instruments. **Strategic policy instruments** are voluntary, non-legally binding agreements or guidelines that provide strategic direction for addressing forest issues. Examples include the European Union (EU) Forest Strategy, the United Nations Non-Legally Binding Instrument on All Types of Forests and the COP26 Declaration on Forests and Land Use. The absence of legally binding obligations often means that enforcement mechanisms are not in place. In contrast, **specific policy instruments**, such as directives or regulations, are legally binding and include formal obligations. Examples include national forestry laws such as the UK Forestry Act and the EU Timber Regulation (EUTR). The EUTR, for instance, legally obliged EU Member States (MS) to combat illegal logging by regulating timber imports. Legally binding agreements typically involve stronger commitments than strategic policy instruments, including enforcement, monitoring, and reporting mechanisms (Arts & Buizer, 2009; McDermott et al., 2010; Rayner et al., 2010) (Fig. 1.1).

The distinction between strategic and specific policy instruments is key to understanding how forest policy frameworks operate at national, regional, and international

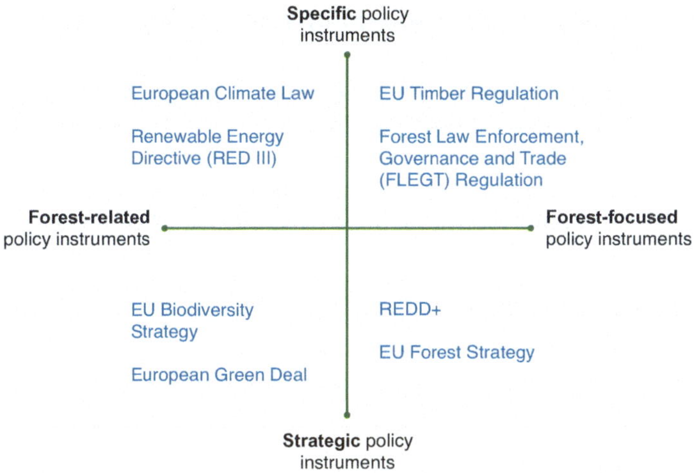

Fig. 1.1 Diversity of EU policy approaches affecting forests. *Source* Own figure

levels. Furthermore, distinctions between policy instruments highlight key trade-offs, such as the trade-off between flexibility and accountability. For instance, as they are non-binding, strategic instruments permit greater adaptability and flexibility but offer no accountability. In contrast, legally binding instruments can ensure compliance, but they are often considered rigid and are commonly associated with more resource-intensive monitoring and reporting. Ultimately, both types of instruments play complementary roles in shaping forest governance.

There are also challenges in applying these definitions. For instance, distinguishing between forest-focused and forest-related policies is challenging due to the interconnected nature of policy systems. Also, the impact of forest-related policies on forest ecosystems may not be immediately apparent (Angelsen, 2017). Another challenge lies in enforcing forest-focused and forest-related policies in countries with weak governance (Sunderlin et al., 2015). The same arguments apply to strategic versus specific policy instruments. For example, while voluntary guidelines or principles, such as Forest Stewardship Council (FSC) certifications, are technically non-binding, they may be treated as quasi-mandatory in some countries and regions (O'Reilly, 2006). Similarly, some specific policy instruments may lack enforcement mechanisms, resulting in a practical impact similar to that of non-binding instruments. The recently adopted EU Nature Restoration Regulation, for example, lacks binding targets (Aggestam, 2024a), which will make it difficult for the European Commission (EC) to enforce its implementation, even though it is a resolution.

1.1.2 Forest Policy in the Context of Transformative Change

Transformative change is another central concept in this book. More specifically, unlike many other ecosystems, forests are uniquely positioned at the intersection of multiple sectors, such as agriculture, energy and public health. However, they are also under pressure from climate change, biodiversity loss and pollution (Aggestam et al., 2021). This multilayered positioning makes the study of forest policy interesting, given the diverse and often competing interests involved, and it also highlights the urgent need for a transformative shift in forest governance to address these interconnected challenges effectively (Sayer et al., 2013). Current policy and governance models must therefore be rethought and restructured to tackle the increasing pressures on forests in a more integrated and adaptive manner (see Box 2).

Box 2 Transformative change

Transformative change refers to a fundamental reconfiguration of forest governance, including its structures, logics and practices, together with shifts in underlying values, norms, and relations between actors (Patterson et al., 2017; Sayer et al., 2013). Rather than incremental adjustments, it involves coordinated changes in how goals and targets are set, who holds authority, and how decisions are justified and implemented. In the EU context, transformative change is essential because incremental policy changes will unlikely be able to address the scale of the **triple planetary crisis** of climate change, biodiversity loss and pollution (De Koning et al., 2023; UNEP, 2021). Addressing this requires structural reforms that challenge dominant productivist and sectoral perspectives, as well as redistribute power and accountability within forest governance (Larson et al., 2021).

In this book, transformative change is approached operationally as a process that unfolds across interconnected dimensions of governance: building capacities, steering transitions, and reconfiguring values and power (Patterson et al., 2017). Rather than viewing these as separate steps, the book treats them as interdependent and connected layers of change. In practical terms, effective transformation in the forest sector depends on the following:

- capacities and infrastructures that enable learning, monitoring, and adaptive management.
- steering mechanisms that align policy targets, instruments, and sectors.
- institutional shifts that embed new norms and rights into the legal and administrative aspects of forest governance.

This perspective highlights the importance of moving beyond siloed policy domains towards cross-sectoral and multi-level frameworks that connect ecological, social, and economic priorities. Examples include landscape approaches, which link production, conservation, and community goals (Sotirov & Arts, 2018; Winkel & Sotirov, 2016), and adaptive governance, which fosters experimentation and feedback when addressing uncertainty (Chaffin et al., 2014; May, 2022; Sousa-Silva et al., 2018). However, persistent transformation ultimately depends on changes in governance, embedding rights-based and justice-oriented commitments in binding instruments and practice. In other words, transformative change involves protecting forests while reimagining their role in European societies by redesigning governance across scales, aligning capacities, steering mechanisms, and value/power relations, so that policy systems can deliver equitable and environmentally robust outcomes (McDermott et al., 2023).

The landscape approach (Box 2) is an example of a governance framework that has evolved to address multifunctional land use across sectors and scales. By considering "forests" as part of a wider natural (or policy) landscape, the landscape approach enables a broader range of interests to be considered within an integrative and adaptive management framework (Sayer et al., 2013). Such flexible, cross-cutting governance systems are better equipped to handle changing conditions, new knowledge, and emerging environmental challenges. Additionally, transformative developments reflect a shift from forest policies and practices focused on products to systems that balance different forest functions, including the management of relevant trade-offs inherent in competing land uses and demands (Puettmann, 2014). Examples of this include policies that incorporate nature-based solutions (NbS), as part of global efforts to combat climate change and biodiversity loss, such as National Biodiversity Strategies and Action Plans (NBSAPs) under the Convention on Biological Diversity (CBD) and Nationally Determined Contributions (NDCs) under the Paris Agreement (Escobedo et al., 2019; Lafortezza et al., 2018), as well as REDD+, which links forest conservation with climate mitigation efforts through financial incentives (Angelsen, 2017).

While approaches such as the landscape approach and NbS represent significant progress and innovation in forest policy, substantial gaps and challenges remain. These gaps are structural and operational, affecting the ability of relevant policy frameworks to facilitate transformative change. For instance, one challenge is integrating competing sectoral demands. Balancing the needs of sectors such as agriculture and energy with those of forests is complex and typically requires cross-sectoral coordination, which is often lacking. Governance structures are often fragmented, and forest policies do not always align with instruments governing land use. This leads to conflicting priorities and target setting that undermine forest governance (Aggestam & Giurca, 2021; Aggestam & Pülzl, 2020).

Additionally, there are gaps in the financial resources and capacity required to implement transformative approaches on a large scale. Effective implementation requires financial commitment and institutional support, both of which are often lacking (Lindenmayer et al., 2016). Another gap is the lack of monitoring and evaluation of these approaches' effectiveness. NbS and landscape approaches often lack robust monitoring frameworks that can assess impact. This makes it difficult to track progress or make necessary policy adjustments. As such, effective monitoring systems are central to ensuring that these types of approaches deliver appropriate long-term benefits (Kremen & Merenlender, 2018). Even more, policy coherence remains a significant challenge (Pülzl et al., 2018). Global frameworks such as the Paris Agreement and the Kunming-Montreal Global Biodiversity Framework (KM-GBF) provide a platform for integrating forest issues into broader policy goals and targets. However, there is often a discrepancy between international commitments and national implementation (Oberthür & Groen, 2018). Many countries struggle to align their national policies with global biodiversity and climate targets, and the enforcement of these commitments remains weak.

Summing up, while many innovative policy frameworks demonstrate significant potential, their success ultimately depends on overcoming context-specific gaps and challenges. This emphasises the need for deeper systemic transformation, stronger governance structures, improved cross-sectoral integration, enhanced financial and technical capacity, and the adoption of more participatory approaches. In other words, addressing these critical gaps and challenges is necessary for the transformative potential of these frameworks to be realised, including the opportunities they offer for meaningful change.

1.1.3 Transformative Change: How This Book Uses the Concept

The approach to transformative change adopted in this book goes beyond the simple introduction of new policy instruments. It covers structural and value-based changes to forest governance, including policy logics, stakeholder configurations, power dynamics, and accountability. In theory, this should enable climate, biodiversity, and other socio-economic objectives to be pursued together. This concept is operationalised through two approaches, which will be presented in more detail in later chapters:

1. The first is the "Pressure-Response-Transformation" approach (see Chap. 3), which explores how external pressures (e.g., climate change, biodiversity loss, and bioenergy targets) generate varied governance responses (e.g., new policy instruments, changing institutional arrangements, and coordination efforts), and the conditions under which these responses lead to transformation rather than incremental policy adjustments. The "Pressure-Response-Transformation" perspective distinguishes between descriptive "pressure to policy response" narratives and actual, systemic, changes in EU forest governance.
2. The second approach is a "frames-in-governance" perspective (see Chap. 5). This approach analyses how policy frames have evolved, coexisted or declined across binding and non-binding EU policy instruments. It also considers how these changes have enabled or limited transformative change. This chapter introduces the concept of "transformative hotspots", representing periods when changes in values and institutional setups align to facilitate a change in governance.

It should also be noted that, as a whole, the book analyses prospects for transformative change through three complementary dimensions, as set out by Patterson et al. (2017). These dimensions are applied across cases in Chaps. 3 and 4:

- Governance **for** transformation: This concerns the capacities and infrastructure that enable system change. In the context of forests, this includes robust monitoring, reporting and verification (e.g., remote sensing, geolocation and interoperable registries), data stewardship (e.g., open standards, auditability and privacy by design), advisory and extension systems for owners and municipalities, and financing and logistics that facilitate learning through experience. While these investments do not determine outcomes on their own, they lower the transaction costs of change and expand the range of viable options.
- Governance **of** transformation: Focuses on how change is governed, including targets, policy mixes, coordination across sectors (e.g., climate, biodiversity, energy, agriculture and trade), and the ordering of instruments from agenda-setting to implementation. In EU forest policy, this involves aligning the Common Agricultural Policy (CAP) measures, the Renewable Energy Directive (RED) sustainability criteria, land use, land-use change and forestry (LULUCF) accounting, restoration obligations, and due diligence mechanisms. These help ensure consistent monitoring and management of trade-offs and conflicts, as well as the use of relevant feedback to address changes.
- Transformations **in** governance: Refer to reconfigurations of values, rights, and power. More specifically, they refer to significant changes that shift what counts as legitimate goals and who gets to decide. Examples include rights-based safeguards (such as participation, standing, access and benefit-sharing), tenure and ownership arrangements that enable communal or municipal stewardship, and justice-oriented distributional commitments that go beyond procedural consultation. This is where value-based changes (e.g., shifting from productivist to ecocentric or justice-oriented frames) become institutionalised.

Crucially, these dimensions interact. More specifically, capacities without steering lead to drift; steering without a change in values or power leads to centred compliance; and attempts at significant reconfiguration will stop without enabling infrastructure and a coherent policy mix. The book uses these dimensions to analyse the current state of EU forest governance, identifying areas where progress has been made, areas where procedures are merely being tightened, and areas where transformation has yet to be achieved.

1.2 Understanding European Forest Policy

In reviewing forest policy, which is the central purpose of this book, it quickly becomes clear that forest governance is decentralised and fragmented across Europe. For example, at the level of the EU, no single policy department (Directorate-General) or executive agency is solely responsible for forests. Furthermore, within MS, competencies are often distributed across ministries and agencies in ways that vary significantly from one country to another. A critical challenge for EU forest governance is the need for a legal mandate that specifically focuses on forest ecosystems. This has led to the EU lacking clear mechanisms or institutions to guide forest policy or represent the interests of the forest-based sector (EP & CoE, 2010; Kleinschmit et al., 2018; Pülzl et al., 2018). This gap stems from the exclusion of forest products (with limited exceptions, such as cork) from EU treaties, leaving forest policy under the jurisdiction of MS, in line with the principles of subsidiarity and proportionality. For example, the Treaty on the Functioning of the European Union (TFEU) does not explicitly cover forests or forestry.

Despite the absence of a common EU forest policy, the EU engages in forest-focused and forest-related policy-making through different policy domains, such as agriculture and rural development, climate action, and energy. These policy areas, in which the EU has explicit competence, influence forest governance directly and indirectly. For example, policy instruments such as Rural Development Programmes (RDPs), carbon accounting frameworks, and international timber trade regulations impact forests by shaping relevant measures (e.g., afforestation), carbon credits, energy efficiency standards and trade practices. Consequently, forest governance within the EU is spread across sectors and levels, affecting forest use and management through many different types of policy instruments (Winkel, 2017; Winkel & Sotirov, 2016; Wolfslehner et al., 2020).

A significant challenge in EU forest policy and governance is the need for coordination across policy areas, mainly to address conflicting or contradictory targets and objectives across sectors. This fragmentation is evident when policies designed to achieve renewable energy, agricultural productivity, and biodiversity conservation conflict with one another. For example, the EU's renewable energy targets prioritise bioenergy, thereby increasing demand for wood-based biomass. While this approach mitigates climate change by reducing reliance on fossil fuels, it can also lead to deforestation and habitat degradation, potentially undermining the restoration goals established in regulations such as the EU Nature Restoration Regulation (Aggestam, 2024a; Jager et al., 2021). Similarly, agricultural subsidies under the Common Agricultural Policy (CAP) incentivise practices that conflict with climate and biodiversity objectives. For example, Rural Development Programmes (RDPs) have historically encouraged the drainage of wetlands and peatlands to increase agricultural land, despite the critical role peatland rewetting plays in carbon sequestration and habitat preservation. The tension between agricultural productivity and climate mitigation underscores some of the complexities behind EU policy-making, leading to fragmented governance and unintended consequences that affect forest use and management (Pe'er et al., 2020; Winter et al., 2014).

Inconsistencies between sectoral policy targets highlight a lack of coherence and integration in EU forest governance. Policies related to agriculture, energy and environmental conservation are often developed in isolation, with limited cross-sectoral alignment. This results in a regulatory environment in which different sectors pursue conflicting goals (Lazdinis et al., 2009, 2019; Winkel & Sotirov, 2016). For instance, although the European Green Deal promotes integrated action to combat climate change and biodiversity loss, the absence of structured coordination mechanisms within EU governance frameworks complicates reconciling conflicting objectives across policy areas (Ellis et al., 2019; Sabatini et al., 2019). This lack of coherence underscores the need for a more integrative approach to forests in EU policy design and implementation.

1.2.1 Is There a de facto EU Forest Policy?

Although there is no clear EU mandate on forests, some argue that the combined effect of various EU policy instruments, whether directly or indirectly influencing forests, constitutes a de facto EU forest policy (Aggestam & Giurca, 2021; Aggestam & Pülzl, 2018, 2020). This suggests that although the EU lacks a unified forest policy, the collective impact of different policy instruments and domains shapes forest use and management across MS. One example of this is the EUTR, a policy instrument specifically designed to prevent illegal logging by regulating timber imports (EP & CoE, 2010). The EUTR sets out requirements for timber operators, thereby promoting SFM and ensuring that only legally sourced timber enters the EU market. Alongside legally binding measures, the EU utilises non-legally binding strategic instruments, such as the EU Forest Strategy (EC, 2021), which sets out long-term goals and principles for forest management across the EU. This strategy is intended to encourage coherence and guide MS in aligning national forest policies with broader EU objectives.

Likewise, policies such as the Common Agricultural Policy (CAP) and the European Climate Law have significant indirect effects on the forest-based sector (EP & CoE, 2021). Through RDPs, the CAP influences forest use by promoting initiatives such as agroforestry, afforestation and rural development that can change forest landscapes. Meanwhile, the European Climate Law aims to achieve climate neutrality by 2050. It is driving efforts in carbon accounting and climate mitigation strategies, with substantial implications for forests given their role in terms of carbon sequestration. By positioning forests as essential carbon sinks, this law indirectly shapes forest management practices, incentivising conservation and reforestation efforts (Cavallin, 2024). Together, these policy instruments demonstrate how the EU's governance approach operates within a complex network of forest-related policies. Even policies that are not explicitly forest-focused, such as the Renewable Energy Directive (RED III) and the Water Framework Directive, affect the forest-based sector by encouraging bioenergy production and regulating water quality in forested catchments (Ellis et al., 2019; Santangeli et al., 2016; Soimakallio et al., 2021). Fundamentally, these instruments demonstrate how the EU influences forests, shaping the conditions under which they are managed and used.

The complexity of EU forest governance is further amplified by the varied nature of the forest-based sector, which spans a wide range of industries. Traditional sectors such as timber and paper and pulp coexist with emerging sectors such as bioplastics and sustainable construction (Aggestam & Giurca, 2022). Each sector has distinctive demands and perspectives on forest resources that must be managed within a broader EU policy framework. Furthermore, this diversity reflects the multifunctional role of forests in the European economy (Borrass et al., 2017; Sutherland & Huttunen,

2018), which is another reason why EU forest governance is shaped by a patchwork of policies and regulations spread across multiple policy domains. This fragmentation makes it difficult to define the scope and content of EU forest governance, primarily because perspectives on what constitutes EU forest policy vary across subsectors and epistemic communities (Haas, 1992; Zito, 2001). For example, the timber industry may prioritise regulations that ensure sustainable logging practices, such as the EUTR. Conversely, the energy sector may focus on policies affecting bioenergy production, such as RED III. Meanwhile, environmental interest groups may emphasise forest policy in the context of the EU Nature Restoration Regulation or the Natura 2000 network. These varied perspectives reflect the many interests within the forest value chain and reveal key challenges in establishing a cohesive forest governance framework (Winkel et al., 2013, 2021; Wolfslehner et al., 2020).

The ongoing debate over whether the EU has a de facto forest policy underscores the need for a common language to understand forest policy. The lack of consensus complicates efforts to characterise, compare, and monitor EU forest policy. In other words, researchers and policymakers must navigate a landscape in which the meaning of forest governance is embedded within different policy domains. While this fragmented understanding reflects the EU's broader policy structures, in which competencies are divided across multiple levels of governance (EU institutions, MS, and national agencies) and sectors, it also implies the absence of a common vocabulary. Additionally, these policy structures tend to operate in isolation, pulling forest management in different directions (Kleinschmit et al., 2018; Lazdinis et al., 2009; Pülzl et al., 2013). In summary, the diversity in our understanding of and attitudes towards forests has led to the conflicting priorities and trade-offs outlined above, emphasising the need for a more harmonised understanding of EU forest policy (Wolfslehner et al., 2020). One possible solution would be to develop a more comprehensive definition of EU forest policy to harmonise objectives across sectors, foster policy coherence, and promote a shared understanding of the multifunctional role of forests.

1.3 What Will Be Covered in This Book?

This book is organised into four sections. The first section is the introduction you are presently reading, as well as an overview of the theory of EU forest governance in practice. The third section looks at cross-cutting dimensions of forest governance. The fourth section provides a summary and synthesis of the content, including future research. The overarching theme is transformative change in European forest governance.

I. **Introduction and Conceptual Framing**
   This Chapter—Forests, Policy and Governance: Framing the transformation challenge.

This introductory chapter (as set out above) introduces some of the key concepts of forests, forest policy and forest governance used in this book. It also defines transformative change as a systemic reconfiguration of capacities, steering mechanisms, and values, and sets out the analytical frameworks that guide parts of the book's analysis of forest policy and governance in Europe.

Chapter 2—Theoretical Perspectives on forest policy and transformative change.

This chapter outlines the evolution of forest policy research, from its technocratic start to its emergence as an interdisciplinary field of governance. It reviews institutional, policy process, discursive, integrative, and reflexive approaches, emphasising how these perspectives have shed light on the interactions between actors, structures, ideas, and learning in forest governance. The chapter situates these perspectives within broader theories of change, such as adaptive governance, sustainability transitions and transformative governance, and argues that they are best understood as complementary. The chapter concludes by proposing an integrative approach centred on four analytical drivers (structure, agency, discourse and learning), offering a conceptual framework for analysing the prospects of transformative change.

II. **Governance in Practice—the EU Case**

Chapter 3—Rethinking EU forest governance: from fragmentation to transformation.

This chapter uses a Pressure–Response–Transformation (PRT) framework to analyse the EU's multi-level, cross-sector forest governance structure. It reviews major forest-relevant instruments (e.g., LULUCF, the Nature Restoration Regulation and the EUDR) and assesses whether recent shifts amount to adaptive layering or structural reorientation. It concludes that transformation is partial and uneven to date.

Chapter 4—Implementation of forest-related policies in the European Union: theoretical background and empirical examples.

Examines the implementation of forest-related policies within the EU, with a focus on the interaction between EU-level legislation and national administrative systems. Using institutional and policy-process perspectives, the chapter analyses how differences in governance arrangements, capacities, and stakeholder configurations influence implementation outcomes across MS. It highlights factors that enable or prevent effective policy delivery, such as coordination challenges, value conflicts, and the complexity of multi-level governance. The chapter offers insights into the practical conditions under which forest policy implementation can support or constrain transformative change.

Chapter 5—Frames in Transition: Prospects for Transformative Change in EU Forest Governance.

Using an AI-based frame analysis of 492 EU documents from 1965 to 2025, this chapter identifies seven recurrent frames (e.g., Policy Architect, Ecological Crisis, Environmental Justice) and maps their diffusion across hard and soft law. The chapter introduces the concept of 'transformative hotspots' (the 1990s, 2000s and 2020s) to demonstrate periods when discursive and institutional shifts align. It also specifies areas where enabling capacities and steering mechanisms have advanced versus where value and power changes remain incomplete.

III. **Cross-Cutting Dimensions of Forest Governance**

Chapter 6—Roots of Ownership: The relations between forest tenure and forest governance.

Examines how forest ownership structures influence management decisions, legitimacy and delivery capacity, and highlights the implications for transformation, such as the use of diversity as a platform for local experimentation and the barriers to scaling up rights-based approaches. The chapter also links tenure to policy uptake and equity, clarifying where soft instruments require support from binding instruments to influence practice.

Chapter 7—The Missing Link: Leveraging communication science insights for forest policy and governance research.

This chapter focuses on how forest topics are communicated and contested and synthesises evidence on media narratives, public perceptions, and participation mechanisms. It demonstrates how communicative practices and epistemic communities influence the knowledge recognised and the issues adopted in policy, identifying levers that connect deliberation to implementation.

Chapter 8—Forest Governance 4.0: The Good, the Bad, and the Ugly.

This chapter analyses how digital infrastructures, such as remote sensing, traceability and platforms, may reconfigure coordination, accountability and power in forest governance. It assesses opportunities such as monitoring, reporting and verification (MRV) and interoperability, as well as risks such as decentralisation and data justice. It also specifies design choices that ensure data-rich governance has a positive impact on equity and compliance in the forest-based sector.

Chapter 9—Measuring what matters: The renewed role of Criteria and Indicators (C&I) in forest policy and governance.

This contribution turns to measurements, focusing on how indicators travel across policy domains and the effects this has on the forest-based sector. It discusses indicator design and integration with EU reporting systems, as well as ways to pair carbon and restoration Key Performance Indicators (KPIs) with territorial and social metrics, ensuring that monitoring supports transformative change.

IV. **Summarising and Concluding Chapters**

Chapter 10—Emerging trends and the future of forest policy research.

This chapter establishes priorities at the science-policy interface, including methods for tracking frame dynamics over time, evaluating policy mixes for transformation, examining ownership-equity linkages, and governing digital forest data. The preceding chapters outline some of the steps needed to advance the transformative research agenda.

Chapter 11—Forests at the crossroads: Navigating pathways to transformative change

The concluding chapter summarises the insights presented throughout the book to evaluate the conditions under which transformative change in forest governance could occur. It identifies cross-cutting patterns, including advances in policy capacities, coordination challenges and uneven progress in value and power shifts, and evaluates their implications for European forest governance. The chapter outlines potential pathways for systemic reconfiguration and identifies areas requiring further research and policy innovation. Finally, it reflects on forests as indicators and drivers of broader socio-ecological transformation, emphasising the importance of integrated and forward-looking governance approaches.

References

Aggestam, F. (2024a). Crosswalking the EU nature restoration regulation and the Kunming-Montreal global biodiversity framework: A forest-centred outlook. *Sustainability, 16*(11), 4863. https://doi.org/10.3390/su16114863

Aggestam, F. (2024b). *European Forest Policy and Governance An Integrative Analytical Framework (1st Edition)*. Routledge.

Aggestam, F., & Giurca, A. (2021). The art of the "Green" deal: Policy pathways for the EU forest strategy. *Forest Policy and Economics, 128*, Article 102456. https://doi.org/10.1016/j.forpol.2021.102456

Aggestam, F., & Giurca, A. (2022). Implementing circular-bioeconomy principles across two value chains of the wood-based sector: A conceptual approach. *Land, 11*(11). https://doi.org/10.3390/land11112037

Aggestam, F., & Pülzl, H. (2018). Coordinating the uncoordinated: The EU forest strategy. *Forests, 9*(3), 125, https://doi.org/10.3390/f9030125

Aggestam, F., & Pülzl, H. (2020). Downloading Europe: A regional comparison in the uptake of the EU forest action plan. *Sustainability, 12*(10). https://doi.org/10.3390/su12103999

Aggestam, F., Prins, K., & Tulsidas, H. (2021). *Natural Resource Nexuses in the ECE Region: Vol. ECE/INF/2020/2*. United Nations Economic Commission for Europe. https://unece.org/sites/default/files/2021-04/2016242_E_web.pdf

Agrawal, A., Chhatre, A., & Hardin, R. (2008). Changing governance of the world's forests. *Science, 320*(5882), 1460–1462. https://doi.org/10.1126/science.1155369

Angelsen, A. (2017). REDD+ as result-based aid: General lessons and bilateral agreements of Norway. *Review of Development Economics, 21*(2), 237–264. https://doi.org/10.1111/rode.12271

Arts, B., & Buizer, M. (2009). Forests, discourses, institutions. *Forest Policy and Economics, 11*(5–6), 340–347. https://doi.org/10.1016/j.forpol.2008.10.004

Arts, B., Behagel, J., Turnhout, E., De Koning, J., & Van Bommel, S. (2014). A practice based approach to forest governance. *Forest Policy and Economics, 49*, 4–11. https://doi.org/10.1016/j.forpol.2014.04.001

Borrass, L., Kleinschmit, D., & Winkel, G. (2017). The "German model" of integrative multifunctional forest management—Analysing the emergence and political evolution of a forest management concept. *Forest Policy Economics, 77*, 16–23. https://doi.org/10.1016/j.forpol.2016.06.028

Bradshaw, C. J. A., Warkentin, I. G., & Sodhi, N. S. (2009). Urgent preservation of boreal carbon stocks and biodiversity. *Trends in Ecology & Evolution, 24*(10), 541–548. https://doi.org/10.1016/j.tree.2009.03.019

Cavallin, E. (2024). Nature restoration and agriculture and forestry: At the opposite side of the fighting ring or compatible after all? An analysis of the proposal and the final agreement on the nature restoration law. *European Energy and Environmental Law Review, 33*(Issue 2), 48–69. https://doi.org/10.54648/EELR2024003

Chaffin, B. C., Gosnell, H., & Cosens, B. A. (2014). A decade of adaptive governance scholarship: Synthesis and future directions. *Ecology and Society, 19*(3), art56. https://doi.org/10.5751/ES-06824-190356

Chazdon, R. L., Brancalion, P. H. S., Laestadius, L., Bennett-Curry, A., Buckingham, K., Kumar, C., Moll-Rocek, J., Vieira, I. C. G., & Wilson, S. J. (2016). When is a forest a forest? Forest concepts and definitions in the era of forest and landscape restoration. *Ambio, 45*(5), 538–550. https://doi.org/10.1007/s13280-016-0772-y

De Koning, S., Boezeman, D., Kaufmann, M., & Visseren-Hamakers, I. J. (2023). Transformative change for biodiversity: A review on the contribution of landscape-oriented partnerships. *Biological Conservation, 277*, Article 109858. https://doi.org/10.1016/j.biocon.2022.109858

EC. (2021). *COM(2021) 572 final: New EU Forest Strategy for 2030*. European Commission.

ECE. (2020). *Who owns our forests? Forest Ownership in the ECE Region*. United Nations Economic Commission for Europe.

Edwards, P., & Kleinschmit, D. (2013). Towards a European forest policy: Conflicting goals. *Forest Policy and Economics, 33*, 87–91.

Ellis, E. C., Pascual, U., & Mertz, O. (2019). Ecosystem services and nature's contribution to people: Negotiating diverse values and trade-offs in land systems. *Current Opinion in Environmental Sustainability, 38*, 86–94. https://doi.org/10.1016/j.cosust.2019.05.001

EP & CoE. (2010). Regulation 995/2010: Laying down the obligations of operators who place timber and timber products on the market. *Official Journal of the European Union*

EP & CoE. (2021). Regulation 2021/1119: Establishing the framework for achieving climate neutrality and amending Regulations (EC) No 401/2009 and (EU) 2018/1999 ('European Climate Law'). *Official Journal of the European Union*

Escobedo, F. J., Giannico, V., Jim, C. Y., Sanesi, G., & Lafortezza, R. (2019). Urban forests, ecosystem services, green infrastructure and nature-based solutions: Nexus or evolving metaphors? *Urban Forestry & Urban Greening, 37*, 3–12. https://doi.org/10.1016/j.ufug.2018.02.011

FAO. (2020). *Global Forest Resources Assessment 2020: Main report*. Food and Agriculture Organization of the United Nations.

FAO. (2022). *The State of the World's Forests 2022. Forest pathways for green recovery and building inclusive, resilient and sustainable economies*. Food and Agriculture Organization of the United Nations. https://www.fao.org/3/cb9360en/cb9360en.pdf

Haas, P. M. (1992). Introduction: Epistemic communities and international policy coordination. *International Organization, 46*(1), 1–35. JSTOR

Jager, H. I., Efroymson, R. A., & McManamay, R. A. (2021). Renewable energy and biological conservation in a changing world. *Biological Conservation, 263*, Article 109354. https://doi.org/10.1016/j.biocon.2021.109354

Kleinschmit, D., Pülzl, H., Secco, L., Sergent, A., & Wallin, I. (2018). Orchestration in political processes: Involvement of experts, citizens, and participatory professionals in forest policy making. *Forest Policy Economins, 89*, 4–15. https://doi.org/10.1016/j.forpol.2017.12.011

Kremen, C., & Merenlender, A. M. (2018). Landscapes that work for biodiversity and people. *Science, 362*(6412), eaau6020. https://doi.org/10.1126/science.aau6020

Lafortezza, R., Chen, J., Van Den Bosch, C. K., & Randrup, T. B. (2018). Nature-based solutions for resilient landscapes and cities. *Environmental Research, 165*, 431–441. https://doi.org/10.1016/j.envres.2017.11.038

Larson, A. M., Mausch, K., Bourne, M., Luttrell, C., Schoneveld, G., Cronkleton, P., Locatelli, B., Catacutan, D., Cerutti, P., Chomba, S., Djoudi, H., Ihalainen, M., Lawry, S., Minang, P., Monterroso, I., Myers, R., Naito, D., Pham, T. T., Reed, J., & Stoian, D. (2021). Hot topics in governance for forests and trees: Towards a (just) transformative research agenda. *Forest Policy and Economics, 131*, Article 102567. https://doi.org/10.1016/j.forpol.2021.102567

Lazdinis, M., Angelstam, P., & Lazdinis, I. (2009). Governing forests of the European Union: Institutional framework for interest representation at the European community level. *Environmental Policy and Governance, 19*(1), 44–56. https://doi.org/10.1002/eet

Lazdinis, M., Angelstam, P., & Pülzl, H. (2019). Towards sustainable forest management in the European Union through polycentric forest governance and an integrated landscape approach. *Landscape Ecology, 34*(7), 1737–1749. https://doi.org/10.1007/s10980-019-00864-1

Lecina-Diaz, J., Alvarez, A., Regos, A., Drapeau, P., Paquette, A., Messier, C., & Retana, J. (2018). The positive carbon stocks–biodiversity relationship in forests: Co-occurrence and drivers across five subclimates. *Ecological Applications, 28*(6), 1481–1493. https://doi.org/10.1002/eap.1749

Lewis, S. L., Edwards, D. P., & Galbraith, D. (2015). Increasing human dominance of tropical forests. *Science, 349*(6250), 827–832. https://doi.org/10.1126/science.aaa9932

Lindenmayer, D., Messier, C., & Sato, C. (2016). Avoiding ecosystem collapse in managed forest ecosystems. *Frontiers in Ecology and the Environment, 14*(10), 561–568. https://doi.org/10.1002/fee.1434

Loisel, J., Gallego-Sala, A. V., Amesbury, M. J., Magnan, G., Anshari, G., Beilman, D. W., Benavides, J. C., Blewett, J., Camill, P., Charman, D. J., Chawchai, S., Hedgpeth, A., Kleinen, T., Korhola, A., Large, D., Mansilla, C. A., Müller, J., Van Bellen, S., West, J. B., & Wu, J. (2021). Expert assessment of future vulnerability of the global peatland carbon sink. *Nature Climate Change, 11*(1), 70–77. https://doi.org/10.1038/s41558-020-00944-0

Lund, H. G. (2002). When Is a forest not a forest? *Journal of Forestry, 100*(8), 21–28. https://doi.org/10.1093/jof/100.8.21

Luyssaert, S., Schulze, E.-D., Börner, A., Knohl, A., Hessenmöller, D., Law, B. E., Ciais, P., & Grace, J. (2008). Old-growth forests as global carbon sinks. *Nature, 455*(7210), 213–215. https://doi.org/10.1038/nature07276

May, C. K. (2022). Complex adaptive governance systems: A framework to understand institutions, organisations, and people in socio-ecological systems. *Socio-Ecological Practice Research, 4*(1), 39–54. https://doi.org/10.1007/s42532-021-00101-7

McDermott, C. L., Montana, J., Bennett, A., Gueiros, C., Hamilton, R., Hirons, M., Maguire-Rajpaul, V. A., Parry, E., & Picot, L. (2023). Transforming land use governance: Global targets without equity miss the mark. *Environmental Policy and Governance, 33*(3), 245–257. https://doi.org/10.1002/eet.2027

McDermott, C., Cashore, B., & Kanowski, P. (2010). *Global Environmental Forest Policies* (0 ed.). Routledge. https://doi.org/10.4324/9781849774925

Millar, C. I., & Stephenson, N. L. (2015). Temperate forest health in an era of emerging megadisturbance. *Science, 349*(6250), 823–826. https://doi.org/10.1126/science.aaa9933

O'Reilly, S. (2006). Governing through markets. forest certification and the emergence of non-state authority * hard choices, soft law: Voluntary standards in global trade, environment and social governance. *Forestry, 79*(5), 610–612. https://doi.org/10.1093/forestry/cpl040

Oberthür, S., & Groen, L. (2018). Explaining goal achievement in international negotiations: The EU and the Paris agreement on climate change. *Journal of European Public Policy, 25*(5), 708–727. https://doi.org/10.1080/13501763.2017.1291708

Patterson, J., Schulz, K., Vervoort, J., Van Der Hel, S., Widerberg, O., Adler, C., Hurlbert, M., Anderton, K., Sethi, M., & Barau, A. (2017). Exploring the governance and politics of transformations towards sustainability. *Environmental Innovation and Societal Transitions, 24*, 1–16. https://doi.org/10.1016/j.eist.2016.09.001

Pauly, M., & Tosteson, J. (2022). Safeguarding natural forests through the voluntary REDD+ scheme. *Nature Plants, 8*(8), 861–866. https://doi.org/10.1038/s41477-022-01208-9

Pe'er, G., Bonn, A., Bruelheide, H., Dieker, P., Eisenhauer, N., Feindt, P. H., Hagedorn, G., Hansjürgens, B., Herzon, I., Lomba, Â., Marquard, E., Moreira, F., Nitsch, H., Oppermann, R., Perino, A., Röder, N., Schleyer, C., Schindler, S., Wolf, C., & Lakner, S. (2020). Action needed for the EU common agricultural policy to address sustainability challenges. *People and Nature*, *2*(2), 305–316. https://doi.org/10.1002/pan3.10080

Puettmann, K. J. (with Coates, K. D., & Messier, C. C.). (2014). *A Critique of Silviculture: Managing for Complexity*. Island Press.
Pülzl, H., Wydra, D., & Hogl, K. (2018). Piecemeal integration: explaining and understanding 60 years of European Union forest policy-making. *Forests, 9*, 719.
Pülzl, H., Hogl, K., Kleinschmit, D., Wydra, D., Arts, B., Mayer, P., Palahí, M., Winkel, G., & Wolfslehner, B. (2013). *European Forest Governance: Issues at Stake and the Way Forward* (No. 2, What Science Can Tell Us). European Forest Institute
Rayner, J., Buck, A., & Katila, P. (2010). Embracing complexity: Meeting the challenges of international forest governance. A global assessment report. *Prepared by the Global Forest Expert Panel on the International Forest Regime.* (No. 28; IUFRO World Series Volume).
Sabatini, F. M., De Andrade, R. B., Paillet, Y., Ódor, P., Bouget, C., Campagnaro, T., Gosselin, F., Janssen, P., Mattioli, W., Nascimbene, J., Sitzia, T., Kuemmerle, T., & Burrascano, S. (2019). Trade-offs between carbon stocks and biodiversity in European temperate forests. *Global Change Biology, 25*, 536–548. https://doi.org/10.1111/gcb.14503
Santangeli, A., Di Minin, E., Toivonen, T., Pogson, M., Hastings, A., Smith, P., & Moilanen, A. (2016). Synergies and trade-offs between renewable energy expansion and biodiversity conservation—A cross-national multifactor analysis. *GCB Bioenergy, 8*(6), 1191–1200. https://doi.org/10.1111/gcbb.12337
Sayer, J., Sunderland, T., Ghazoul, J., Pfund, J.-L., Sheil, D., Meijaard, E., Venter, M., Boedhihartono, A. K., Day, M., Garcia, C., Van Oosten, C., & Buck, L. E. (2013). Ten principles for a landscape approach to reconciling agriculture, conservation, and other competing land uses. *Proceedings of the National Academy of Sciences, 110*(21), 8349–8356. https://doi.org/10.1073/pnas.1210595110
Sikor, T., & Lund, C. (2009). Access and property: A question of power and authority. *Development and Change, 40*(1), 1–22. https://doi.org/10.1111/j.1467-7660.2009.01503.x
Soimakallio, S., Kalliokoski, T., Lehtonen, A., & Salminen, O. (2021). On the trade-offs and synergies between forest carbon sequestration and substitution. *Mitigation and Adaptation Strategies for Global Change, 26*(1), 4. https://doi.org/10.1007/s11027-021-09942-9
Sotirov, M., & Arts, B. (2018). Integrated Forest Governance in Europe: An introduction to the special issue on forest policy integration and integrated forest management. *Land Use Policy, 79*, 960–967. https://doi.org/10.1016/j.landusepol.2018.03.042
Sousa-Silva, R., Verbist, B., Lomba, A., Valent, P., Suškevičs, M., Picard, O., Hoogstra-Klein, M. A., Cosofret, V.-C., Bouriaud, L., Ponette, Q., Verheyen, K., & Muys, B. (2018). Adapting forest management to climate change in Europe: Linking perceptions to adaptive responses. *Forest Policy Econmics, 90*, 22–30. https://doi.org/10.1016/j.forpol.2018.01.004.
Sunderlin, W. D., Sills, E. O., Duchelle, A. E., Ekaputri, A. D., Kweka, D., Toniolo, M. A., Ball, S., Doggart, N., Pratama, C. D., Padilla, J. T., Enright, A., & Otsyina, R. M. (2015). REDD+ at a critical juncture: Assessing the limits of polycentric governance for achieving climate change mitigation. *The International Forestry Review, 17*(4), 400–413. JSTOR
Sutherland, L.-A., & Huttunen, S. (2018). Linking practices of multifunctional forestry to policy objectives: Case studies in Finland and the UK. *Forest Policy Econmics, 86*, 35–44. https://doi.org/10.1016/j.forpol.2017.10.019
Turetsky, M. R., Benscoter, B., Page, S., Rein, G., Van Der Werf, G. R., & Watts, A. (2015). Global vulnerability of peatlands to fire and carbon loss. *Nature Geoscience, 8*(1), 11–14. https://doi.org/10.1038/ngeo2325
UNEP. (2021). *Making Peace with Nature: A scientific blueprint to tackle the climate, biodiversity and pollution emergencies.* United Nations Environment Programme. https://www.unep.org/resources/making-peace-nature
Winkel, G., Sotirov, M., & Moseley, C. (2021). Forest environmental frontiers around the globe: Old patterns and new trends in forest governance. *Ambio, 50*(12), 2129–2137. https://doi.org/10.1007/s13280-021-01647-2

Winkel, G., & Sotirov, M. (2016). Whose integration is this? European forest policy between the gospel of coordination, institutional competition, and a new spirit of integration. *Environment and Planning C, 34*(3), 496–514, https://doi.org/10.1068/c1356j

Winkel, G., Aggestam, F., Sotirov, M., & Weiss, G. (2013). *Forest Policy in the European Union* (European Forest Governance: Issues at Stake and the Way Forward, pp. 52–63).

Winkel, G. (2017). *Towards a sustainable European forest-based bioeconomy- assessment and the way forward. What Science Can Tell Us 8*. European Forest Institute. https://www.efi.int/publications-bank.

Winter, S., Borrass, L., Geitzenauer, M., Blondet, M., Breibeck, R., Weiss, G., & Winkel, G. (2014). The impact of Natura 2000 on forest management: A socio-ecological analysis in the continental region of the European Union. *Biodiversity and Conservation, 23*(14), 3451–3482. https://doi.org/10.1007/s10531-014-0822-3

Wolfslehner, B., Pülzl, H., Kleinschmit, D., Aggestam, F., Winkel, G., Candel, J., Eckerberg, K., Feindt, P., McDermott, C., Secco, L., Sotirov, M., Lackner, M., & Lazya Roux, J. (2020). *European forest governance post-2020* (From Science to Policy 10). European Forest Institute. https://doi.org/10.36333/fs10

Zito, A. R. (2001). Epistemic communities, collective entrepreneurship and European integration. *Journal of European Public Policy, 8*, 585–603.

Chapter 2
Theoretical Perspectives on Forest Policy and Transformative Change

Filip Aggestam

Abstract This chapter considers how theoretical approaches to forest policy and governance research can help to facilitate transformative change. It starts by outlining the evolution of forest policy research, from its technocratic and sector-specific origins to its current interdisciplinary status, shaped by political science, sociology, institutional theory, and sustainability research. The chapter reviews key theoretical perspectives, including institutional and policy process frameworks, constructivist and discursive approaches, and integrative and reflexive perspectives. It demonstrates how each has expanded our understanding of forest governance. These approaches are then placed within broader theories of change, such as adaptive governance, sustainability transitions and transformative governance. These theories provide insights into how forest governance systems react to complexity, uncertainty, and conflicting values. The chapter argues that these perspectives should be viewed as complementary rather than competing. Furthermore, the chapter outlines the need for an integrative framework to analyse transformative forest governance, organised around four drivers: structure, agency, discourse, and learning.

2.1 Introduction

Forest policy research initially emerged from a technocratic field that focused on sustainable timber production, principally from a state-led forest management perspective. In other words, back in the day, the main focus was to ensure that there was enough timber available. Since that time, forest policy has evolved into an interdisciplinary research domain that now covers a broad range of socio-economic and environmental factors shaping forest use and governance. In other words, forests are now recognised as part of a much broader policy system in which competing values, interests, and different forms of knowledge interact and shape forest use (Aggestam, 2025; Art & Visseren-Hamakers, 2012; Gordeeva et al., 2025; Pülzl et al., 2018).

F. Aggestam (✉)
ORBIS Green Policy and Governance, Forest Policy Research Network—European Forest Institute (FPRN-EFI), Vienna, Austria
e-mail: aggestam@orbis-gpg.com

F. Aggestam (ed.), *Governing Forests Through Transformative Change*, Advances in Global Change Research 70, https://doi.org/10.1007/978-3-032-19072-7_2

The reframing of forest policy has furthermore broadened the analytical scope, going beyond issues of resource use and institutional design to cover legitimacy, participation, justice, and sustainability (Arts & Buizer, 2009; Winkel, 2014). This gradual expansion, however, raises important questions. How should different forest policy processes be understood across sectors and levels? Moreover, relevant to this book, how can forest policy facilitate or prevent transformative change?

Forest policy research has increasingly become associated with global narratives and frameworks that address societal challenges such as biodiversity loss, climate change, and energy production. For example, international policy instruments such as the Paris Agreement, the Kunming-Montreal Global Biodiversity Framework, and the 2030 Agenda for Sustainable Development address forests as relevant subjects and as agents of change (Aggestam, 2024; O'Brien et al., 2025). These developments have expanded the role of forest policy, increasing its cross-sectoral relevance and potential contribution to transformative change. Even more, the broadening of forest policy research has involved epistemic developments, with forest policy researchers increasingly incorporating perspectives from disciplines such as political science, sociology, and environmental governance. This has significantly increased the number of approaches and frameworks that policy researchers use. These changes have also resulted in a more diverse theoretical landscape, generating new insights with regards to how forest policy problems are defined, how actors interact, and how changes occur.

Arguably, the increasingly diverse theoretical landscape and approaches reflect the diversification of the forest-based sector itself. Compared to the past, when timber ruled the day, forest-related products are today used across many levels (e.g., state agencies, private companies, and local communities) and in many more sectors (e.g., plastics and textiles). This implies that any decision affecting forest use is also shaped by many more interests and policy frameworks, ranging from national regulations to international trade, climate, and biodiversity-related instruments. Forest governance has in turn shifted from a more hierarchically (top-down) controlled system towards a system that is increasingly characterised by negotiated outcomes and multiple use (Borrass et al., 2017; Winkel et al., 2022). In other words, policy-making is no longer a linear process (Folke, 2006; Voß et al., 2006). Against this background, transformative change (as a concept) is of interest as it provides an integrative perspective that can connect governance, sustainability transition, and forest policy (Patterson et al., 2017; Scoones & Stirling, 2020). For example, taking transformative change seriously could prompt us to reconsider land tenure and property rights as well as production and consumption systems. It could help us reconsider the numerous socio-economic and environmental objectives that are imposed on forests today.

Building on these developments, this chapter explores how theories used in forest policy research can provide conceptual foundations for understanding transformative change. It is argued that "transformation" can and should be better understood through theoretical pluralism, in which insights from different frameworks and perspectives (e.g., institutional, discursive, and adaptive) can improve how forests are managed and used. The following subsections will begin by reviewing some classical theoretical perspectives and approaches used in forest policy research.

2.2 Tracing the Origins of Forest Policy

To understand forest policy, it is important to know its origins. As set out above, historically, forest policy emerged from the need to ensure a predictable timber supply, most often coupled with the belief that managing forest resources was the state's responsibility. The initial formulations of forest policy were therefore normative, seeking to define the principles of "good forest management" and focusing on how the state could ensure its timber supply. One such principle was "nachhaltige Nutzung" (sustained yield forestry), introduced by Hans Carl von Carlowitz in 1713, which over time became a central pillar of early forest policy. This outlook framed forests as a renewable and finite resources that require long-term (and rational) management by the state (Carlowitz, 2013).

Within this tradition, European forest policy (particularly in the German-speaking context of Forstpolitik) developed as a sub-discipline of forestry. It focused primarily on the legal, economic, and administrative instruments affecting timber supply and forest ownership. Consequently, early forest policy research was more prescriptive than analytical, emphasising improvements in policy design (e.g., optimising forest laws, pricing mechanisms, and management standards), rather than developing theories surrounding forest governance (Arts & Buizer, 2009; Burns et al., 2017). As such, forest policy analysis remained technocratic for a long time, assuming that social and environmental problems could be solved through expertise, regulation, and state control.

Throughout the mid-twentieth century, forest policy remained largely sector-specific, focusing on forestry while paying limited attention to other sectors. The dominant assumption was that forests should be separated from social and political pressures and governed independently. In practice, however, forests were increasingly influenced by external factors such as the expansion of agricultural production, industrialisation, urbanisation, and global trade. This created a tension between the traditional, sector-specific, perspectives of forestry and the cross-sectoral nature of land use change. These developments gradually laid the groundwork for more integrative approaches to forest governance (Borrass et al., 2017; Winkel & Sotirov, 2016).

Despite its limitations, early forest policy research provided some of the conceptual foundations for later theoretical developments. For example, the concept of public interest in forestry predates ongoing discussions around ecosystem services and public goods (Winkel, 2014; Winkel et al., 2022). Similarly, the emphasis on long-term planning can be seen as a precursor to discussions around sustainability and intergenerational equity (Arts & Buizer, 2009; McDermott et al., 2023). Nevertheless, these contributions were embedded within a dominant narrative that depoliticised forest use and governance. Forest policy was framed as a technical system to be managed rather than as a political landscape that competing values could shape. This means that early forest policy analysis often overlooked issues such as power, legitimacy, and social justice. In a nutshell, it was assumed that rational design and expert knowledge would make sure that outcomes were optimised.

Starting with the modern environmental movements in the 1960s, 1970s, and beyond, the more technocratic approach to forest policy began to face significant criticism (Scott, 2020). Researchers increasingly highlighted the social and environmental costs of state-centred forest policy, including failures of top-down reforestation and conservation programmes, the displacement of local communities, and the loss of traditional knowledge. In response, social forestry and community-based forest management emerged, emphasising participation and livelihood benefits together with conservation. These movements introduced more process-oriented analytical approaches, treating forest governance as a dynamic and multi-actor process (Agrawal et al., 2008; Ostrom, 1990).

By the late twentieth century, forest policy researchers had begun to include multiple interdisciplinary approaches. Aside from inherent sectoral changes, these developments reflected broader epistemological shifts in systems thinking and in participatory and adaptive approaches (Folke, 2006; Voß et al., 2006). As noted at the onset of this chapter, forest policy research entered a phase of theoretical and practical diversification, with researchers increasingly drawing on theories and perspectives from political science, sociology and institutional economics to explain how forest policies were being formulated, implemented, and changed (Agrawal et al., 2008; Art & Visseren-Hamakers, 2012; Krott, 2005).

2.3 Institutional and Discursive Perspectives

Institutional and discursive perspectives have been central to the evolution of forest policy research. These perspectives offer different but complementary explanations of how forest policies emerge, how actor constellations form, how power is used, and how certain perspectives become dominant in forest governance.

2.3.1 Institutions and the Forest Policy Process

The introduction of new forms of institutional analysis in the late 1980s and 1990s led to a significant expansion of forest policy research. For instance, researchers began exploring in more detail the rules, values, and policy-making processes that influence forest management and governance. Essentially, forest policy was no longer considered simply an extension of a regulation, but rather an outcome of interactions within (or between) governance systems. Institutional theory provided the conceptual basis for this change. Rational-choice institutionalism, for instance, considers strategic behaviour under institutional constraints, historical institutionalism describes policy stability and change through path dependencies and critical junctures, and sociological institutionalism explains how norms and ideas shape actors' perceptions of legitimacy (Baumgartner & Mahoney, 2008; Mahoney & Thelen, 2009).

More specifically, in forest policy research, the Advocacy Coalition Framework has been widely used to analyse policy change. This model describes the policy-making process as a competition between actor coalitions that share belief systems and act to influence a policy (sub)system (Sabatier, 1988; Weible, 2006; Weible & Sabatier, 2018). Changes can occur through learning or through external shocks that affect existing coalitions. Similarly, the Multiple Streams Framework and Punctuated Equilibrium Theory emphasise the conditions for policy change. For example, these models demonstrate how "windows of opportunity"[1] occur when the political, problem, and policy streams converge or when long periods of stability are disrupted by a crisis (Baumgartner & Jones, 2009; Baumgartner & Mahoney, 2008; Hoefer, 2022; Kingdon, 1984).

In Europe, these frameworks have been used to analyse the EU Forest Strategy, the Natura 2000 initiative and the Forest Europe processes. This work has revealed how institutional layering and cross-sectoral coordination have shaped the coherence of policies relating to forests, climate, and biodiversity (Pülzl et al., 2013; Winkel & Sotirov, 2016; Winkel et al., 2015). In North America, the Advocacy Coalition Framework and Punctuated Equilibrium Theory have been used to analyse public land management, wildfire policy, and certification standards (Cashore et al., 2004; Elliott & Schlaepfer, 2001). Together, these studies have demonstrated the long-term impact of belief systems, learning, and institutional inaction on forest use and policy-making. In effect, it highlights the importance of the structural dynamics of forest policy systems and processes.

Polycentric governance[2] has also become an important concept in understanding how forest-relevant institutions interact across levels and over time (Thiel, 2023). Polycentric governance represents a system with layered policy-making (e.g., from national to local), arguably, making it more resilient and adaptive (e.g., as policies can be adjusted based on local conditions through learning and experimentation). While the Advocacy Coalition Framework frames learning as changes in beliefs based on new information, other institutional approaches emphasise organisational and social learning through feedback and experimentation (Hall, 1993; Henly-Shepard et al., 2015). Researchers have adopted these ideas to analyse how forest governance reforms have evolved, covering topics such as how certification schemes, participatory monitoring, and forest partnerships adapt to local conditions (e.g., through cycles of trial and error). This means that learning became an analytical concept, linking institutional design to a broader discussion around adaptive and reflexive governance.

Nevertheless, despite its analytical power, institutional frameworks also have their drawbacks. For instance, it has been argued that institutional approaches underestimate the informal and creative factors that characterise forest governance in practice. Hybrid governance arrangements (e.g., community-based forestry) have,

[1] A "window of opportunity" implies a brief period during which conditions align to allow significant policy change.

[2] A governance system with multiple interacting centres of power across scales, enabling decentralised coordination and learning.

for example, challenged the hierarchical perspective of formal institutional models (Agrawal et al., 2008; Cleaver, 2017). This emphasises the importance of context, agency, and cultural meaning-making in understanding how institutions operate. More specifically, many institutional frameworks tend to take a functionalist or apolitical approach, focusing on structures and rules and treating the institution as a neutral space that is unaffected by power and ideology. However, forest governance is shaped by institutional capacities, discursive authority, and formal rules in equal measure (Krott & Giessen, 2014). Furthermore, work on reducing emissions from deforestation and forest degradation (REDD+) has demonstrated the importance of underlying power imbalances and socio-economic inequalities (Angelsen, 2017). In other words, there is much more to the forest governance puzzle than the institutions.

2.3.2 The Rise of Constructivism, Discourse, and Frame Analysis

By the late 1990s and early 2000s, forest policy research had adopted interpretive and critical perspectives that considered the role of language, meaning-making, and legitimacy in policy-making, moving beyond institutions (Dryzek, 2005; Fischer, 2003; Hajer, 1997). These developments reflect a wider constructivist shift,[3] whereby forest governance was understood not only as a response to a problem, but also as a process through which problems are framed and created. This means that forest policy also defines what constitutes a "problem" for the forest-based sector. These developments were supported by discursive institutionalism and frame analysis, which conceptualise policies as narratives, frames, or interpretive schemes through which actors make sense of complex problems and mobilise support (Rein & Schön, 1996; Schmidt, 2008). Forest policy researchers began using these approaches to investigate how ideas, beliefs, and perspectives shape forest governance (Aggestam, 2015; Aggestam et al., 2025; Elomina & Pülzl, 2021; Lindahl, 2008).

One key message from this area of research is that changes in discourse or framing can create opportunities for, or prevent, policy change. Often in tandem with changes in power relationships. These dynamics have been demonstrated in Europe and globally. Within the EU, for instance, the emergence of the bioeconomy narrative allowed forest-based industries to present themselves as climate actors, thereby influencing the allocation of research funding (Giurca & Späth, 2017). In the context of tropical forests, the emergence of the REDD+ discourse reframed deforestation as a climate mitigation issue, attracting new investments while centralising control over land and carbon rights (Angelsen, 2017; Angelsen et al., 2018). These examples demonstrate that changing narratives affects the interests that are prioritised as well as how costs and benefits are distributed.

---

[3] Constructivism argues that policy problems and solutions are socially constructed, e.g., through discourse and shared meaning-making.

Constructivist approaches have transformed forest policy research by broadening our understanding of what "knowledge" means. Rather than relying on quantitative indicators or institutional metrics, constructivist approaches have encouraged researchers to examine how knowledge is used and produced, including the co-production of forest-related expertise that in turn informs (or shapes) forest management. This epistemic shift has revealed that governance outcomes cannot be fully understood without analysing the power relations embedded in knowledge systems (e.g., who can communicate, whose knowledge is used, and how is evidence interpreted). Methodologically, these developments have led to the adoption of qualitative and interpretive approaches, such as discourse analysis, ethnography, and participatory observation, and more recently, computational text analysis. These methods focus on meaning-making, narrative construction, and the practices through which forest governance is implemented in specific contexts.

Overall, constructivism has achieved far more than simply criticising the institutional perspective. By conceptualising forest governance as a meaning-making process and a struggle over legitimacy, constructivism has demonstrated that transformation also involves cognitive and normative changes as well as institutional reforms. It has demonstrated that power operates through formal structures, such as institutions, and through knowledge and the framing of policy problems and desirable futures.

2.3.3 Moving Towards Integration

By 2010, forest policy research had become increasingly diverse in the theories and approaches used. This diversification established the foundation for new forms of analytical integration, mainly because forest governance was now perceived as increasingly complex and cross-sectoral. In short, it was recognised that forest governance could not be understood through a single lens and that forest policy researchers needed frameworks to capture the complex interactions between institutions, actors, and ideas (Folke, 2006; Voß et al., 2006). For instance, institutional theories emphasise the impact of rules and resources on behaviour, whereas actor-centred approaches reveal the strategies, power relationships, and learning processes that facilitate policy change. Additionally, discursive approaches highlight the effect of meaning-making and legitimacy on policy-making. These analytical dimensions (e.g., structure, agency, and discourse) are distinct yet empirically connected, each revealing a different aspect of the governance system (Aggestam, 2025; Art & Visseren-Hamakers, 2012; Krott & Giessen, 2014).

Integrative research treats different theoretical frameworks as complementary heuristics, or perspectives, that can be combined depending on the problem being analysed (Aggestam, 2025). For instance, analysing the transition to a forest-based bioeconomy might involve using institutional theory to examine how policies have evolved, discourse analysis to investigate changing narratives associated with the bioeconomy, and network theory to understand the relationships between actors (e.g., public–private partnerships) during the transition. This problem-oriented approach provides an analytical framework that offers multiple and complementary explanations while moving away from universal explanatory models.

One important feature of the shift towards "integration" is the adoption of reflexivity[4] as a governance factor and a research principle. For instance, reflexivity has emerged in forest policy as countries have tried to formalise policy learning and adaptive management. Instruments, such as Criteria and Indicators (C&I) (see Chap. 9), national forest programmes, and forest-relevant stakeholder platforms have been redesigned to monitor performance, facilitate dialogue, and enable adaptation. The Forest Europe[5] process, for instance, incorporated learning cycles into its ministerial commitments, while national C&I initiatives in countries such as Canada and Finland evolved from mainly acting as reporting tools to facilitating stakeholder engagement (Baycheva-Merger & Wolfslehner, 2016; Lier et al., 2021; Linser et al., 2018). These developments have established that forest governance is dynamic, knowledge-intensive, and value-laden, and that learning and participation are important factors when considering policy change.

Reflexivity also signifies a methodological shift, as researchers are no longer considered impartial observers. This implies that the framing of research questions, the selection of indicators (or data), and the communication of findings influence policy discourse, outcomes, and the analytical process itself. In practice, the distinction between analysis and practice has become blurred, leading to collaborative efforts to produce knowledge across epistemic communities, including scientists, policymakers, and practitioners (Jasanoff, 2004; Wyborn & Bixler, 2013). Methods used for this include collaborative modelling and participatory scenario building. Furthermore, integrative approaches serve a normative function by making explicit the assumptions, trade-offs, and value-based decisions inherent in the research process and the use of theory.

Finally, the groundwork behind integrative and reflexive approaches provided the foundation for transformation-oriented perspectives. For example, as governance challenges became more complex and uncertain, researchers began to ask not only how governance systems adapt to new pressures, but also how they could be deliberately realigned towards sustainability and other societal goals (Patterson et al., 2017; Scoones & Stirling, 2020). In forest policy research, these developments aligned with the emergence of the bioeconomy discourse, transition agendas, and ecosystem restoration agendas, each of which frames governance as a process of intentional societal change (Briers et al., 2025; Chazdon et al., 2016; Giurca & Späth, 2017; Winkel, 2017).

---

[4] Reflexivity relates to the capacity of actors, institutions, and researchers to question underlying assumptions, learn from feedback, and adjust strategies in light of new knowledge.

[5] See https://foresteurope.org.

2.4 Theories-of-Change and Transformation

One focus area for forest policy research has been to explain how and why changes occur in governance and management systems. Theoretical and conceptual frameworks, such as the Advocacy Coalition Framework, policy network analysis and institutional approaches, have provided valuable insights into governance regime dynamics, policy learning, value change and adaptation. However, debates surrounding forest use, governance, and sustainability require us to extend this analytical scope. In order to address current and emerging challenges, it is necessary to engage with a broader range of theories-of-change, approaches that consider not only incremental adjustments within existing policy systems, but also the conditions under which systemic and transformative change becomes possible:

- Classical policy-process theories understand change as incremental and negotiated within existing institutional structures. Path dependence, learning, coalition dynamics and occasional policy windows shape these structures. This approach is useful for explaining gradual shifts in forest governance, but it is limited in its ability to account for deeper transformations in values, power relations, and governance (Baumgartner & Jones, 2009; Hall, 1993; Mahoney & Thelen, 2009).
- Adaptive governance approaches, emerging from resilience thinking and systems theory, conceptualise change as arising from adaptability, feedback, and iterative learning, commonly found within polycentric and participatory governance arrangements. These perspectives have been applied to forest management, such as through community forestry and cross-scale coordination, to explain how systems cope with uncertainty and adjust through monitoring, experimentation, and partnerships. While adaptive governance is valuable for understanding resilience, it offers few tools for analysing the intentional or normative dimensions of transformation. This can include addressing whose interests guide adaptation and how larger structural changes may occur (Blythe et al., 2021; Folke, 2006).
- Sustainability transition frameworks, grounded in socio-technical systems and innovation studies, conceptualise change as a systemic, multi-level process in which innovations interact with dominant (policy) regimes within a broader landscape of pressures. When applied to the forest-based sector, these approaches have been used, for example, to analyse the development of new wood-based materials and digital forest innovations (see Chaps. 7 and 8). This demonstrates how technological and organisational breakthroughs can reshape value chains and governance arrangements. While offering tools for understanding co-evolutionary dynamics, transitions theory has also been criticised for insufficient attention to politics, power, and justice. Developments, such as transformative innovation policy (Haddad et al., 2022) and transformative social innovation (Avelino et al., 2019; Pamplona et al., 2024), explicitly incorporate agency, experimentation, and value-based objectives, positioning these perspectives within ongoing debates on sustainable transformation (Kemp & Loorbach, 1 C.E.Loorbach, 2010; Geels, 2002, 2021; Turnheim et al., 2015)

Drawing on insights from both adaptive and transition theories, transformative governance incorporates value-based (or normative), reflexive, and justice-oriented aspects of change. It defines governance as the deliberate reconfiguration of institutions, discourses, and power relations towards sustainability, supported by anticipatory, adaptive, and transformative capacities. Recent work has linked this perspective with theory-of-change arguments, treating "transformation" as an analytical lens and a strategic process involving explicit causal assumptions, actor mapping, and iterative learning towards a defined objective. This approach emphasises the enabling function of law and institutional design, which operate as both constraints and mechanisms for embedding reflexivity, adaptability, and procedural justice. When applied to forest-related policies, for example, transformative governance provides a framework for analysing ecosystem restoration, rights-based conservation, and sustainable transitions. In a nutshell, it emphasises that governance is a political and moral process that is shaped by competing perspectives (Blythe et al., 2021; Chaffin et al., 2014; Korhonen-Kurki et al., 2025; Patterson et al., 2017; Scoones & Stirling, 2020).

2.4.1 Insights for Forest Policy Research

Together, adaptive governance, sustainability transitions and transformative governance offer complementary perspectives on how changes can occur within complex governance systems. For instance, adaptive governance highlights the importance of learning and feedback processes in enabling responsiveness within existing institutional structures. Meanwhile, sustainability transition frameworks explain how innovations, policy regimes, and broader landscape pressures interact to change, or reconfigure, socio-technical systems. Building on these insights, transformative governance highlights the normative and political dimensions of change, as well as the deliberate reshaping of institutions, discourses and power relations. Rather than representing competing paradigms, these perspectives can be best understood as complementary layers of analysis (or as analytically nested). While adaptive governance focuses on incremental adjustment, transition theory demonstrates how new configurations emerge and stabilise. Meanwhile, transformative governance provides the value-based and reflexive framework needed to guide policy change.

In the context of forest policy research, this layered understanding offers a foundation for theorising how system innovations could facilitate transformative change. Furthermore, integrating theory-of-change arguments strengthens this framework by adding an operational dimension. Theory-of-change approaches typically involve explicitly formulating causal pathways, underlying assumptions, and actor roles. This enables researchers and practitioners to connect analysis with strategic action and iteratively test proposals for policy change. This is particularly relevant in forest governance debates, as there is a need for clarity regarding how to realise sustainability and resilience-related objectives in practice. Ultimately, theories-of-change may help forest policy researchers understand and explain the institutional, discursive, and material transformations needed to shape forest governance. In other words, to understand how governance systems evolve (or how they might be deliberately reoriented toward sustainability), forest policy research should adopt an integrative theoretical perspective that connects these complementary perspectives and approaches.

2.4.2 *Transformative Forest Governance*

Calls for integration, whether within forest policy research or across other domains, are nothing new. Many researchers and policy processes have emphasised the limitations of siloed and linear analytical approaches when addressing complex governance problems. For example, systems theorists and researchers in the fields of ecosystems and resilience noted the need to connect the ecological, social, and ethical dimensions of resource management during the early twentieth century (Boulding, 1966; Odum, 1953; Singer, 1998; Zeide, 1998). International initiatives, such as the 1972 Stockholm Conference[6] and the 1971 Man and the Biosphere (MAB) Programme,[7] helped to institutionalise calls for interdisciplinary and cross-sectoral approaches. Another example is the 1987 Brundtland Report,[8] which framed sustainability as an integrative process that requires integrating environmental, social, and economic considerations into natural resource management. These examples highlight long-standing efforts to integrate policy and research. Furthermore, while they highlight that individual theoretical frameworks can provide valuable insights into specific aspects of forest governance, they rarely align precisely with the actual (or real life) policy processes and problems.

Rather than attempting to combine multiple theories into a single, universal model, an integrative approach, as understood in this book, implies using multiple perspectives to analyse and understand complex governance problems. This approach reflects developments in sustainability science, where integration is increasingly recognised as an iterative learning process involving boundary work[9] rather than the development of fixed analytical structures (Miller & Wyborn, 2020; Wyborn & Bixler, 2013). Within forest governance, such boundary work can mean linking structural analyses of institutions and policy instruments with interpretive approaches that review discourses and values. It also implies embedding the work within reflexive social learning processes. This creates an analytical space to understand how governance systems change over time, particularly in terms of transformative change.

6 See https://legal.un.org/avl/ha/dunche/dunche.html.

7 See https://www.unesco.org/en/mab.

8 See https://www.brundtland.co.za.

9 Boundary work involves defining, negotiating, or bridging the boundaries between different knowledge systems, disciplines, or governance areas to enable collaboration and learning.

2.4.3 Structure, Agency, Discourse, and Learning

An integrative framework for analysing transformative forest governance could be organised around the following four interdependent factors: structure, agency, discourse, and learning. These factors (arguably) represent the main drivers behind systemic change:

- **Structure** relates to the institutional and material factors that enable and restrict action. These can include laws, property rights, and policy instruments. From a transformative change perspective, structures are objects and mediums for change. In other words, they stabilise governance systems while also setting the conditions for innovation and reform. For instance, structural inflexibility can lead to path dependency and lock-in (Mahoney & Thelen, 2009), whereas adaptive legal frameworks can encourage flexibility and experimentation (Korhonen-Kurki et al., 2025; Ostrom, 2010). Recognising the dual role of structures is important for identifying transformative leverage points within existing governance structures.
- **Agency** relates to the capacity of actors (e.g., individuals, organisations, or networks) to influence governance processes and outcomes. For instance, policy network and coalition frameworks demonstrate how agency is established through collaborations, negotiations, and learning (Rhodes, 1997; Sabatier, 1988). However, transformative governance extends this idea by emphasising distributed and collective agency, in which change can emerge from the interactions of different actors (Loorbach, 2010). Agency, therefore, reflects both strategic action and reflexivity. This means having the capacity not only to influence policy but also to reimagine governance.
- **Discourse**, as a driver, addresses the value-based aspects of governance, such as how problems are defined, which values are being prioritised, and how legitimacy is constructed. In this sense, language and narratives are not secondary, but an important aspect of governance. In the context of transformation, for example, discursive changes (such as the emergence of narratives concerning ecosystem restoration, climate justice, and the bioeconomy) can undermine dominant paradigms and create openings for alternative policy pathways (Hajer, 1997; Schmidt, 2008). Power is, as such, expressed not only through institutional power but also through meaning-making and narrative framing of governance challenges.
- **Learning** can be seen as the glue that connects the drivers of systemic change. In other words, it is the mechanism through which actors, structures, and discourses may evolve together (or not) over time. Theories of policy learning and adaptive governance emphasise feedback and iteration, whereas transformative learning implies shifts in values and perspectives. This effectively means that learning is both a methodological and a normative aspect of transformation (Folke, 2006; Patterson et al., 2017).

These four dimensions provide a conceptual foundation for analysing governance change and methodological guidance for structuring integrative research (Box 1).

Box 1. Relating the drivers of systemic change to governance FOR, OF, and IN transformation

As outlined in the introduction to this book, several chapters utilise Patterson et al.'s (2017) distinction between governance *for*, *of*, and *in* transformation. The four analytical drivers (structure, agency, discourse and learning) can be understood as complementary dimensions of these modes of governance. Together, they can clarify different aspects of governance and how they relate to the prospects of transformative change:

Structure → Governance *of* Transformation

Institutional and material structures, such as laws, property rights and strategies, shape how transformation processes are coordinated and regulated. These structures act as stabilising factors, but they also create opportunities for reorientation and reform (*see related chapters*: Rethinking EU Forest Governance (This Chapter) and Implementation of forest-related policies in the European Union (Chap. 3)).

Agency → Governance *for* Transformation

Actors and networks set the framework conditions, such as the capacities, strategies, and visions, needed to initiate and support transformative change. Different forms of agency across public, private, and civil society actors can, in turn, enable or prevent transformative change (*see related chapters*: Roots of Ownership (Chap. 6), The Missing Link (Chap. 7), and Forest Governance 4.0 (Chap. 8)).

Discourse → Governance *in* transformation

Discourses shape how problems are framed, which values gain importance, and which futures become legitimate. Shifts in narratives can change expectations, identities, and power relations, thereby influencing the direction and depth of governance transformations (*see related chapter*: Frames in transition (Chap. 5)).

Learning → Connecting mechanism across governance modes

Learning links structure, agency and discourse over time. It explains how governance systems absorb feedback, reinterpret knowledge, and adjust strategies. Transformative learning, in particular, enables deeper shifts in perspectives, institutional logics and norms (*see related chapters*: Measuring what matters (Chap. 10)).

Understanding how the four drivers correspond to governance *for*, *of*, and *in* transformation reveals the different ways in which forest governance can evolve. Importantly, it also highlights that transformation does not emerge from a single process, but rather from an interplay between structural factors, actor strategies, shifting narratives and iterative learning.

2.4.4 Moving Towards a Conceptual Framework for Research on Transformative Forest Governance

These insights can be synthesised within an integrative framework for transformative forest governance (Fig. 2.1). This combines the four analytical dimensions of transformation (Box 1) with the procedural stages of an integrative analysis, as derived from policy studies and interdisciplinary research design (Aggestam, 2025; Dunn,

2017; Spies & Alff, 2020). Together, these components demonstrate how governance systems and research processes can co-evolve through learning.

Learning and reflexivity are at the heart of the framework, driving the iterative reassessment of assumptions, perspectives, and objectives. The three principal drivers surrounding this core are agency (actors and networks), discourse (ideas and legitimacy) and structure (institutions and material systems), representing the arenas in which governance change takes place. These drivers interact through continuous feedback loops, facilitated by problem framing, methodological integration, and evaluation.

The outer ring of the framework represents the procedural stages of an integrative analysis, conceptualised as an iterative sequence ranging from initial problem identification to synthesis and communication, which feeds into reflexive learning over time. In this way, the framework operates at the analytical and methodological levels. Analytically, it captures the co-evolutionary drivers of structure, agency, and discourse that support policy change. Methodologically, it provides a heuristic for designing, implementing, and evaluating research or policy interventions that follow transformative governance principles.

Adopting a more integrated approach based on an integrative framework would have several implications for the future of forest policy research, which has long been oriented towards explanation and diagnosis:

First a range of different methods are needed, including a combination of comparative case studies and longitudinal analyses alongside interpretive and participatory approaches, to capture the interactions between structural factors, discursive changes, and behavioural patterns (Aggestam, 2025; Dunn, 2017). This reflects the emphasis on multiple analytical perspectives interacting through learning processes.

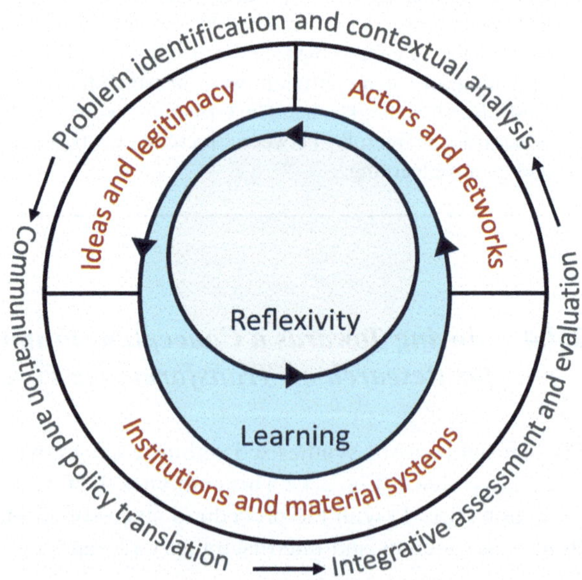

Fig. 2.1 An integrative framework for research on transformative change. *Source*: own figure

Second, integration requires spaces for reflection and learning. Boundary organisations, research-policy interfaces, and collaborative platforms, for example, can provide spaces where different stakeholders can exchange ideas, negotiate meanings, and engage in iterative sense-making (Guston, 2001; Turnhout et al., 2020). These spaces facilitate the feedback loops needed to promote co-production across scientific, policy, and practitioner communities.

Third, the framework implies the need to reframe research practices as collective learning. Rather than explaining governance problems and outcomes, transformative research contributes to re-evaluating objectives, assumptions, and approaches (Jasanoff, 2004; Miller & Wyborn, 2020). This involves embedding reflexivity within the research design, engaging with alternative knowledge systems and actively promoting the capacities required to support transformative governance.

Taken together, these implications highlight that integrative research represents a shift in epistemic orientation and research practice. It is not simply a matter of combining methods or perspectives. However, it also needs to be recognised that, while these types of designs may offer richer analytical insight, they require substantial time, resources, and coordination (Bianchi et al., 2021; Large, 2024). In other words, full integration may not always be feasible, particularly in traditional project-based research settings. Nevertheless, smaller-scale or ad hoc mechanisms (e.g., facilitated workshops or multi-actor dialogues) can be used to support reflexive learning.

2.4.5 *Transformative Research*

Transformative research acknowledges that no single theory (or method) can fully capture the complexity of a policy system. Instead, it adopts a pluralistic approach to learning and adaptation. This means that the focus shifts from how governance systems change to how research practices should evolve. In this sense, research becomes part of the system it studies, influencing how problems are framed, how knowledge is mobilised and used, and how potential futures are defined.

Pluralism furthermore provides the foundation for methodological innovation. By using a variety of approaches, researchers can more effectively map governance patterns, trace the evolution of discourse, and analyse collaborative networks. This integration of mixed methods reflects the iterative processes of synthesis and learning underlying transformative governance.

- **Reflexivity as part of the research design**

One important implication of transformative research is to acknowledge that the researcher is embedded in the governance system; he or she is not a detached observer. Their perspectives, conceptual frameworks, methodological choices and dissemination strategies influence how governance problems are defined and which interventions are considered legitimate or feasible (Jasanoff, 2004). Reflexivity should therefore involve systematically reflecting on how research practices may affect perspectives and power relations (Fazey et al., 2006a, b).

Moreover, incorporating reflexivity requires specific design features. For example, these include transparent theoretical assumptions, participatory co-definition of research questions, and iterative feedback loops that enable policy learning. Tools such as theory-of-change mapping (Korhonen-Kurki et al., 2025) and participatory foresight (Rutting et al., 2022) could help make these feedback loops visible and help transform a research project into a "learning system" that evolves. Reflexivity can in this way transform the research process, facilitating learning and adaptive governance.

- **Value-based engagement**

It also needs to be recognised that transformation is inherently value-based. This raises important and difficult questions about what constitutes "desirable change", which value and knowledge systems should be prioritised, and how costs and benefits should be distributed (Schlosberg, 2007). In forest governance, justice concerns can cover many dimensions, from the distribution of rights and resources to procedural fairness in decision-making processes (Art & Visseren-Hamakers, 2012).

A transformative research agenda must therefore explicitly address values and questions of justice. This implies considering epistemic and procedural elements, such as who is included in research and governance processes, which perspectives are legitimised and how deliberations develop (Avelino et al., 2019). Such engagements align with emerging discussions around transformative justice (Blythe et al., 2021) and imply that normative engagement is integral to transformative research.

- **Co-production**

Co-production is another defining feature of transformative research. It establishes direct links between knowledge production and forest governance, zooming in on the exchange between researchers, policy-makers, practitioners, and relevant communities (Fazey et al., 2006a, b; Wyborn & Bixler, 2013). In the forest-based sector, for example, co-production could be participatory indicator development, community-based monitoring, or participatory foresight workshops (Rutting et al., 2022). When aligned with theory-of-change perspectives, co-production enables researchers to test their underlying assumptions about how change is expected to occur iteratively (Korhonen-Kurki et al., 2025).

This ultimately reconfigures the research cycle itself. Each stage of problem framing, integration, and communication generates feedback that informs future analytical (or governance) decisions, thereby embedding learning directly into the research design. In this way, co-production transforms research projects into reflexive and adaptive systems.

- **Bridging theory and practice**

Moving beyond traditional policy analysis, the transformative research agenda requires us to reconsider the relationship between theory and practice. Rather than treating theory as an explanatory device and practice as the domain of policymakers, transformative approaches emphasise co-evolution. This is particularly relevant given that researchers are increasingly involved in policy deliberation, design, and implementation processes, including within forest governance systems (Meadowcroft, 2009; Clark et al., 2016). Consequently, there is a demand for tools that facilitate interactions between different groups of actors and knowledge systems (e.g., participatory foresight workshops, design labs, and learning platforms) to better analyse the constraints and opportunities for transformative change (Howlett & Rayner, 2013; Meadows, 1999).

2.5 Conclusion

Forest policy and governance research has evolved from a technocratic field concerned mainly with regulating timber production to a complex and interdisciplinary domain that now addresses the multiple (political, socio-economic, and environmental) dimensions of forest use. The diversification of theoretical perspectives, including institutional, network, discursive, adaptive and critical approaches, has broadened our understanding of how forest governance systems function and change over time. However, it is clear that forest governance is increasingly influenced by changing global policy priorities, cross-sectoral pressures and demands, as well as increasing socio-ecological complexity. Addressing these challenges will require new theoretical and methodological tools that can analyse and explain systemic and transformative pathways towards policy change.

Transformation-oriented perspectives, such as adaptive governance, sustainability transitions, and transformative governance, provide this broader lens. These perspectives shift analytical attention from incremental adjustments within existing systems to the structural, discursive, and value-based changes needed for major policy change (whether transformative or not). When combined with theory-of-change arguments, these approaches demonstrate the assumptions and values embedded in governance interventions, clarifying how agency, knowledge production, and institutional design can enable or prevent systemic transformation. Moreover, taken together, these approaches highlight the added value of theoretical pluralism, organised around the analytical drivers of structure, agency, discourse, and learning.

References

Aggestam, F. (2015). Framing the ecosystem concept through a longitudinal study of developments in science and policy. *Conservation Biology, 29*(4), 1052–1064.

Aggestam, F. (2024). Crosswalking the EU nature restoration regulation and the Kunming-Montreal global biodiversity framework: A forest-centred outlook. *Sustainability, 16*(11), 4863. https://doi.org/10.3390/su16114863

Aggestam, F., Trummer, J., & Wilkes-Allemann, J. (2025). Sixty years of EU forest policy discourse: An AI-based frame analysis of policy documents. *Forest Policy and Economics, Under review*.

Aggestam, F. (2025). *European Forest Policy and Governance: An Integrative Analytical Framework* (1st ed.). Routledge. https://doi.org/10.4324/9781003198987
Agrawal, A., Chhatre, A., & Hardin, R. (2008). Changing Governance of the world's forests. *Science, 320*(5882), 1460–1462. https://doi.org/10.1126/science.1155369
Angelsen, A. (2017). REDD+ as result-based aid: General lessons and bilateral agreements of Norway. *Review of Development Economics, 21*(2), 237–264. https://doi.org/10.1111/rode.12271
Angelsen, A., Martius, C., De Sy, V., Duchelle, A. E., Larson, A. M., Pham, T. T., & (Eds.). (2018). *Transforming REDD+: Lessons and new directions.* Center for International Forestry Research (CIFOR). https://doi.org/10.17528/cifor/007045
Art, B., & Visseren-Hamakers, I. (2012). Forest governance: A state of the art review. In B. Arts, S. Van Bommel, M. Ros-Tonen, & G. Verschoor (Eds.), *Forest-people interfaces* (pp. 241–257). Wageningen Academic Publishers. https://doi.org/10.3920/978-90-8686-749-3_15
Arts, B., & Buizer, M. (2009). Forests, discourses, institutions. *Forest Policy and Economics, 11*(5–6), 340–347. https://doi.org/10.1016/j.forpol.2008.10.004
Avelino, F., Wittmayer, J. M., Pel, B., Weaver, P., Dumitru, A., Haxeltine, A., Kemp, R., Jørgensen, M. S., Bauler, T., Ruijsink, S., & O'Riordan, T. (2019). Transformative social innovation and (dis)empowerment. *Technological Forecasting and Social Change, 145*, 195–206. https://doi.org/10.1016/j.techfore.2017.05.002
Baumgartner, F. R., & Jones, B. D. (2009). *Agendas and instability in American politics* (2. ed). The University of Chicago Press.
Baumgartner, F. R., & Mahoney, C. (2008). The two faces of framing: Individual-level framing and collective issue definition in the European Union. *European Union Politics, 9*, 435–449. https://doi.org/10.1177/1465116508093492
Baycheva-Merger, T., & Wolfslehner, B. (2016). Evaluating the implementation of the Pan-European criteria and indicators for sustainable forest management—A SWOT analysis. *Ecological Indicators, 60*, 1192–1199. https://doi.org/10.1016/j.ecolind.2015.09.009
Bianchi, C., Nasi, G., & Rivenbark, W. C. (2021). Implementing collaborative governance: Models, experiences, and challenges. *Public Management Review, 23*(11), 1581–1589. https://doi.org/10.1080/14719037.2021.1878777
Blythe, J. L., Armitage, D., Bennett, N. J., Silver, J. J., & Song, A. M. (2021). The politics of ocean governance transformations. *Frontiers in Marine Science, 8*, Article 634718. https://doi.org/10.3389/fmars.2021.634718
Borrass, L., Kleinschmit, D., & Winkel, G. (2017). The "German model" of integrative multifunctional forest management—Analysing the emergence and political evolution of a forest management concept. *Forest Policy Economics, 77*, 16–23. https://doi.org/10.1016/j.forpol.2016.06.028
Boulding, E. K. (1966). The Economics of the Coming Spaceship Earth. In H. Jarrett (Ed.), *Environmental Quality in a Growing Economy.* Johns Hopkins University Press.
Briers, S., Ackermann, A., Živojinović, I., Linser, S., Rinn, R., De Arano, I. M., Klapper, J., Wallius, V., Kriján, M. A., Orbe, L. B., Dominguez, M. G., Koivula, S., Van Langenhove, G., & Wieland, S. (2025). Pathways to bioeconomy development: A multi-regional perspective from Europe. *Ambio.* https://doi.org/10.1007/s13280-025-02297-4
Burns, S. L., Brukas, V., Krott, M., & Giessen, L. (2017). Trends in European forest policy research. *Forest Policy and Economics, 83*, 179–180. https://doi.org/10.1016/j.forpol.2017.08.013
Carlowitz, H. C. (2013). *Sylvicultura oeconomica The Future of Nature: Documents of Global Change* (L. Robin, S. Sörlin, & P. Warde, Eds., pp. 67–77). Yale University Press. https://doi.org/10.12987/9780300188479-009
Cashore, B., Auld, G., & Newsom, D. (2004). *Governing Through Markets: Forest Certification and the Emergence of Non-State Authority.* Yale University Press, JSTOR. www.jstor.org/stable/j.ctt1npqtr

Chaffin, B. C., Gosnell, H., & Cosens, B. A. (2014). A decade of adaptive governance scholarship: Synthesis and future directions. *Ecology and Society, 19*(3), art56. https://doi.org/10.5751/ES-06824-190356

Chazdon, R. L., Brancalion, P. H. S., Laestadius, L., Bennett-Curry, A., Buckingham, K., Kumar, C., Moll-Rocek, J., Vieira, I. C. G., & Wilson, S. J. (2016). When is a forest a forest? Forest concepts and definitions in the era of forest and landscape restoration. *Ambio, 45*(5), 538–550. https://doi.org/10.1007/s13280-016-0772-y

Clark, W. C., van Kerkhoff, L., Lebel, L., & Gallopin, G. C. (2016). Crafting usable knowledge for sustainable development. *The Proceedings of the National Academy of Sciences, 113*(17) 4570–4578. https://doi.org/10.1073/pnas.1601266113

Cleaver, F. (2017). *Development Through Bricolage: Rethinking Institutions for Natural Resource Management* (1st ed.). Routledge. https://doi.org/10.4324/9781315094915

Dryzek, J. S. (2005). *The Politics of the Earth: Environmental Discourses* (2nd ed.). Oxford University Press.

Dunn, W. N. (2017). *Public Policy Analysis: An Integrated Approach* (6th ed.). Routledge. https://doi.org/10.4324/9781315181226

Elliott, C., & Schlaepfer, R. (2001). Understanding forest certification using the advocacy coalition framework. *Forest Policy and Economics, 2*(3–4), 257–266. https://doi.org/10.1016/S1389-9341(01)00043-0

Elomina, J., & Pülzl, H. (2021). How are forests framed? An analysis of EU forest policy. *Forest Policy and Economics, 127*, Article 102448. https://doi.org/10.1016/j.forpol.2021.102448

Fazey, I., Fazey, A. J., Salisbury, J. G., & Lindenmayer, D. B. (2006a). The nature and role of experiential knowledge for environmental conservation. *Environmental Conservation, 33*, 1–10.

Fazey, I., Proust, K., Newell, B., Johnson, B., & Fazey, J. A. (2006b). Eliciting the implicit knowledge and perceptions of on-ground conservation managers of the Macquarie marshes. *Ecology and Society, 11*(1), 25.

Fischer, F. (2003). *Reframing Public Policy: Discursive Politics and Deliberative Practices* (1st ed.). Oxford University PressOxford. https://doi.org/10.1093/019924264X.001.0001

Folke, C. (2006). Resilience: The emergence of a perspective for social–ecological systems analyses. *Global Environmental Change, 16*(3), 253–267. https://doi.org/10.1016/j.gloenvcha.2006.04.002

Geels, F. W. (2002). Technological transitions as evolutionary reconfiguration processes: A multi-level perspective and a case-study. *Research Policy, 31*(8–9), 1257–1274. https://doi.org/10.1016/S0048-7333(02)00062-8

Geels, F. W. (2021). From leadership to followership: A suggestion for interdisciplinary theorising of mainstream actor reorientation in sustainability transitions. *Environmental Innovation and Societal Transitions, 41*, 45–48. https://doi.org/10.1016/j.eist.2021.10.021

Giurca, A., & Späth, P. (2017). A forest-based bioeconomy for Germany? Strengths, weaknesses and policy options for lignocellulosic biorefineries. *Journal of Cleaner Production, 153*, 51–62. https://doi.org/10.1016/j.jclepro.2017.03.156

Gordeeva, E., Wolfslehner, B., & Weber, N. (2025). 25 years of EU forest policy—An analysis. *Forests, 16*(2), 256. https://doi.org/10.3390/f16020256

Guston, D. H. (2001). Boundary organisations in environmental policy and science: An introduction. *Science, Technology, & Human Values, 26*(4), 399–408. JSTOR.

Haddad, C. R., Nakić, V., Bergek, A., & Hellsmark, H. (2022). Transformative innovation policy: A systematic review. *Environmental Innovation and Societal Transitions, 43*, 14–40. https://doi.org/10.1016/j.eist.2022.03.002

Hajer, M. (1997). *The Politics of Environmental Discourse: Ecological Modernisation and the Policy Process*. Oxford University Press.

Hall, P. A. (1993). Policy paradigms, social learning, and the state: The case of economic policymaking in Britain. *Comparative Politics, 25*(3), 275. https://doi.org/10.2307/422246

Henly-Shepard, S., Gray, S. A., & Cox, L. J. (2015). The use of participatory modeling to promote social learning and facilitate community disaster planning. *Environmental Science & Policy, 45*, 109–122. https://doi.org/10.1016/j.envsci.2014.10.004

Hoefer, R. (2022). The multiple streams framework: Understanding and applying the problems, policies, and politics approach. *Journal of Policy Practice and Research, 3*(1), 1–5. https://doi.org/10.1007/s42972-022-00049-2

Howlett, M., & Rayner, J. (2013). Patching versus packaging in policy formulation: Assessing policy portfolio design. *Politics and Governance, 1*(2), 170–182.

Jasanoff, S. (Ed.). (2004). *States of Knowledge* (0 ed.). Routledge. https://doi.org/10.4324/9780203413845

Kemp, R., & Loorbach, D. (1 C.E.). Transition Management: A Reflexive Governance Approach. *Reflexive Governance for Sustainable Development.* Cheltenham, UK and Northampton, MA, USA: Edward Elgar, pp. 103–130.

Kingdon, J. W. (1984). *Agendas.* Alternatives and Public Policies.

Korhonen-Kurki, K., D'Amato, D., Belinskij, A., Lazarevic, D., Leskinen, P., Nylén, E.-J., Pappila, M., Penttilä, O., Pitzén, S., Pykäläinen, N., Turunen, T., & Vikström, S. (2025). Transformative governance: Exploring theory of change and the role of the law. *Earth System Governance, 23*, Article 100230. https://doi.org/10.1016/j.esg.2024.100230

Krott, M. (2005). Forest policy analysis. *Forest Policy Analysis, 7*, 1–323. https://doi.org/10.1007/1-4020-3485-7

Krott, M., & Giessen, L. (2014). Learning from practices—Implications of the "practice based approach" for forest and environmental policy research. *Forest Policy and Economics, 49*, 12–16. https://doi.org/10.1016/j.forpol.2013.04.013

Large, D. (2024). The limits of co-production: Linking regulatory capacity to co-production of authoritative knowledge for environmental policy. *Science and Public Policy, 51*(5), 978–991. https://doi.org/10.1093/scipol/scae038

Lier, M., Köhl, M., Korhonen, K. T., Linser, S., & Prins, K. (2021). Forest relevant targets in EU policy instruments—Can progress be measured by the pan-European criteria and indicators for sustainable forest management? *Forest Policy and Economics, 128*, Article 102481. https://doi.org/10.1016/j.forpol.2021.102481

Lindahl, B. K. (2008). Frame analysis, place perceptions and the politics of natural resource management—exploring a forest policy controversy in Sweden [Doctoral Thesis, Swedish University of Agricultural Sciences]. In *Department of Urban and Rural Development.* http://pub.epsilon.slu.se/1810/1/Karins_Acta_Thesis_080829_slutversion.pdf

Linser, S., Wolfslehner, B., Bridge, S. R. J., Gritten, D., Johnson, S., Payn, T., Prins, K., Raši, R., & Robertson, G. (2018). 25 years of criteria and indicators for sustainable forest management: How intergovernmental C&I processes have made a difference. *Forests, 9*(9), 578.

Loorbach, D. (2010). Transition management for sustainable development: A prescriptive, complexity-based governance framework. *Governance, 23*(1), 161–183. https://doi.org/10.1111/j.1468-0491.2009.01471.x

Mahoney, J., & Thelen, K. (Eds.). (2009). A theory of gradual institutional change. In *Explaining Institutional Change* (1st ed., pp. 1–37). Cambridge University Press. https://doi.org/10.1017/CBO9780511806414.003

McDermott, C. L., Montana, J., Bennett, A., Gueiros, C., Hamilton, R., Hirons, M., Maguire-Rajpaul, V. A., Parry, E., & Picot, L. (2023). Transforming land use governance: Global targets without equity miss the mark. *Environmental Policy and Governance, 33*(3), 245–257. https://doi.org/10.1002/eet.2027

Meadows, D. (1999). *Leverage Points: Places to Intervene in a System.* The Sustainability Institute. https://donellameadows.org/archives/leverage-points-places-to-intervene-in-a-system/

Meadowcroft, J. (2009). What about the politics? Sustainable development, transition management, and long term energy transitions. *Policy Sciences, 42*, 323–340. https://doi.org/10.1007/s11077-009-9097-z

Miller, C. A., & Wyborn, C. (2020). Co-production in global sustainability: Histories and theories. *Environmental Science & Policy, 113*, 88–95. https://doi.org/10.1016/j.envsci.2018.01.016

O'Brien, K., Garibaldi, L., Agrawal, A., Bennett, E., Biggs, R., Calderón Contreras, R., Carr, E. R., Frantzeskaki, N., Gosnell, H., Gurung, J., Lambertucci, S. A., Leventon, J., Chuan, L., Reyes García, V., Shannon, L., Villasante, S., Wickson, F., Zinngrebe, Y., Périanin, L., & Zaccagnini, M. E. (2025). *IPBES Transformative Change Assessment: Summary for Policymakers* (Version v.10.1.0). Zenodo. https://doi.org/10.5281/ZENODO.11382230

Odum, E. (1953). *Fundamentals of ecology*. Saunders.

Ostrom, E. (2010). Beyond markets and states: Polycentric governance of complex economic systems. *American Economic Review, 100*(3), 641–672. https://doi.org/10.1257/aer.100.3.641

Ostrom, E. (1990). *Governing the Commons: The Evolution of Institutions for Collective Action*. Cambridge University Press.

Pamplona, L., Estellita Lins, M., Xavier, A., & Almeida, M. (2024). Transformative social innovation as a guideline to enhance the sustainable development goals' framework. *Sustainability, 16*(16), 7114. https://doi.org/10.3390/su16167114

Patterson, J., Schulz, K., Vervoort, J., Van Der Hel, S., Widerberg, O., Adler, C., Hurlbert, M., Anderton, K., Sethi, M., & Barau, A. (2017). Exploring the governance and politics of transformations towards sustainability. *Environmental Innovation and Societal Transitions, 24*, 1–16. https://doi.org/10.1016/j.eist.2016.09.001

Pülzl, H., Wydra, D., & Hogl, K. (2018). Piecemeal integration: Explaining and understanding 60 years of European Union forest policy-making. *Forests, 9*, 719.

Pülzl, H., Hogl, K., Kleinschmit, D., Wydra, D., Arts, B., Mayer, P., Palahi, M., Winkel, G., & Wolfslehner, B. (2013). *European Forest Governance: Issues at Stake and the Way Forward. What Science Can Tell Us 2*. European Forest Institute. https://www.efi.int/publications-bank

Rein, M., & Schön, D. A. (1996). Frame-critical policy analysis and frame-reflective policy practice. *Knowledge and Policy, 9*(1), 85–104.

Rhodes, R. A. W. (1997). *Understanding Governance: Policy Networks, Governance, Reflexivity and Accountability*.

Rutting, L., Vervoort, J., Mees, H., & Driessen, P. (2022). Strengthening foresight for governance of social-ecological systems: An interdisciplinary perspective. *Futures, 141*, Article 102988. https://doi.org/10.1016/j.futures.2022.102988

Sabatier, P. A. (1988). An advocacy coalition framework of policy change and the role of policy-oriented learning therein. *Policy Sciences, 21*(2), 129–168. https://doi.org/10.1007/BF00136406

Schlosberg, D. (2007). *Defining Environmental Justice: Theories, Movements, and Nature* (1st ed.). Oxford University PressOxford. https://doi.org/10.1093/acprof:oso/9780199286294.001.0001

Schmidt, V. A. (2008). Discursive institutionalism: The explanatory power of ideas and discourse. *Annual Review of Political Science, 11*(1), 303–326. https://doi.org/10.1146/annurev.polisci.11.060606.135342

Scoones, I., & Stirling, A. (2020). *The Politics of Uncertainty: Challenges of Transformation*. Routledge.

Scott, J. C. (2020). Seeing like a state: How certain schemes to improve the human condition have failed. *Yale University Press*. https://doi.org/10.2307/j.ctvxkn7ds

Singer, P. (1998). *The Expanding Circle: Ethics and sociobiology*. Clarendon Press.

Spies, M., & Alff, H. (2020). Assemblages and complex adaptive systems: A conceptual crossroads for integrative research? *Geography Compass, 14*(10), Article e12534. https://doi.org/10.1111/gec3.12534

Thiel, A. (2023). Polycentric Governing and Polycentric Governance. In F. Gadinger & J. A. Scholte (Eds.), *Polycentrism* (1st ed., pp. 98–120). Oxford University PressOxford. https://doi.org/10.1093/oso/9780192866837.003.0005

Turnheim, B., Berkhout, F., Geels, F., Hof, A., McMeekin, A., Nykvist, B., & Van Vuuren, D. (2015). Evaluating sustainability transitions pathways: Bridging analytical approaches to address governance challenges. *Global Environmental Change, 35*, 239–253. https://doi.org/10.1016/j.gloenvcha.2015.08.010

Turnhout, E., Metze, T., Wyborn, C., Klenk, N., & Louder, E. (2020). The politics of co-production: Participation, power, and transformation. *Current Opinion in Environmental Sustainability, 42*, 15–21. https://doi.org/10.1016/j.cosust.2019.11.009

Voß, J.-P., Bauknecht, D., & Kemp, R. (2006). Reflexive governance for sustainable development. In *Reflexive Governance for Sustainable Development*. https://doi.org/10.4337/9781847200266

Weible, C. M. (2006). An advocacy coalition framework approach to stakeholder analysis: Understanding the political context of California marine protected area policy. *Journal of Public Administration Research and Theory, 17*(1), 95–117. https://doi.org/10.1093/jopart/muj015

Weible, M. C., & Sabatier, P. A. (2018). *Theories of the Policy Process: Vol. 4th Edition*. Routledge.

Winkel, G. (2014). When the pendulum doesn't find its center: Environmental narratives, strategies, and forest policy change in the US Pacific Northwest. *Global Environmental Change, 27*, 84–95. https://doi.org/10.1016/j.gloenvcha.2014.04.009

Winkel, G., Blondet, M., Borras, L., Frei, T., Geitzenauer, M., Gruppe, A., Jump, A., De Koning, J., Sotirov, M., Weiss, G., Winter, S., & Turnhout, E. (2015). The implementation of Natura 2000 in forests: A trans- and interdisciplinary assessment of challenges and choices. *Environmental Science and Policy, 52*, 23–32.

Winkel, G., Lovrić, M., Muys, B., Katila, P., Lundhede, T., Pecurul, M., Pettenella, D., Pipart, N., Plieninger, T., Prokofieva, I., Parra, C., Pülzl, H., Roitsch, D., Roux, J.-L., Thorsen, B. J., Tyrväinen, L., Torralba, M., Vacik, H., Weiss, G., & Wunder, S. (2022). Governing Europe's forests for multiple ecosystem services: Opportunities, challenges, and policy options. *Forest Policy and Economics, 145*, Article 102849. https://doi.org/10.1016/j.forpol.2022.102849

Winkel, G., & Sotirov, M. (2016). Whose integration is this? European forest policy between the gospel of coordination, institutional competition, and a new spirit of integration. *Environment Planning C, 34*(3), 496–514, https://doi.org/10.1068/c1356j

Winkel, G. (2017). *Towards a sustainable European forest-based bioeconomy- assessment and the way forward. What Science Can Tell Us 8*. European Forest Institute. https://www.efi.int/publications-bank.

Wyborn, C., & Bixler, R. P. (2013). Collaboration and nested environmental governance: Scale dependency, scale framing, and cross-scale interactions in collaborative conservation. *Journal of Environmental Management, 123*, 58–67. https://doi.org/10.1016/j.jenvman.2013.03.014

Zeide, B. (1998). Another look at Leopold's land ethic. *Journal of Forestry, 96*, 13–19.

Part II
Governance in Practice—The EU Case

Chapter 3
Rethinking EU Forest Governance—From Fragmentation to Transformation

Filip Aggestam

Abstract EU forest governance is increasingly shaped by interacting pressures relating to climate change, biodiversity loss, and the push for a sustainable bioeconomy. Together with the EU's institutional complexity and the principle of subsidiarity, these pressures have resulted in a fragmented forest governance system that is characterised by overlapping mandates and sectoral misalignment. This chapter uses a novel Pressure–Response–Transformation (PRT) framework to demonstrate EU forest governance works and to determine whether recent policy changes indicate adaptive adjustments or significant structural changes in forest governance. Based on an extensive review of EU forest-related policies, including the EU Forest Strategy, the LULUCF Regulation, the Nature Restoration Regulation, and the EU Deforestation-Free Regulation, the chapter examines governance responses across three critical areas: climate change, biodiversity, and the bioeconomy. It identifies persistent challenges in aligning strategic and specific policy instruments, integrating actor coalitions, and overcoming institutional inaction. While the analysis reveals emerging signs of discursive and regulatory shifts, such as new monitoring frameworks and legally binding restoration targets, it concludes that transformation remains partial and uneven. Ultimately, the chapter argues that a more coordinated and reflexive governance approach will be necessary to navigate the trade-offs between sustainability and economic development. The chapter offers insights into the conditions under which fragmented adaptation could evolve into coordinated transformation, thereby contributing to the broader debate around sustainability transition and integrative forest policy.

F. Aggestam (✉)
ORBIS Green Policy and Governance, Forest Policy Research Network—European Forest Institute (FPRN-EFI), Vienna, Austria
e-mail: aggestam@orbis-gpg.com

F. Aggestam (ed.), *Governing Forests Through Transformative Change*, Advances in Global Change Research 70, https://doi.org/10.1007/978-3-032-19072-7_3

3.1 EU Forest Governance Through a Transformative Lens

European Union (EU) forest governance operates under increasing and often competing interests and pressures, including climate change, biodiversity loss, and the transition towards a circular bioeconomy. A growing body of literature has, in fact, examined the structure, development, and challenges of EU forest governance, providing valuable insights into its fragmented nature and cross-sectoral complexities. Recent reviews include Aggestam (2025), who provides an integrative analytical framework for understanding the transformation of forest policy across EU levels, and the recent European Forest Institute's (EFI) report on EU wood-based policy (Aggestam et al., 2025), which provides a comprehensive synthesis of policy dynamics between the forest-based sector and the bioeconomy. These works analyse the evolving governance structures and the pressures shaping the implementation of EU forest policies. Other notable contributions include reviews by Pülzl et al. (2013), Winkel (2017), Wolfslehner et al. (2020) and Stubenrauch et al. (2022), which highlight the long-standing tension between voluntary coordination mechanisms and binding regulatory instruments.

These studies demonstrate the limitations of soft policy instruments, ongoing policy incoherence, and the underutilisation of integrative tools for sustainable forest management (SFM) and bioeconomy integration. Despite lacking a common forest policy (due to the subsidiarity principle), the EU has become an area where these pressures interact, are addressed, and mediated, through different policy responses. Readers seeking more details on specific policies, implementation gaps and cross-sectoral linkages shaping the forest-based sectors in Europe are encouraged to read the sources listed above.

This chapter sets out to examine whether these pressures are facilitating transformative change in EU forest governance. It employs a novel take on the **pressure–state–response** (PSR) framework (see Box 3.1), where the focus is on transformation in an adapted **pressure–response–transformation (PRT)** approach. Rather than addressing governance change as linear or incremental, the PRT approach sets out to determine whether current responses to external pressures reflect systemic reorientation[1] or adaptive layering.[2] Specifically, it examines:

- the nature and scale of the pressures,
- the governance responses in terms of instruments, institutions, and actor configurations, and
- the extent to which these responses signal transformation (Biermann et al., 2012; Feindt & Weiland, 2018; Pahl-Wostl, 2009).

---

[1] Systemic reorientation refers to a shift in the logic, objectives, and institutional structure of a governance system, often involving a redefinition of policy paradigms and the realignment of institutional rules and actor constellations (Hall, 1993; Streeck & Thelen, 2005).

[2] Adaptive layering represents the process by which new policy instruments (or goals) are added to existing policy frameworks without replacing them, potentially improving flexibility but also risking fragmentation and incoherence (Howlett & Rayner, 2007; Oberthür & Stokke, 2011).

Box 3.1. PRT Versus PSR Frameworks

The pressure–state–response (PSR) framework, developed by the OECD and UNEP, has been widely used for environmental monitoring and reporting. It focuses on:

- **Pressure**: Human activities or natural events that represent stress on the environment.
- **State**: The condition or trends of environmental systems (e.g., forest cover, biodiversity indices).
- **Response**: Efforts to prevent, compensate, or adapt to changes in environmental quality.

While PSR is useful for tracking environmental outcomes, it tends to be descriptive and static, focusing on status indicators rather than institutional or policy change. In contrast, the **pressure–response–transformation (PRT) approach**, adapted from sustainability transitions and governance research, shifts the emphasis:

- **Pressure**: Systemic and external drivers (e.g., climate crises, biodiversity loss, bioeconomic factors).
- **Response**: Governance adaptations, including policy instruments, institutional change, and actor reconfiguration.
- **Transformation**: Structural and normative shifts in governance systems that change policy instruments, institutional arrangements, and sectoral priorities.

PRT is, therefore, more suited to analysing whether policy responses represent systemic change rather than incremental developments. It sets out to highlight the conditions under which fragmented adaptation might evolve into coordinated transformation (Aggestam, 2025; Feindt & Weiland, 2018; Pahl-Wostl, 2009).

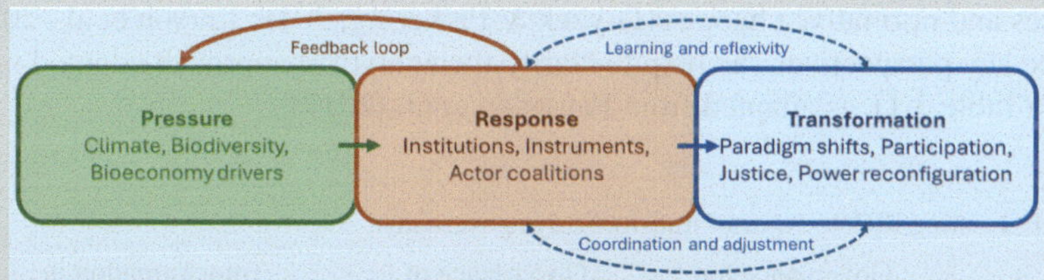

Although the PRT framework distinguishes between pressures and responses, it is important to recognise that policies designed in response to external drivers can themselves become sources of new governance pressure. For example, climate regulations, such as the EU Climate Law, can create new conflicts or trade-offs (e.g., relating to land use or carbon accounting), thereby putting extra pressure on forest governance systems. This reflects the iterative nature of transformation and the feedback loops in complex policy environments.

In addition to offering a novel analytical approach, this chapter aims to introduce how EU forest governance works, including its institutional architecture, policy instruments and coordination mechanisms. To this end, it begins by outlining the conceptual basis for governing transformation and the PRT approach, before analysing the key pressures shaping EU forest governance today (Sect. 3.2). Section 3.3 focuses on specific governance responses, including instruments such as the EU Forest Strategy and EUDR, to examine how institutional dynamics respond to these pressures. Section 3.4 synthesises these findings to assess whether EU forest governance is undergoing a transformation. Section 3.5 then explores the conditions needed to navigate such a transformation more deliberately.

3.1.1 Conceptualising Transformation in Forest Governance

In the context of EU forest governance, transformation in governance refers to fundamental changes in policymaking structures, logic, and practice (Olsson et al., 2014; Patterson et al., 2017; Voß & Bornemann, 2011). For EU forest governance, this would imply moving away from fragmented, sector-specific, governance towards more integrated and cross-sectoral coordination that aligns with sustainability objectives.

Transformative governance is, in this sense, not limited to the design of new policy instruments or institutions. It also requires a normative change, e.g., rethinking how forests are valued and governed, and a reconfiguration of actor networks, knowledge systems, and accountability mechanisms. As such, transformation covers a wide range of factors ranging from institutional change, social learning, participatory processes and normative changes (Dryzek & Pickering, 2018; Larson et al., 2021).

Three key perspectives can help to clarify the analytical approach adopted in this chapter (Table 3.1), as adapted from Patterson et al. (2017):

Table 3.1 Linking PRT to types of transformative governance

	Governance for transformation	Governance of transformation	Transformation in governance
Pressure	Framing the problem	Recognising urgency	Challenging dominant narratives
Response	Coordinating actors, enabling tools	Steering mechanisms and new instruments	Emerging institutional shifts
Transformation	Empowering actors, building capacity	Deliberate system change	Paradigm shifts, equity, and justice

1. **Governance for transformation**: creating conditions that support and enable systemic change.
2. **Governance of transformation**: coordinating and managing the transformation process itself.
3. **Transformations in governance**: changes within the governance system to allow for transformative change.

This chapter focuses on the second and third perspectives. It explores whether EU forest governance is actively steering a transformative agenda (e.g., through the Green Deal or new regulatory instruments) or if internal institutional and policy changes are occurring in response to environmental pressures. Elements of "governance for transformation" are also considered, such as when considering forest monitoring and cross-sectoral coordination.

Key characteristics of transformative governance in the forest-based sector include (Larson et al., 2021):

- **A paradigm shift**: A redefinition of policy goals and problem framings, e.g., moving from productive forest models to ecosystem-based multifunctionality.
- **Multi-level coordination**: Stronger vertical and horisontal interlinkages between EU institutions, EU Member States (EU MS) and non-state actors.
- **Policy coherence**: Aligning forest-relevant objectives across policy domains (e.g., biodiversity, climate, agriculture, energy and trade).
- **Institutional innovation**: New or reconfigured governance arrangements that, for example, address transboundary and cross-sectoral challenges.
- **Reflexivity and learning**: The capacity to provide continuous feedback, engage in deliberation, and make adjustments based on evolving knowledge and values.
- **Justice, participation, and accountability**: Ensuring inclusive policymaking processes, transparency, and consideration of distributional impacts, particularly when balancing policy objectives.

These characteristics contrast with adaptive governance, which tends to operate within existing institutional and normative frameworks. While adaptive governance focuses on incremental change and flexibility, transformation requires deeper structural changes, including the redistribution and reconfiguration of power and values, if needed (Feindt & Weiland, 2018; Meijerink & Huitema, 2010; Pahl-Wostl, 2009).

The EU is an interesting case study for these dynamics. More specifically, EU forest governance has been characterised by fragmentation, the prevalence of strategic policy instruments, and the lack of a common EU forest policy for a long time (Aggestam, 2025; Pülzl et al., 2018). However, recent developments, such as the Nature Restoration Regulation (EP & CoE, 2024a, 2024b), forest-carbon integration under LULUCF (Regulation, 2018; Urrutia et al., 2021) and efforts to revise coordination mechanisms (EC, 2019, 2021b), represents a potential shift.

This chapter uses the PRT framework (see Box 3.1 and Table 3.1) to assess whether recent responses to climate-, biodiversity-, and bioeconomy-related pressures reflect adaptive adjustments or more significant transformations. In doing so, it contributes to a wider debate around sustainability transitions by asking not only whether a transformation is occurring, but also what kind of transformation is emerging. The chapter also provides a structured overview of the key policy instruments, institutional developments, and stakeholder groups that are shaping EU forest governance today.

3.2 Structural Pressures that Shape and Define EU Forest Governance

Using the PRT framework, the next sections examine key pressures that are shaping EU forest governance. 1. climate change and carbon accounting (see Sect. 2.1), 2. biodiversity conservation and ecological integrity (see Sect. 2.2), and 3. the bioeconomy, circularity, and innovation (see Sect. 2.3). These interconnected pressures interact across policy domains and actor coalitions, often generating trade-offs and increasing governance tensions (Biermann et al., 2012; Feindt & Weiland, 2018). Rather than describing each pressure in isolation, the analysis highlights how these pressures have led to regulatory mandates, shifting discourses, and institutional challenges, requiring adaptive or transformative governance responses (Pahl-Wostl, 2009).

Section 3.2 focuses on these structural pressures and their origins in environmental, economic and strategic developments. Section 3.3 turns to the governance structures that have evolved at the EU level. These responses include specific EU policy instruments and coordination mechanisms that address forests. Moreover, it should be re-acknowledged that certain policy responses can generate new sources of pressure, namely, by creating implementation burdens or reinforcing institutional misalignments. This feedback dynamic reflects the non-linear and iterative nature of governance transformation. Each section identifies relevant EU policy instruments, evaluates the governance responses they have generated and considers whether these responses indicate a shift towards institutional realignment or continued policy fragmentation (Aggestam & Pülzl, 2020).

3.2.1 Climate Change and Carbon Accounting

The climate crisis is one of the most significant and transformative structural pressures facing Europe. Since the adoption of the European Green Deal in 2019 and the subsequent Fit-for-55 package, climate imperatives have become mainstream across EU policy areas. This has resulted in the adoption of legally binding targets for reducing greenhouse gas (GHG) emissions and the elevation of forests' role as carbon sinks and providers of renewable resources. This dual framing has arguably resulted in political momentum as well as complex governance tensions, particularly between afforestation and SFM for mitigation purposes and the use of bioenergy as part of the clean energy transition (Aggestam & Giurca, 2021; Lindner et al., 2017; Winkel, 2017).

3.2.1.1 Pressure: Climate Commitments and Emission Targets

Climate change is reflected in EU governance through multiple legislative targets and initiatives. The European Climate Law (EP & CoE, 2021b) sets out the goal of achieving climate neutrality by 2050, establishing interim targets such as reducing net GHG emissions by 55% by 2030. Forests are central to this agenda. Under the revised Land Use, Land Use Change and Forestry (LULUCF) Regulation (EP & CoE, 2023b), EU MS must increase their carbon sinks to remove 310 Mt CO_2e by 2030 collectively. This legal pressure is reinforced by accounting innovations, such as forest reference levels, net removal benchmarks, and an EU-wide GHG inventory system, which tracks progress.

3.2.1.2 Response: Governance Realignments and Monitoring Innovation

Policy responses to climate change pressures have focused significantly on expanding and institutionalising monitoring regimes. Carbon accounting has evolved into a core pillar of forest-related governance, supported by the development of digital monitoring tools, enhanced satellite-based reporting and the proposal for a common EU Forest Observation Framework (EC, 2023b). The Renewable Energy Directive III (EP & CoE, 2023a), a specific policy instrument, strengthens the connection between climate and forests by setting binding targets for renewable energy that rely on forest biomass. The directive includes sustainability and GHG-saving criteria for forest biomass as well as provisions to phase out support for electricity-only installations using forest biomass by 2026.

Other key instruments include the Effort Sharing Regulation (EP, 2018; EP & CoE, 2023c), a specific policy instrument that sets national emission reduction targets, and the Emissions Trading System (ETS) Directive (EP & CoE, 2018), which is currently under revision and is being expanded to include emissions from land use sectors. Together, these instruments work with the LULUCF and the Renewable Energy Directive III (RED III) frameworks to incorporate forest-related carbon dynamics into the broader climate policy mechanisms.

These responses indicate an increasing integration of forest and climate policies. However, these are often through technical and procedural innovations rather than deeper institutional restructuring. Furthermore, it should be noted that the tension between maintaining high forest carbon stocks and harvesting for biomass energy remains unresolved. Forests are caught between two competing demands as they are protected for their climate mitigation function and exploited as a renewable energy source (see Box 3.2).

Box 3.2. Balancing Carbon Sinks and Harvest

Efforts to enhance EU forest-based climate mitigation is increasingly faced with an important trade-off: Should forests be protected as carbon sinks or managed as renewable sources of biomass for energy and materials? Under the LULUCF Regulation, EU MS are under pressure to increase net carbon removals. At the same time, RED III promotes the use of biomass to meet renewable energy targets. This dual mandate represents conflicting incentives: (1) maximising biomass extraction would reduce long-term sink potential, while (2) maximising sinks would reduce the amount of available renewable feedstock. This challenge is complicated by methodological uncertainty in carbon accounting, such as temporal carbon debt and substitution effects as well as the lack of harmonised sustainability thresholds across EU MS (Ellison et al., 2011; Schulze et al., 2012; Searchinger et al., 2022). A transformative governance approach would require resolving these trade-offs (e.g., through integrated planning) as well as confronting the different approaches to what forest climate mitigation should achieve, such as immediate emissions offsets versus long-term carbon storage.

3.2.1.3 Transformation: Signs of Reframing but not Restructuring

The shift in framing from viewing forests primarily as timber suppliers to viewing forests as multifunctional ecosystems is a significant discursive shift. This reflects the growing importance of ecosystem services, particularly carbon sequestration, in EU forest governance. However, institutional changes remain limited. The lack of a common EU forest policy may arguably prevent coherence as governance remains spread across different policy domains.

Multiple studies have highlighted this misalignment. Aggestam and Giurca (2021), for example, argue that the Green Deal has reinforced existing sectoral silos by layering climate ambitions onto already fragmented forest governance structures. Similarly, recent evaluations have highlighted that, although the regulatory push around RED III and LULUCF is significant, it lacks cross-sectoral coordination with biodiversity and rural development policy objectives (Aggestam et al., 2025; Searchinger et al., 2018; Söderberg & Eckerberg, 2013).

Thus, while governance responses to climate change have introduced new policy instruments and frameworks, they largely reflect adaptive integration rather than transformative restructuring. The challenge ahead lies not only in improving forest-based carbon accounting but also in balancing competing objectives, such as mitigation, bioenergy and restoration, within a (hopefully) coherent governance system.

3.2.2 Biodiversity Conservation and Ecological Integrity

Biodiversity loss has emerged as another defining challenge for forest governance. The EU's commitment to halting and reversing biodiversity loss and ecosystem degradation, as set out in the EU Biodiversity Strategy for 2030 (EC, 2020c) and further operationalised through the Nature Restoration Regulation (EP & CoE, 2024a, 2024b), has introduced a regulatory pressure that establishes forests as a key component for ecosystem recovery. This pressure emphasises the ecological role of forests. It takes EU forest governance further into an area of legally binding instruments, contrasting with earlier approaches that have mainly been based on strategic and soft (non-binding) coordination.

3.2.2.1 Pressure: Legal Mandates and Ecological Decline

The biodiversity crisis has resulted in a shift from strategic policy instruments to specific policy instruments with formal obligations. While non-binding, the EU Biodiversity Strategy for 2030 sets ambitious targets, including the legal protection of 30% of the EU's land area and the restoration of at least 20% of degraded ecosystems by 2030. These targets are based on legislative frameworks such as the Birds Directive (EP & CoE, 2009) and the Habitats Directive (EP & CoE, 1992), which together form the basis of the Natura 2000 network, Europe's largest coordinated network of protected areas, including forest ecosystems.

The adoption of the Nature Restoration Law in 2024, a specific policy instrument, further increases this pressure. It is the first major EU legislation to mandate restoration rather than just conservation. It requires the EU MS to submit national restoration plans and align them with climate and biodiversity strategies. They must also monitor specific forest-related indicators, including deadwood, forest connectivity, carbon stocks and species diversity.

Other relevant instruments include the LIFE Programme (EP & CoE, 2021a), a strategic policy instrument supporting biodiversity and restoration projects, and the EU Taxonomy Regulation (EP & CoE, 2020), which establishes technical screening criteria linking sustainable finance to biodiversity performance. Provisions under the Common Agricultural Policy (CAP), particularly through rural development funds, indirectly influence forest biodiversity by providing incentives for agroforestry, afforestation, and forest environmental services (EP & CoE, 2021c, 2021d).

3.2.2.2 Response: Regulatory Turn and Policy Integration

Governance responses have included a regulatory shift towards enforceable obligations and measurable outcomes. The Nature Restoration Law and the Biodiversity Strategy signal the EU's intent to move beyond habitat protection towards active ecosystem recovery. These responses build on and align with the broader agenda of the European Green Deal (EC, 2019), aiming to restore ecosystem services and achieve climate neutrality.

However, implementing these legal frameworks faces several barriers (see Box 3.3). As noted in a recent EFI report, there are trade-offs between restoration and bioenergy production across EU MS (Aggestam et al., 2025). Forest owners and those involved in the bioeconomy have expressed concerns about land-use restrictions and conflicting targets. This reflects broader tensions between ecosystem-based approaches and the extraction of forest resources.

Even more, while the European Forest Strategy for 2030 acknowledges biodiversity concerns, it primarily addresses them through afforestation and reforestation goals. Critics have argued that this approach may lead to oversimplified ecological solutions and potentially reinforce monocultures, missing opportunities for more ecologically oriented restoration efforts (EC, 2021a, 2021b).

Box 3.3. The Spatial Mismatch

EU biodiversity policy principally operates through spatial targets, such as protecting 30% of EU land and restoring 20% of degraded ecosystems by 2030. However, these targets often overlook the spatial and ecological complexity of forest landscapes. Forest biodiversity is unevenly distributed, with many high-biodiversity areas lying outside designated Natura 2000 sites. Meanwhile, restoration obligations under the Nature Restoration Law focus on easily monitored metrics (e.g., deadwood and canopy cover) rather than ecological functionality. This spatial mismatch can lead to counterproductive outcomes, such as restoring degraded forests in one location while allowing continued degradation in other areas. To address this challenge, better ecological mapping, dynamic prioritisation tools, and integrated spatial planning across the domains of biodiversity, climate and forestry are required (Hermoso et al., 2022; Luyssaert et al., 2018; Strzelecka et al., 2021).

3.2.2.3 Transformation: Ambitious Mandates, Ambivalent Integration

From a transformational perspective, biodiversity policy represents a partial yet significant shift in approach. The Nature Restoration Regulation introduce legally binding, time-bound targets, essentially elevating biodiversity restoration to a central EU priority. However, governance structures remain fragmented and implementation mechanisms depend heavily on national discretion and priorities (e.g., the restoration target is an EU-wide target rather than a national target).

Aggestam and Pülzl (2020) have furthermore demonstrated that biodiversity targets commonly conflict with economic and climate-driven forest-related objectives, resulting in incoherent or weak responses. While the recent regulatory advances may signal a normative reframing of forests as a biodiversity-rich ecosystem rather than as an economic asset, institutional realignment remains limited.

The emergence of monitoring indicators and enforcement mechanisms suggests the potential for deeper change. However, the true measure of transformation will be whether biodiversity is consistently prioritised in land-use decisions across EU MS and whether new forms of coordination emerge to balance ecosystem integrity with climate and bioeconomy objectives.

3.2.3 Bioeconomy, Circularity and Innovation

The EU's transition to a sustainable and circular bioeconomy has placed yet one more pressure on forest governance, particularly in framing wood and forest biomass as key resources for decarbonisation, enhancing resource efficiency, and developing green industrial policies. This pressure is being promoted through strategic policy instruments such as the EU Bioeconomy Strategy (EC, 2012, 2018), the Circular Economy Action Plan (EC, 2015, 2020a) and the updated EU Industrial Strategy (EC, 2020b). These instruments have placed forest-based resources in a central position with regards to the EUs ecological transition. However, these developments have also intensified cross-sectoral demands on forests, highlighting the limitations of current governance structures in integrating socio-economic and environmental policy objectives.

3.2.3.1 Pressure: Strategic Demands for Wood-Based Innovations

Forests are now framed as a provider of renewable materials that can replace fossil-based materials in sectors such as construction, packaging, textiles, and energy. For instance, the EU Bioeconomy Strategy emphasises the sustainable mobilisation of biomass and innovation in value chains, from wood-based bioplastics to engineered timber products. This vision is reinforced by the Circular Economy Action Plan, which prioritises longer-lived products, material recovery, and life cycle thinking (EC, 2018, 2020a).

Furthermore, the Green Deal industrial strategy has increased interest in wood as a "future-proof" material for sustainable manufacturing and construction. Forest-based solutions are now considered not only climate-neutral but also climate-beneficial, capable of delivering carbon storage and substitution benefits simultaneously (EC, 2019).

3.2.3.2 Response: Policy Fragmentation and Innovation Platforms

Examples of governance responses include targeted R&I funding (such as Horizon Europe), the establishment of sectoral platforms (such as the Circular Bio-based

Europe Joint Undertaking), and the inclusion of sustainability criteria within broader regulatory frameworks (such as the EU Taxonomy Regulation and RED III). However, these responses often occur in isolation, with limited coordination across value chains. For example, the Single-Use Plastics Directive (EC & CoE, 2019), a specific policy instrument aimed at reducing marine litter, has impacted forest-based sectors by classifying cellulose-based bioplastics as conventional plastics, thereby undermining the sector's potential role in the bioeconomy. The EU Ecolabel Regulation (EC & CoE, 2010a, 2010b) and the Registration, Evaluation, Authorisation and Restriction of Chemicals (REACH) Regulation (EC & CoE, 2006), both specific policy instruments, impose specific sustainability, health, and safety requirements on wood-based products. Moreover, the Construction Products Regulation (EC & CoE, 2024a, 2024b), also a specific policy instrument, sets standards for the marketing of wood in the construction sector.

Value-chain fragmentation remains a significant barrier (Aggestam, 2025; Aggestam et al., 2025). The various policy regimes that govern product safety, industrial emissions, consumer protection and waste management create inconsistent expectations for forest-based industries. Furthermore, the EU forest strategies tend to underplay the industrial and commercial implications of the bioeconomy transition (Aggestam & Pülzl, 2018). Although the new EU Forest Strategy references bioeconomy ambitions, these lack a clear governance framework or integrated monitoring mechanisms, which limits their transformative potential.

Finally, a growing number of instruments are shaping the forest-based bioeconomy, particularly on the demand side. These include Green Public Procurement (GPP) and standardisation mechanisms (EC, 2008). GPP criteria in construction and public purchasing are increasingly referencing sustainability criteria for forest-based materials. However, implementation remains inconsistent and voluntary across EU MS, which limits its transformative potential (Krieger & Zipperer, 2022; Yu & Morotomi, 2022). Furthermore, standardisation processes under the European Committee for Standardization (CEN)[3] and the International Organization for Standardization (ISO)[4] establish technical standards for wood products that have an impact on market uptake. These standards remain secondary to governance discussions regardless of their regulatory implications (Ludvig & Weiss, 2013; Ludvig et al., 2021).

[3] CEN develops voluntary technical standards across various sectors, including forestry, wood products, and sustainability criteria. EU MS often adopt CEN standards, which can influence market access, safety requirements, and sustainability claims, even though these standards are not legally binding. In forest-related sectors, CEN has been involved in setting norms for wood durability, fire resistance, environmental labelling, and performance in construction (Ludvig & Weiss, 2013).

[4] ISO develops globally recognised technical standards through international consensus. In the forest-based sector, ISO standards influence sustainability certification (e.g. ISO 38200 for wood traceability), environmental management (e.g. ISO 14001), and life cycle assessments. While not enforceable, ISO standards are often integrated into national regulations or procurement frameworks and function as de facto benchmarks for industry compliance and product credibility (Cashore et al., 2004).

3.2.3.3 Transformation: Enabling the Bioeconomy Within Structural Constraints

The discourse around the bioeconomy introduces enabling conditions for transformation, considering its emphasis on innovation, systemic change, and material substitution. However, current responses tend to be adaptive adjustments within existing policy frameworks rather than genuine transformations. As Aggestam and Giurca (2022) argue, the political relevance of bioeconomy policies varies by actor and sector, often reinforcing the interests of those already in power and creating an uneven distribution of influence on policy implementation.

Taking on a transformation lens highlights the need to reconcile forest-based bioeconomy strategies with biodiversity, climate, and rural development objectives. This requires overcoming fragmentation, aligning incentives and developing cross-sectoral governance arrangements that can ensure long-term ecological integrity (see Box 3.4). In summary, while the bioeconomy introduces promising frameworks and investments, it risks becoming a technocratic solution unless it is embedded within broader governance innovations. Whether the bioeconomy becomes a driver of transformation or simply another pathway towards new contradictory policy objectives will depend on the EU's ability to integrate policy domains, redistribute influence between actors, and institutionalise sustainability safeguards along different value chain (Aggestam & Giurca, 2021; Aggestam et al., 2021; Winkel, 2017).

Box 3.4. Fragmented Accountability

Despite its intention to be integrative, the bioeconomy remains governed by uncoordinated accountability mechanisms. The sustainability criteria under RED III, the classification rules under the Taxonomy Regulation, and the plastic bans under the Single-Use Plastics Directive reflect different logics: climate mitigation, green finance, and pollution prevention, respectively. Without harmonised oversight, these instruments send inconsistent messages to forest-based industries, ultimately, undermining the legitimacy and effectiveness of these policy instruments (Kleinschmit et al., 2025; Winkel, 2017; Wolfslehner et al., 2020). In fact, as Töller et al. (2021) argue, the bioeconomy has functioned primarily as a conceptual umbrella rather than as a coherent governance framework, yielding limited results. In other words, instead of encouraging integrated action, bioeconomy strategies have largely absorbed existing sectoral objectives, thereby reinforcing fragmentation and conflicting logics across policy domains.

3.3 Institutional and Policy Responses in EU Forest Governance

The governance responses to the pressures shaping European forest policy have become increasingly complex and multilayered. This has been described in the literature as "policy layering" (Howlett & Rayner, 2007) and "institutional bricolage" (Cleaver, 2017), reflecting the accumulation of policy instruments and the lack of clear coordination between relevant institutional actors (Bauer & Knill, 2014). For instance, work by Hogl et al. (2012) and Peters (2018) suggests that such failures in coordination are commonly rooted in competing policy paradigms and sectoral interests, which can create barriers for integration and transformation. Within the EU, the difficulties of designing coherent governance frameworks are further increased by overlapping responsibilities between Directorates-General (DGs), differing national forest priorities, and the subsidiarity principle (see Box 3.5). Rather than replacing existing instruments or frameworks, new ones are added incrementally, resulting in an institutional space characterised by overlapping mandates, regulatory inconsistencies, and unclear implementation pathways.

Box 3.5. The Architecture of EU Forest Governance

Forest governance in the EU is shaped by a complex institutional framework that combines national competencies with an expanding range of EU policies that are relevant to forests. Although forest policy remains the responsibility of EU MS under the principle of subsidiarity, the European Commission plays an important coordinating role across sectors such as climate, the environment, agriculture and trade. Several DGs (including DG AGRI, DG ENVI, DG CLIMA and DG GROW) contribute to the development and implementation of policies that affect forests, directly or indirectly, often with limited horizontal coordination. The Standing Forestry Committee (SFC) has historically served as the primary consultative body between the Commission and EU MS, particularly regarding rural development and agricultural policy. Additional governance forums, such as the Civil Dialogue Group on Forestry and Cork, and inter-service groups within the Commission, are intended to facilitate stakeholder engagement and internal coherence. However, numerous studies have demonstrated that institutional fragmentation, competing policy priorities, and unclear mandates continue to limit the effectiveness of EU-level coordination of forest governance (for more information on EU forest-related policy instruments, see Pülzl et al. (2013, 2018) Wolfslehner et al. (2020) and Aggestam (2025)).

EU forest governance has led to the proliferation of policies, strategies, and institutional mechanisms. More specifically, the trends and patterns associated with policy-making in the EU's forest-related policy domains contribute to persistent fragmentation and raise important questions regarding the EU's capacity to develop a more coordinated, coherent, and transformational governance framework. Having this background in mind, this section will consider whether recent governance responses, covering strategic and specific policy instruments, reflect more than adaptive change and represent meaningful steps towards integrated forest governance, producing what the literature would describe as policy layering or institutional bricolage (Cleaver, 2017; Howlett & Rayner, 2007). Drawing on recent developments, such as the evolution of the EU Forest Strategies, the debate around revising the Standing Forestry Committee (SFC), and the adoption of new regulations like the EU Deforestation-Free Regulation (EUDR), this section will examine whether current governance responses are addressing underlying fragmentation or reinforcing it.

3.3.1 Strategic Instruments and the Evolution of the EU Forest Strategy

The EU Forest Strategy has become one of the main policy instruments for soft coordination on forest governance in the EU. Since its initial adoption in 1998 (EC, 1998), the Strategy has provided a shared, albeit non-binding, policy instrument for aligning forest-related priorities across EU MS (Pelli et al., 2012). However, its effectiveness has been limited by the principle of subsidiarity, which restricts EU competence in forestry, and by the absence of binding enforcement mechanisms.

The second Forest Strategy, adopted in 2013, aimed to offer a more integrated approach by acknowledging the cross-sectoral complexity of forest-related policies (EC, 2013; Wolfslehner et al., 2019). Nevertheless, Aggestam and Pülzl (2018) observe that the second iteration of the Strategy struggled to gain traction among EU MS and failed to address conflicts and trade-offs between different EU policy domains affecting the forest-based sector.

The latest and third iteration, the EU Forest Strategy for 2030, was introduced under the European Green Deal and set out a more ambitious agenda (Aggestam & Giurca, 2021; EC, 2021b). It sets out goals, including planting 3 billion trees by 2030, enhancing forest resilience and biodiversity, and supporting sustainable forest-based industries. Notably, it also proposes establishing a common EU Forest Monitoring Framework and revising the SFM indicators.

Even though the Strategy has evolved significantly, it remains a strategic policy instrument that relies on voluntary commitments and cooperation between EU MS. At the same time, it signals a stronger commitment to integration and a longer-term vision, however, it lacks the institutional setup for driving structural change. Moreover, its emphasis on afforestation has generated debate relating to the ecological soundness and social implications of large-scale tree-planting initiatives (Holl & Brancalion, 2020; Pörtner et al., 2021).

Box 3.6. The Missed Opportunity of the EU Forest Strategy?

Despite its ambitious goals and close alignment with the European Green Deal, the new EU Forest Strategy for 2030 still faces structural limitations. Previous evaluations have highlighted uneven uptake among EU MS, and the Strategy's non-binding nature offers limited leverage in driving policy implementation (Pelli et al., 2012; Wolfslehner et al., 2019). Furthermore, the Strategy's vision is often disconnected from the legally binding instruments of other policy domains affecting forests, such as the Renewable Energy Directive and the Common Agricultural Policy, which undermines policy coherence. Without dedicated funding, robust monitoring, or enforcement tools, the latest strategy will likely repeat the fate of its predecessors - influential in terms of rhetoric but marginal in terms of impact (Aggestam & Pülzl, 2018, 2020; Wolfslehner et al., 2020).

3.3.2 Institutional Restructuring: Reforming the Standing Forestry Committee

Established in 1989, the Standing Forestry Committee (SFC) has long served as a forum for dialogue between the European Commission and EU MS on forestry issues. However, its mandate was historically focused on supporting forestry measures under rural development and agricultural policy, with limited scope for addressing broader cross-sectoral priorities, such as biodiversity and climate change.

In line with the new EU Forest Strategy for 2030, the European Commission proposed reforming and renaming the SFC to create the Standing Forest and Forestry Expert Group (SFFEG). This reform aimed to modernise EU forest governance by broadening the group's mandate and updating its composition. Under the proposed decision (EC, 2023a), the new expert group would reinforce cooperation and multi-disciplinary exchange between the Commission and EU MS, provide advice on forest-relevant Union policy initiatives, including legislation and implementation under the EU Forest Strategy, include representatives from a wider range of national authorities responsible for climate, biodiversity, forest health, rural development and the bioeconomy, and foster coherence across EU policy domains as well as facilitate the exchange of good practices.

This proposal represents a transition from a narrower agricultural coordination body to a more comprehensive, cross-sectoral expert group, in line with the EU's broader Green Deal objectives. However, the European Economic and Social Committee has expressed concerns about the risk of diluting forest-specific expertise, the exclusion of civil society, and the potential administrative burden of expanded monitoring responsibilities (EESC, 2024). The renaming and restructuring have also generated debate regarding subsidiarity and the Commission's role in coordinating forest data collection and planning.

Although the reform has not been finalised, it demonstrates a wider initiative (or interest) to establish more integrated forest governance. Whether the proposed expert group will become an effective coordination mechanism or remain an advisory forum with limited impact depends on how it is implemented, its inclusivity, and its capacity to address and balance different policy objectives and stakeholder perspectives.

3.3.3 Specific Policy Instruments: The EU Deforestation-Free Regulation

Adopting the EU Deforestation-Free Regulation (EUDR) represents a significant step towards binding forest-related legislation (EP & CoE, 2023d). Replacing the EU Timber Regulation (EP & CoE, 2010a, 2010b), the EUDR requires companies to prove that commodities such as wood, soy, cocoa and coffee placed on the EU market are deforestation-free. The regulation applies to imports and domestically produced goods, introducing traceability requirements, risk assessment obligations, and potential sanctions for non-compliance.

The timber regulation and the EUDR mark a significant shift towards the use of specific and legally binding policy instruments in EU forest governance, with implications for trade policy, private sector compliance, and forest monitoring. However, it should be noted that the EUDR focuses on global deforestation rather than internal EU forest governance. As such, although it introduces legal obligations and enforcement capacity, its alignment with broader EU forest and biodiversity objectives is only partial.

3.3.4 Fragmentation, Uptake, and Sectoral (Mis)alignment

The challenges described below reflect broader critiques of fragmented forest governance at the EU level (Elomina & Pülzl, 2021; Pülzl et al., 2013; Winkel et al., 2013, 2022), where forest-focused and forest-related policies frequently operate in isolation from one another. It has been argued that this disconnect limits synergies and leads to contradictory outcomes, such as between afforestation subsidies under rural development programmes and biodiversity targets under Natura 2000. Literature on policy integration suggests that, without stronger coordination mechanisms, the EUs integrative ambitions are unlikely to be realised (Candel & Biesbroek, 2016).

These governance developments highlight persistent problems:

1. *The balance between strategic and specific policy instruments remains unresolved.* While strategies such as the EU Forest Strategy promote integrative thinking, they lack effectiveness. Hard policy instruments such as the EUDR have a greater impact, but they often operate outside the traditional silos associated with forest-focused policies.
2. *Policy uptake varies by sector*. Forest-based industries tend to prioritise economic and product-based regulations (e.g., RED III and the Construction Products Regulation), whereas environmental stakeholders focus on biodiversity and conservation frameworks. This asymmetry contributes to sectoral misalignment and undermines cross-policy coherence.
3. *Governance fragmentation remains a systemic barrier to transformation.* Despite calls for an EU Forest Policy, competencies remain scattered, coordination mechanisms are underdeveloped, and interests among actors and EU MS vary significantly. Neither the Inter-Service Group on Forestry nor the Civil Dialogue Group on Forestry and Cork have compensated for the lack of binding coordination mechanisms.

3.3.5 Governance Drift or Emerging Realignment?

From a PRT perspective, these governance responses show little movement from adaptation to transformation. Although there are signs of institutional innovation and reframing (e.g., viewing forests from a climate and biodiversity perspective), the underlying governance structures remain unchanged. The EU Forest Strategy continues to rely on voluntary alignment, the SFC reform remains unresolved, and the most robust instruments (e.g., EUDR) operate in policy domains next to, rather than within, core areas affecting EU forest governance. This implies that responses to rising pressures remain fragmented, reflecting a mixture of policy layering, sectoral insulation, and limited experimentation. Whether these tensions can be resolved in favour of more integrated and transformative forest governance depends on the political will to overcome institutional inaction to develop new forms of cross-sectoral coordination.

Research on sustainability transitions emphasises the significance of having policy environments, actor coalitions, and experimental spaces that facilitate institutional learning and innovation (Geels, 2002, 2021; Kemp & Loorbach, 2011; Loorbach, 2010). Without such conditions, governance reform will likely remain incremental and limited by existing power structures. As Jordan and Lenschow (2010) argue, environmental policy integration in the EU has often fallen short of expectations due to a lack of institutional ownership and enforcement tools as well as resistance from entrenched sectoral interests. These challenges remain particularly relevant in the context of EU forest governance, where integrating policies across biodiversity, climate, agriculture, and industry requires a political commitment and, more importantly, operational mechanisms that support coordination and accountability.

3.4 From Pressure to Transformation

The preceding sections have outlined some of the EU policy responses to climate change, biodiversity loss and the bioeconomy as they relate to forest governance. These responses include an expanding range of strategic and specific policy instruments, institutional restructuring efforts, such as reforming the SFC, and a greater focus on monitoring, data transparency, and cross-sectoral integration. This section considers whether these developments represent more than just adaptation or whether they represent a transformative shift in forest governance.

- From a PRT perspective, transformation is not only understood in terms of a change in tools or actors but as a structural reconfiguration of governance paradigms, institutions, and actor coalitions. Several indicators suggest the potential for such a shift. These include the adoption of legally binding frameworks, such as the LULUCF Regulation, the EUDR, and the Nature Restoration Regulation, as well as increasing efforts to integrate biodiversity and climate considerations into forest planning and monitoring. Additionally, proposed changes to forest data governance are being made through the forest monitoring framework. Arguably, these developments suggest a more systemic, multi-objective governance model, however, the transformative effects of these changes remain uneven.
- The literature on institutional change highlights that transformation is rarely linear. Instead, it often emerges through processes such as institutional layering, policy drift, and discursive reframing (Hall, 1993; Howlett & Rayner, 2007; Streeck & Thelen, 2005). In the EU context, new policy instruments are often introduced alongside older ones rather than replacing them, resulting in complex and contradictory governance structures. The coexistence of multiple forest-relevant frameworks across policy domains exemplifies this issue. Despite efforts to align forest-related policy instruments with climate and biodiversity goals, many governance structures remain siloed, with weak mechanisms for cross-sectoral integration (Aggestam, 2025; Pülzl et al., 2018).
- The tension between voluntary strategic instruments and binding legal mandates reinforces this pattern. Strategic instruments, such as the EU Forest Strategy and the Bioeconomy Strategy, promote integrative narratives and serve as relevant agenda-setting functions, however, their non-binding nature limits their enforcement and uptake by EU MS and institutions. On the other hand, specific instruments such as the EUDR, RED III, and the Nature Restoration Regulation have greater regulatory force, yet, they are frequently implemented within sector-specific silos, thereby reducing their ability to address the complex interdependencies between climate, biodiversity, and forest management (Winkel et al., 2021, 2022).
- Actor constellations remain largely state-centric, with limited engagement from non-state stakeholders in EU forest governance processes. Although some participatory mechanisms exist, such as the Civil Dialogue Group on Forestry and Cork, they often lack formal decision-making power or coordination mandates. This is problematic because transformative change usually requires pluralistic and

networked governance arrangements that integrate a wider range of perspectives and knowledge systems (Pahl-Wostl, 2009; Turnhout et al., 2016).

- Changes in the discourse around forests, such as the increased focus on ecosystem services, nature-based solutions, and forest resilience, suggest a shift in how forests are being framed. However, as Hall (1993) and Sabatier (2007) have demonstrated, discursive changes must be accompanied by changes in policy instruments and institutional configurations in order to represent a paradigm shift. While the increased coupling of climate and biodiversity objectives in policy language is promising, it remains uncertain whether this will translate into integrated implementation.

The current trajectory for EU forest governance indicates an interplay between incremental changes and some steps towards transformation. Although enabling conditions are in place, such as a possible new monitoring framework, expanded legislative mandates and public awareness, barriers remain. These include institutional inaction, legal fragmentation, insufficient enforcement capacity, and the EU MS reluctance to relinquish control over forests and forestry. It is important to recognise these ambiguities. For example, while adaptive governance, characterised by flexible, learning-based approaches, can deliver pragmatic short-term solutions, it does not inherently resolve or address deeper structural barriers or power asymmetries. Moreover, transformative governance would involve institutional realignment and normative reorientation (e.g., realigning forest governance with long-term sustainability goals and ecological limits). As such, it remains a question as to whether EU forest governance will move in this direction, which will ultimately depend on institutional reform, mobilisation of actor coalitions, strengthened deliberative processes and the ability to link strategic vision with operational coherence (Dryzek & Pickering, 2018; Voß & Bornemann, 2011). The coming years will be crucial in determining whether EU forest governance consolidates around this more ambitious vision or maintains a conservative response to increased ecological pressures. However, looking backwards, it is likely we will have more of the same (Pelli et al., 2012; Pülzl et al., 2018; Wolfslehner et al., 2019).

3.5 Navigating the Transformation

This chapter set out to demonstrate some of the key policy instruments that characterise EU forest governance and to highlight that the forest policy domain is in a state of flux, exhibiting signs of innovation and realignment while remaining characterised by long-standing institutional arrangements. This final section will briefly consider how these barriers could be addressed and the conditions necessary for a transformation. One key entry point for structural change resides in consolidating and using recent innovations, such as the proposed EU Forest Monitoring Framework (even though this initiative appears to be heading towards withdrawal at the time of writing).

If paired with institutional mandates and participatory processes, this policy instrument could become more than a technical solution and serve as an integrating node between different sectors. For example, studies have shown that robust monitoring systems can support learning, feedback, and accountability, but only if indicators are embedded within deliberative and adaptive governance structures (Gupta et al., 2010; Van Der Heijden, 2011).

The configuration of actor coalitions is also crucial. Current EU forest governance tends to prioritise formal and MS-centred representation, often sidelining civil society, scientific institutions and subnational actors. However, sustainability transition research emphasises the importance of polycentric governance arrangements that can adapt across levels and domains (Jordan et al., 2018; Ostrom, 2010). In practice, this suggests that transformation will require new mechanisms for building connections between stakeholder groups as well as for co-steering the design and implementation of forest-relevant policies.

Additionally, there needs to be a willingness to experiment with institutional structures. The EU has achieved some success in other sectors through experimental governance models, such as the Open Method of Coordination and iterative target-setting, where goals are agreed upon centrally. However, implementation is flexible and complemented by reporting, benchmarking, and peer learning (Sabel & Zeitlin, 2012). Applying this approach to EU forest governance could facilitate innovation in monitoring, restoration planning, and ecosystem services valuation while still aligning with Union-wide targets and objectives. At the same time, structural barriers across policy domains must also be addressed. There are well-documented misalignments between CAP subsidies, energy market incentives and biodiversity directives (Pe'er et al., 2019; Winkel et al., 2022). Without coordinated changes to land use, attempts to transform forest governance will remain incomplete and susceptible to policy conflicts. As Haas (2004) notes, epistemic communities and evidence-based policymaking can help to address such misalignments and conflicts, but only where institutional learning is matched with political willingness.

Finally, paradigm shifts require more than institutional design, it depends on debate and social learning. Forest governance is shaped by underlying narratives surrounding forest use, including who should manage the forests and which trade-offs are considered acceptable. Shifts in dominant problem framings, such as from productivism to ecosystem-based paradigms, are already evident in policy documents. However, making these shifts institutional depends on creating spaces for value conflicts, negotiations, and plural knowledge systems (Dryzek & Pickering, 2018; Turnhout et al., 2016). Thus, navigating this transition will require the development of capacities for this type of reflexive governance. These capacities would help adjust existing systems, question assumptions, and enable collective sense-making. The tools for this change, such as participatory foresight, experimental platforms, and integrative monitoring, are already in use. However, the question still remains as to whether any actual change will occur.

References

Aggestam, F. (2025). *European Forest Policy and Governance: An integrative analytical framework* (1st ed.). Routledge. https://doi.org/10.4324/9781003198987

Aggestam, F., & Giurca, A. (2021). The art of the "Green" deal: Policy pathways for the EU forest strategy. *Forest Policy and Economics, 128*, 102456. https://doi.org/10.1016/j.forpol.2021.102456

Aggestam, F., & Giurca, A. (2022). Implementing circular-bioeconomy principles across two value chains of the wood-based sector: A conceptual approach. *Land*, *11*(11). https://doi.org/10.3390/land11112037

Aggestam, F., Kacprzak, A., & Wollnik, R. (2021). *Circularity concepts in forest-based industries: Vol. ECE/TIM/SP/49*. United Nations Economic Commission for Europe.

Aggestam, F., & Pülzl, H. (2018). Coordinating the uncoordinated: The EU forest strategy. *Forests, 9*(3), 125. https://doi.org/10.3390/f9030125

Aggestam, F., & Pülzl, H. (2020). Downloading Europe: A regional comparison in the uptake of the EU forest action plan. *Sustainability*, *12*(10). https://doi.org/10.3390/su12103999

Aggestam, F., Weiss, G., Elomina, J., & Pülzl, H. (2025). *Taking a bird's eye view of EU wood-based policy: Untangling policy, institutional, and actor frameworks affecting the wood-based sectors* (knowledge to action). European Forest Institute. https://doi.org/10.36333/k2a07

Bauer, M. W., & Knill, C. (2014). A conceptual framework for the comparative analysis of policy change: Measurement, explanation and strategies of policy dismantling. *Journal of Comparative Policy Analysis: Research and Practice, 16*(1), 28–44. https://doi.org/10.1080/13876988.2014.885186

Biermann, F., Abbott, K., Andresen, S., Bäckstrand, K., Bernstein, S., Betsill, M. M., Bulkeley, H., Cashore, B., Clapp, J., Folke, C., Gupta, A., Gupta, J., Haas, P. M., Jordan, A., Kanie, N., Kluvánková-Oravská, T., Lebel, L., Liverman, D., Meadowcroft, J., Mitchell, R. B., Newell, P., Oberthür, S., Olsson, L., Pattberg, P., Sánchez-Rodríguez, R., Schroeder, H., Underdal, A., Camargo Vieira, S., Vogel, C., Young, O. R., Brock, A., Zondervan, R. (2012). Navigating the anthropocene: Improving earth system governance. *Science*, *335*(6074), 1306–1307. https://doi.org/10.1126/science.1217255

Candel, J. J. L., & Biesbroek, R. (2016). Toward a processual understanding of policy integration. *Policy Sciences, 49*(3), 211–231. https://doi.org/10.1007/s11077-016-9248-y

Cashore, B., Auld, G., & Newsom, D. (2004). *Governing through markets: Forest certification and the emergence of non-state authority*. Yale University Press; JSTOR. www.jstor.org/stable/j.ctt1npqtr

Cleaver, F. (2017). *Development through bricolage: Rethinking institutions for natural resource management* (1st ed.). Routledge. https://doi.org/10.4324/9781315094915

Dryzek, J. S., & Pickering, J. (2018). *The politics of the anthropocene* (1st ed.). Oxford University Press. https://doi.org/10.1093/oso/9780198809616.001.0001

EC. (1998). *COM 1998/649 final: A forestry strategy for the European Union*. European Commission.

EC. (2008). *COM(2008) 400 final: Public procurement for a better environment*. European Commission.

EC. (2012). *COM(2012) 60 final: Innovating for sustainable growth: A bioeconomy for Europe*. European Commission.

EC. (2013). *COM(2013) 659 final: A new EU forest strategy: For forests and the forest-based sector*. European Commission. http://ec.europa.eu/agriculture/forest/strategy/communication_en.pdf

EC. (2015). *COM(2015) 614: Closing the loop—An EU action plan for the circular economy*. European Commission.

EC. (2018). *COM(2018) 673 final: A sustainable bioeconomy for Europe: Strengthening the connection between economy, society and the environment*. European Commission.

EC. (2019). *COM(2019) 640 final: The European Green Deal*. European Commission. Retrieved April 08, 2025, from https://eur-lex.europa.eu/legal-content/EN/TXT/?uri=uriserv:OJ.C_.2020.104.01.0003.01.ENG

EC. (2020a). *COM(2020) 98 final: A new circular economy action plan: For a cleaner and more competitive Europe*. European Commission.

EC. (2020b). *COM(2020) 102 final: A new industrial strategy for Europe*. European Commission.

EC. (2020c). *COM(2020) 380 final: EU biodiversity strategy for 2030: Bringing nature back into our lives*. European Commission. Retrieved April 08, 2025, from https://eur-lex.europa.eu/legal-content/EN/TXT/?uri=CELEX:52020DC0380

EC. (2021a). *COM(2021) 572 final: Annex to the new EU forest strategy for 2030: Roadmap of the commission's action to implement the pledge to plant 3 billion additional trees by 2030 in the EU*. European Commission.

EC. (2021b). *COM(2021) 572 final: New EU forest strategy for 2030*. European Commission.

EC. (2023a). *COM(2023) 727: Proposal for a decision of the European parliament and of the council amending council decision 89/367/EEC setting up a standing forestry committee*. European Commission.

EC. (2023b). *COM(2023) 728 final: Proposal for a regulation on a monitoring framework for resilient European forests*. European Commission.

EC & CoE. (2006). Regulation 1907/2006: Registration, evaluation, authorisation and restriction of chemicals (REACH), establishing a European Chemicals Agency, amending directive 1999/45/EC and repealing council regulation (EEC) No 793/93 and commission regulation (EC) No 1488/94 as well as council directive 76/769/EEC and commission directives 91/155/EEC, 93/67/EEC, 93/105/EC and 2000/21/EC. In *OJ L 396*. Official Journal of the European Communities.

EC & CoE. (2010). *Regulation 66/2010: EU ecolabel*. Official Journal of the European Union.

EC & CoE. (2019). *Directive 2019/904: Reduction of the impact of certain plastic products on the environment*. Official Journal of the European Union.

EC & CoE. (2024). *Regulation (EU) 2024/3110: Laying down harmonised rules for the marketing of construction products and repealing regulation (EU) No 305/2011*. Official Journal of the European Union.

EESC. (2024). *C/2024/3380: Opinion of the European economic and social committee—Proposal for a regulation of the European parliament and of the council on a monitoring framework for resilient European forests—Proposal for a decision of the European parliament and of the council amending council decision 89/367/EEC setting up a standing forestry committee*. Official Journal of the European Union.

Ellison, D., Lundblad, M., & Petersson, H. (2011). Carbon accounting and the climate politics of forestry. *Environmental Science & Policy, 14*(8), 1062–1078. https://doi.org/10.1016/j.envsci.2011.07.001

Elomina, J., & Pülzl, H. (2021). How are forests framed? An analysis of EU forest policy. *Forest Policy and Economics, 127*, 102448. https://doi.org/10.1016/j.forpol.2021.102448

EP. (2018). *Regulation 2018/842: Binding annual greenhouse gas emission reductions by member states from 2021 to 2030 contributing to climate action to meet commitments under the Paris Agreement and amending Regulation (EU) No 525/2013*. Official Journal of the European Union. https://eur-lex.europa.eu/legal-content/EN/TXT/?uri=uriserv:OJ.L_.2018.156.01.0026.01.ENG

EP & CoE. (1992). *Directive 92/43/EEC: The conservation of natural habitats and wild fauna and flora*. Official Journal of the European Communities. https://eur-lex.europa.eu/legal-content/EN/TXT/?uri=CELEX%3A31992L0043

EP & CoE. (2009). *Directive 2009/147/EC: The conservation of wild birds*. Official Journal of the European Union.

EP & CoE. (2010). *Regulation 995/2010: Laying down the obligations of operators who place timber and timber products on the market*. Official Journal of the European Union.

EP & CoE. (2018). *Directive 2018/410: Amending Directive 2003/87/EC to enhance cost-effective emission reductions and low-carbon investments, and Decision (EU) 2015/1814*. Official Journal of the European Union.

EP & CoE. (2020). *Regulation 2020/852: Establishment of a framework to facilitate sustainable investment, and amending Regulation (EU) 2019/208*. Official Journal of the European Union.

EP & CoE. (2021a). *Regulation 2021/783: Establishing a programme for the environment and climate action (LIFE), and repealing Regulation (EU) No 1293/2013*. Official Journal of the European Union.

EP & CoE. (2021b). *Regulation 2021/1119: Establishing the framework for achieving climate neutrality and amending Regulations (EC) No 401/2009 and (EU) 2018/1999 ('European Climate Law')*. Official Journal of the European Union.

EP & CoE. (2021c). *Regulation 2021/2115: Establishing rules on support for strategic plans to be drawn up by member states under the common agricultural policy (CAP Strategic Plans) and financed by the European agricultural guarantee fund (EAGF) and by the European agricultural fund for rural development (EAFRD) and repealing Regulations (EU) No 1305/2013 and (EU) No 1307/2013*. Official Journal of the European Union.

EP & CoE. (2021d). *Regulation 2021/2116: Financing, management and monitoring of the common agricultural policy and repealing Regulation (EU) No 1306/2013*. Official Journal of the European Union.

EP & CoE. (2023a). *Directive 2023/2413: Amending Directive (EU) 2018/2001, Regulation (EU) 2018/1999 and Directive 98/70/EC as regards the promotion of energy from renewable sources, and repealing Council Directive (EU) 2015/652*. Official Journal of the European Union.

EP & CoE. (2023b). *Regulation 2023/839: Amending Regulation (EU) 2018/841 as regards the scope, simplifying the reporting and compliance rules, and setting out the targets of the member states for 2030, and Regulation (EU) 2018/1999 as regards improvement in monitoring, reporting, tracking of progress and review*. Official Journal of the European Union. https://eur-lex.europa.eu/eli/reg/2023/839/oj

EP & CoE. (2023c). *Regulation 2023/857: Amending Regulation (EU) 2018/842 on binding annual greenhouse gas emission reductions by member states from 2021 to 2030 contributing to climate action to meet commitments under the Paris Agreement, and Regulation (EU) 2018/1999*. Official Journal of the European Union.

EP & CoE. (2023d). *Regulation 2023/1115: Making available on the Union market and the export from the Union of certain commodities and products associated with deforestation and forest degradation and repealing Regulation (EU) No 995/2010*. Official Journal of the European Union.

EP & CoE. (2024). *Regulation (EU) 2024/1991: Nature restoration and amending Regulation (EU) 2022/869*. Official Journal of the European Union. https://eur-lex.europa.eu/eli/reg/2024/1991/oj

Feindt, P. H., & Weiland, S. (2018). Reflexive governance: Exploring the concept and assessing its critical potential for sustainable development. Introduction to the special issue. *Journal of Environmental Policy & Planning, 20*(6), 661–674. https://doi.org/10.1080/1523908X.2018.1532562

Geels, F. W. (2002). Technological transitions as evolutionary reconfiguration processes: A multi-level perspective and a case-study. *Research Policy, 31*(8–9), 1257–1274. https://doi.org/10.1016/S0048-7333(02)00062-8

Geels, F. W. (2021). From leadership to followership: A suggestion for interdisciplinary theorising of mainstream actor reorientation in sustainability transitions. *Environmental Innovation and Societal Transitions, 41*, 45–48. https://doi.org/10.1016/j.eist.2021.10.021

Gupta, J., Termeer, C., Klostermann, J., Meijerink, S., Van Den Brink, M., Jong, P., Nooteboom, S., & Bergsma, E. (2010). The adaptive capacity wheel: A method to assess the inherent characteristics of institutions to enable the adaptive capacity of society. *Environmental Science & Policy, 13*(6), 459–471. https://doi.org/10.1016/j.envsci.2010.05.006

Haas, P. M. (2004). When does power listen to truth? A constructivist approach to the policy process. *Journal of European Public Policy, 11*(4), 569–592. https://doi.org/10.1080/1350176042000248034

Hall, P. A. (1993). Policy paradigms, social learning, and the state: The case of economic policymaking in Britain. *Comparative Politics, 25*(3), 275. https://doi.org/10.2307/422246

Hermoso, V., Carvalho, S. B., Giakoumi, S., Goldsborough, D., Katsanevakis, S., Leontiou, S., Markantonatou, V., Rumes, B., Vogiatzakis, I. N., & Yates, K. L. (2022). The EU biodiversity strategy for 2030: Opportunities and challenges on the path towards biodiversity recovery. *Environmental Science & Policy, 127*, 263–271. https://doi.org/10.1016/j.envsci.2021.10.028

Hogl, K., Kvarda, E., Nordbeck, R., & Pregernig, M. (2012). *Environmental governance: The challenge of legitimacy and effectiveness*. Edward Elgar Publishing.

Holl, K. D., & Brancalion, P. H. S. (2020). Tree planting is not a simple solution. *Science, 368*(6491), 580–581. https://doi.org/10.1126/science.aba8232

Howlett, M., & Rayner, J. (2007). Design principles for policy mixes: Cohesion and coherence in 'New Governance Arrangements.' *Policy and Society, 26*(4), 1–18. https://doi.org/10.1016/S1449-4035(07)70118-2

Jordan, A., Huitema, D., Van Asselt, H., & Forster, J. (Eds.). (2018). *Governing climate change: Polycentricity in action?* (1st ed.). Cambridge University Press. https://doi.org/10.1017/9781108284646

Jordan, A., & Lenschow, A. (2010). Environmental policy integration: A state of the art review. *Environmental Policy and Governance, 20*(3), 147–158. https://doi.org/10.1002/eet.539

Kemp, R., & Loorbach, D. (2011). Transition management: A reflexive governance approach. *Reflexive governance for sustainable development* (pp. 103–130). Edward Elgar.

Kleinschmit, D., Giurca, A., Lehmann, R., Rodríguez, F., & Sinaga, H. (2025). Bioeconomy governance in the global south: State of the art and the way forward. *Forest Policy and Economics, 171*, 103403. https://doi.org/10.1016/j.forpol.2024.103403

Krieger, B., & Zipperer, V. (2022). Does green public procurement trigger environmental innovations? *Research Policy, 51*(6), 104516. https://doi.org/10.1016/j.respol.2022.104516

Larson, A. M., Mausch, K., Bourne, M., Luttrell, C., Schoneveld, G., Cronkleton, P., Locatelli, B., Catacutan, D., Cerutti, P., Chomba, S., Djoudi, H., Ihalainen, M., Lawry, S., Minang, P., Monterroso, I., Myers, R., Naito, D., Pham, T. T., Reed, J., Barletti, J. P. S., Sola, P., & Stoian, D. (2021). Hot topics in governance for forests and trees: Towards a (just) transformative research agenda. *Forest Policy and Economics, 131*, 102567. https://doi.org/10.1016/j.forpol.2021.102567

Lindner, M., Hanewinkel, M., & Nabuurs, G.-J. (2017). How can a forest-based bioeconomy contribute to climate change adaptation and mitigation? In G. Winkel (Ed.), *Towards a sustainable European forest-based bioeconomy—Assessment and the way forward*. European Forest Institute.

Loorbach, D. (2010). Transition management for sustainable development: A prescriptive, complexity-based governance framework. *Governance, 23*(1), 161–183. https://doi.org/10.1111/j.1468-0491.2009.01471.x

Ludvig, A., Braun, M., Hesser, F., Ranacher, L., Fritz, D., Gschwantner, T., Jandl, R., Kindermann, G., Ledermann, T., Pölz, W., Schadauer, K., Schmid, B. F., Schmid, C., Schwarzbauer, P., Weiss, G., Wolfslehner, B., & Weiss, P. (2021). Comparing policy options for carbon efficiency in the wood value chain: Evidence from Austria. *Journal of Cleaner Production, 292*, 125985. https://doi.org/10.1016/j.jclepro.2021.125985

Ludvig, A., & Weiss, G. (2013). Governing carbon efficiency. The international regime of standards in wooden construction. *Osterreichische Zeitschrift für Politikwissenschaft, 42*, 329–342. https://doi.org/10.15203/ozp.133.vol42iss3

Luyssaert, S., Marie, G., Valade, A., Chen, Y.-Y., Njakou Djomo, S., Ryder, J., Otto, J., Naudts, K., Lansø, A. S., Ghattas, J., & McGrath, M. J. (2018). Trade-offs in using European forests to meet climate objectives. *Nature, 562*(7726), 259–262. https://doi.org/10.1038/s41586-018-0577-1

Meijerink, S., & Huitema, D. (2010). Policy entrepreneurs and change strategies: Lessons from sixteen case studies of water transitions around the globe. *Ecology and Society, 15*(2), art21. https://doi.org/10.5751/ES-03509-150221

Oberthür, S., & Stokke, O. S. (Eds.). (2011). *Managing institutional complexity: Regime interplay and global environmental change.* The MIT Press. https://doi.org/10.7551/mitpress/9780262015912.001.0001

Olsson, P., Galaz, V., & Boonstra, W. J. (2014). Sustainability transformations: A resilience perspective. *Ecology and Society, 19*(4), art1. https://doi.org/10.5751/ES-06799-190401

Ostrom, E. (2010). Beyond markets and states: Polycentric governance of complex economic systems. *American Economic Review, 100*(3), 641–672. https://doi.org/10.1257/aer.100.3.641

Pahl-Wostl, C. (2009). A conceptual framework for analysing adaptive capacity and multi-level learning processes in resource governance regimes. *Global Environmental Change, 19*(3), 354–365. https://doi.org/10.1016/j.gloenvcha.2009.06.001

Patterson, J., Schulz, K., Vervoort, J., Van Der Hel, S., Widerberg, O., Adler, C., Hurlbert, M., Anderton, K., Sethi, M., & Barau, A. (2017). Exploring the governance and politics of transformations towards sustainability. *Environmental Innovation and Societal Transitions, 24*, 1–16. https://doi.org/10.1016/j.eist.2016.09.001

Pe'er, G., Zinngrebe, Y., Moreira, F., Sirami, C., Schindler, S., Müller, R., Bontzorlos, V., Clough, D., Bezák, P., Bonn, A., Hansjürgens, B., Lomba, A., Möckel, S., Passoni, G., Schleyer, C., Schmidt, J., & Lakner, S. (2019). A greener path for the EU Common Agricultural Policy. *Science, 365*(6452), 449–451. https://doi.org/10.1126/science.aax3146

Pelli, P., Aggestam, F., Weiss, G., Inhaizer, H., Keenleyside, C., Gantioler, S., Boglio, D., & Poláková, J. (2012). *Ex-post evaluation of the EU Forest Action Plan.* http://ec.europa.eu/agriculture/evaluation/market-and-income-reports/forest-action-plan-2012_en.htm. https://doi.org/10.13140/RG.2.1.4025.6163

Peters, B. G. (2018). *The politics of bureaucracy: An introduction to comparative public administration* (7th ed.). Routledge. https://doi.org/10.4324/9781315813653

Pörtner, H.-O., Scholes, R. J., Agard, J., Archer, E., Arneth, A., Bai, X., Barnes, D., Burrows, M., Chan, L., Cheung, W. L. (William), Diamond, S., Donatti, C., Duarte, C., Eisenhauer, N., Foden, W., Gasalla, M. A., Handa, C., Hickler, T., Hoegh-Guldberg, O., … Ngo, H. (2021). *Scientific outcome of the IPBES-IPCC co-sponsored workshop on biodiversity and climate change* (version 5). Zenodo. https://doi.org/10.5281/ZENODO.4659158

Pülzl, H., Hogl, K., Kleinschmit, D., Wydra, D., Arts, B., Mayer, P., Palahi, M., Winkel, G., & Wolfslehner, B. (2013). *European Forest Governance: Issues at stake and the way forward. What science can tell us 2.* European Forest Institute. https://www.efi.int/publications-bank

Pülzl, H., Wydra, D., & Hogl, K. (2018). Piecemeal integration: Explaining and understanding 60 years of European Union forest policy-making. *Forests, 9*, 719.

Regulation. (2018). *Inclusion of greenhouse gas emissions and removals from land use, land use change and forestry in the 2030 climate and energy framework, and amending Regulation (EU) No 525/2013 and Decision No 529/2013/EU.* Official Journal of the European Union.

Sabatier, P. A. (Ed.) (2007). *Theories of the policy process* (2nd ed.). Westview Press

Sabel, C. F., & Zeitlin, J. (2012). *Experimentalist governance.* Oxford University Press. https://doi.org/10.1093/oxfordhb/9780199560530.013.0012

Schulze, E., Körner, C., Law, B. E., Haberl, H., & Luyssaert, S. (2012). Large-scale bioenergy from additional harvest of forest biomass is neither sustainable nor greenhouse gas neutral. *GCB Bioenergy, 4*(6), 611–616. https://doi.org/10.1111/j.1757-1707.2012.01169.x

Searchinger, T. D., Beringer, T., Holtsmark, B., Kammen, D., Lambin, E. F., Lucht, W., Raven, P., & Van Ypersele, J.-P. (2018). Europe's renewable energy directive poised to harm global forests. *Nature Communications, 9*(1), 3741. https://doi.org/10.1038/s41467-018-06175-4

Searchinger, T., James, O., Dumas, P., Kastner, T., & Wirsenius, S. (2022). EU climate plan sacrifices carbon storage and biodiversity for bioenergy. *Nature, 612*(7938), 27–30. https://doi.org/10.1038/d41586-022-04133-1

Söderberg, C., & Eckerberg, K. (2013). Rising policy conflicts in Europe over bioenergy and forestry. *Forest Policy and Economics, 33*, 112–119. https://doi.org/10.1016/j.forpol.2012.09.015

Streeck, W., & Thelen, K. (Eds.). (2005). *Beyond continuity: Institutional change in advanced political economies*. Oxford University Press. https://doi.org/10.1093/oso/9780199280452.001.0001

Strzelecka, M., Rechciński, M., Tusznio, J., Akhshik, A., & Grodzińska-Jurczak, M. (2021). Environmental justice in Natura 2000 conservation conflicts: The case for resident empowerment. *Land Use Policy, 107*, 105494. https://doi.org/10.1016/j.landusepol.2021.105494

Stubenrauch, J., Garske, B., Ekardt, F., & Hagemann, K. (2022). European forest governance: Status Quo and optimising options with regard to the Paris climate target. *Sustainability, 14*(7), 4365. https://doi.org/10.3390/su14074365

Töller, A. E., Vogelpohl, T., Beer, K., & Böcher, M. (2021). Is bioeconomy policy a policy field? A conceptual framework and findings on the European Union and Germany. *Journal of Environmental Policy & Planning, 23*(2), 152–164. https://doi.org/10.1080/1523908X.2021.1893163

Turnhout, E., Dewulf, A., & Hulme, M. (2016). What does policy-relevant global environmental knowledge do? The cases of climate and biodiversity. *Current Opinion in Environmental Sustainability, 18*, 65–72. https://doi.org/10.1016/j.cosust.2015.09.004

Urrutia, C., Herold, A., & Gores, S. (2021). *2030 climate target plan: Review of land use, land use change and forestry (LULUCF) regulation* (ENVI workshop proceedings). European Parliament.

Van Der Heijden, J. (2011). Institutional layering: A Review of the use of the concept. *Politics, 31*(1), 9–18. https://doi.org/10.1111/j.1467-9256.2010.01397.x

Voß, J.-P., & Bornemann, B. (2011). The politics of reflexive governance. *Ecology and Society, 16*(2). JSTOR. http://www.jstor.org/stable/26268901

Winkel, G. (2017). *Towards a sustainable European forest-based bioeconomy—assessment and the way forward. What science can tell us 8*. European Forest Institute. https://www.efi.int/publications-bank

Winkel, G., Aggestam, F., Sotirov, M., & Weiss, G. (2013). *Forest policy in the European Union* (European forest governance: Issues at stake and the way forward, pp. 52–63).

Winkel, G., Lovrić, M., Muys, B., Katila, P., Lundhede, T., Pecurul, M., Pettenella, D., Pipart, N., Plieninger, T., Prokofieva, I., Parra, C., Pülzl, H., Roitsch, D., Roux, J.-L., Thorsen, B. J., Tyrväinen, L., Torralba, M., Vacik, H., Weiss, G., & Wunder, S. (2022). Governing Europe's forests for multiple ecosystem services: Opportunities, challenges, and policy options. *Forest Policy and Economics, 145*, 102849. https://doi.org/10.1016/j.forpol.2022.102849

Winkel, G., Sotirov, M., & Moseley, C. (2021). Forest environmental frontiers around the globe: Old patterns and new trends in forest governance. *Ambio, 50*(12), 2129–2137. https://doi.org/10.1007/s13280-021-01647-2

Wolfslehner, B., Aggestam, F., Pulzl, H., Hendriks, K., Lindner, M., Sotirov, M., Kulikova, E., Secco, L., Lovric, M., Nabuurs, G.-J., Hurmekoski, E., Pettenella, D., Arets, E., Baulenas Serra, E., Bozzano, M., Derks, J., Gatto, P., Linser, S., Masiero, M., … Giessen, L. (2019). *Study on progress in implementing the EU Forest Strategy. Final report*. Publications Office of the European Union. https://doi.org/10.2762/160685

Wolfslehner, B., Pülzl, H., Kleinschmit, D., Aggestam, F., Winkel, G., Candel, J., Eckerberg, K., Feindt, P., McDermott, C., Secco, L., Sotirov, M., Lackner, M., & Lazya Roux, J. (2020). *European forest governance post-2020* (From Science to Policy 10). European Forest Institute. https://doi.org/10.36333/fs10

Yu, C., & Morotomi, T. (2022). Impacts of green public procurement on eco-innovation: Evidence from EU countries. *Global Public Policy and Governance, 2*(2), 154–174. https://doi.org/10.1007/s43508-022-00041-2

Chapter 4
Implementation of Forest-related Policies in the European Union: Theoretical Background and Empirical Examples

Aron Buzogány, Daniela Kleinschmit, and Helga Pülzl

Abstract This chapter examines the implementation of forest-related policies in the European Union (EU) and the reasons for the significant variations in outcomes among EU Member States (MS). It contextualises EU forest policy within the broader field of implementation research, which has long debated how to assess compliance, as well as the respective merits of top-down, bottom-up and hybrid approaches. The chapter also considers the role of ideas and interests in shaping policy delivery. After outlining the core assumptions of each implementation perspective, the chapter illustrates their relevance using forest-related examples, including Natura 2000, the EU Timber Regulation, and emerging instruments such as the Nature Restoration Law and the Deforestation-Free Products Regulation. It demonstrates that EU forest policy research has traditionally favoured a top-down approach, focusing on legal compliance and enforcement, while studies of bottom-up street-level practice remain rare. Hybrid approaches are increasingly being used to capture how EU rules are adapted to domestic contexts. It argues that forest policy implementation research must now consider differentiated national capacities, the growing importance of non-state actors, and the simultaneous and sometimes contradictory introduction of multiple forest-related regulations. Strengthening context-sensitive, actor-centred, and multi-level analyses is essential for understanding how the EU's expanding regulatory framework can deliver effective and equitable forest governance.

A. Buzogány (✉) · H. Pülzl
BOKU University, Vienna, Austria
e-mail: aron.buzogany@boku.ac.at

D. Kleinschmit
Albert-Ludwigs-Universität Freiburg, Freiburg, Germany

H. Pülzl
European Forest Institute (EFI), Joensuu, Finland

F. Aggestam (ed.), *Governing Forests Through Transformative Change*, Advances in Global Change Research 70, https://doi.org/10.1007/978-3-032-19072-7_4

4.1 Introduction

This chapter focuses on the implementation of European forest policies in the European Union's Member States. Although studies of the implementation of forest policy in the US, such as Kaufman's "Forest Ranger" (1960), have become classics not only of administrative and organisational studies as well as early policy implementation research, similar attention has not been given to the implementation of forest policy in Europe. This is surprising given that adjacent fields, such as environmental policy, are among the most important ones for studying implementation in the EU's 27 Member States (MS) of the European Union (EU). Adopting a multidisciplinary approach that draws on disciplines such as public administration and international relations, research on Europeanisation—understood as the analysis of EU-triggered changes at a domestic level—has been the key conceptual approach applied in this field. The European Commission's activism has sparked interest in compliance with EU environmental policies, influencing MS sectoral policies and causing numerous implementation problems. This has made the policy field the one with the largest non-implementation gap (Börzel, 2003). While this gap has narrowed over the years (Börzel & Buzogány, 2019), the research field has advanced considerably in terms of both sophistication and geographic scope. It now includes not only EU MS but also candidate states and non-EU states that are affected directly or indirectly by EU forest policies (Overdevest & Zeitlin, 2018; Partzsch, 2021; Partzsch et al., 2023).

One reason for the considerable growth of the implementation literature in the EU is the disagreement among scholars about what exactly to should be measured. The field of measuring compliance is a field rife with methodological flaws, with different approaches yielding varying assessments of member-state performance. At the same time, the literature acknowledges only legally binding instruments as requiring follow-up on implementation, thus indirectly questioning the importance of non-legally binding strategies. This is based on the assumption that automatically legally binding instruments are inherently more important than non-legally binding ones. Formal and legalistic approaches typically rely on European Commission data, including the analysis of formal letters and infringement procedures issued by the Commission. Expert evaluations and case studies provide a more nuanced, albeit less comprehensive understanding of implementation realities. The temporal dimension also matters: whether researchers focus on cumulative implementation deficits over time or on specific year "snapshots" can substantially affect conclusions about member state compliance. Scholars have increasingly recognised that transposition is only one step in the broader implementation process (Falkner et al., 2007; Versluis, 2007). Qualitative studies have revealed significant discrepancies between the "law on the books" and "law in practice," as policy-makers fail to comply with both EU and national laws designed to enforce EU directives.

In the field of forest policy, most scholarly attention has focused on the governance of EU forest policies, while the systematic analysis of their implementation has been overlooked. These forest policy studies primarily contribute detailed empirical accounts, rather than in generating new theoretical insights that advance the field, although more recent work that also explores more theoretical arguments (Roux et al., 2025). This is consistent with Arts (2012) assertion that early forest policy research was atheoretical. More recent studies increasingly emphasise the importance of comparing policy implementation in different national contexts (Aggestam & Pülzl, 2020; Pecurul-Botines et al., 2025). This is especially important in the context of forestry, where the failure to implement EU forest-related policy instruments in MS can lead to a greater reliance on regulatory approaches. Indeed, over the last 5–10 years, a significant number of EU regulations have been introduced that are directly applicable in MS (e.g., the Regulation on Deforestation-Free Products (EUDR) and the Nature Restoration Regulation (NRL)), eliminating the need for individual transposition. However, no one has yet systematically examined how the implementation literature is taken up in forest policy.

The primary aim in this chapter is to make the implementation literature accessible to a wider forest-related audience by providing a general understanding of the key pillars of the policy implementation literature and illustrating how these concepts have been applied in forest policy research within the EU. The chapter is structured to reflect important developments in the study of policy implementation by focusing on the top-down, bottom-up, and hybrid perspectives, which have emerged sequentially. Each section begins with a brief description of the given perspective on implementation studies, followed by examples from the forest-related research area (for details, see Buzogány & Pülzl, 2024; Pülzl & Treib, 2006). The chapter then discuss the roles of ideas and interests in forest policy-related implementation studies and conclude by summarising and highlighting potential new pathways in forest policy implementation in Europe.

4.2 The Top-Down Perspective

The top-down approach to policy implementation emerged from the realisation that the "great expectations" of policy-makers often encounter practical and organisational barriers when transitioning from political decision-making to administrative action. Pressman and Wildavsky (1973) identified imprecise orders and a lack of control over lower levels of the organisation as major contributors to implementation failures. This perspective assumes a clear division between politics and administration, with bureaucratic actors responisble for faithfully executing policies designed by their political superiors.

Historically, the top-down approach has been most closely aligned with classical Weberian and Wilsonian theories of public administration theories, as well as the central planning ethos of the New Deal and the planning euphoria of the 1970s. These frameworks view governments as the primary actors in the policy process, equipped with both the mandate and capacity to design policies for the public good. From this perspective, the role of policy analysts is to provide actionable advice that helps governments to overcome implementation challenges. Similarly, top-down implementation focuses on hierarchical control and clear attribution of responsibilities.

Pressman and Wildavsky (1973) argued that successful implementation depends on bureaucratic procedures ensuring adherence to policy goals. Sabatier and Mazmanian (1979) identified six conditions for effective implementation: clear objectives, a robust causal theory, a sound legal framework, committed officials, supportive interest groups, and favourable socioeconomic conditions. From an EU perspective, the first wave of implementation studies in the 1990s focused primarily on analysing whether centrally agreed policy objectives were being met at national and regional levels within the MS, and from a comparative perspective (Pülzl & Treib, 2006).

Taken together, the top-down perspective has several advantages, particularly in providing practical advice to policy practitioners in various fields. Nevertheless, it also has significant limitations. It tends to overemphasise administrative processes and neglect the role of politics in shaping implementation outcomes. This focus on centralised control often fails to account for complex situations in which local implementers and community actors play a role and can significantly hinder implementation.

4.2.1 Examples of Forest-related Studies

Top-down approaches to implementation are frequently adopted in forest policy research at the EU level. One of the policy areas most often studied in relation to this hierarchical understanding of policy implementation is Natura 2000. A major reason for the particular scientific interest in this policy area may be the recognition that the national implementation of Natura 2000 varies across the EU (Galiana et al., 2013; Hogl et al., 2016). Many studies that take a top-down approach focus on particular aspects of the Natura 2000 Directive, such as its impact on the management of beech forests in MS (Winter et al., 2014), or the different ways in which compensation and restrictions for private forest owners are implemented differently (Sarvašová et al., 2019).

The focus is partly geo-political basis, for example, when considering whether post-communist countries might face more substantial or least different implementation challenges to long-term EU MS with distinct political cultures (Křenová & Kindlmann, 2015). Articles that adopt a top-down implementation perspective frequently refer to existing institutions with implementation responsibilities (and power). Apart from national and partly sub-national governments, the central actors identified in these studies are the European Commission and the European Court of Justice. The latter fosters implementation through compliance and enforcement actions, under the threat of infringement procedures, and through its legal authority which should provide clarity of meaning. However, some actors, such as those in environmental coalitions, claim that top-down implementation has been ineffective due to the resistance of forest and land use actors. Consequently, they have adopted a more cooperative approache (Sotirov et al., 2021), the top-down regulatory approach continued to inform their actions, repeating legal procedures to enforce the protection of Natura 2000 sites in forests (Borrass, 2014; Sotirov et al., 2015a, 2015b). The limited effectiveness of top-down implementation has also been observed in reviews of other policies, such as the EU Timber Regulation. Here, authors argue that it was the "dissatisfied member states and interest groups from the forestry and land use coalitions" weakened its implementation (Sotirov et al., 2021: 2165).

In summary, the top-down approach to policy implementation emphasises hierarchical control and clear bureaucratic procedures to ensure that policy goals are met. However, it tends to overlook the complexities and political dynamics that influence implementation outcomes. In EU forest policy research, and specifically in studies focusing on the implementation of Natura 2000, this approach has been both employed and criticised for its limited effectiveness. National differences, resistance from local stakeholders, and enforcement challenges reveal the shortcomings of purely top-down regulatory strategies.

4.3 The Bottom-Up Perspective

The bottom-up approach emerged as a critique of the top-down perspective, emphasising the socio-organisational dynamics of policy implementation. It starts from the perceptions and actions of local implementers and and those affected by the policy. It challenges the state-centric and hierarchical focus of the top-down approach, arguing that vague policy goals, competing aims, and complex networks of actors hinder the selection of optimal policy options (Hill & Hupe, 2009). From a bottom-up perspective, effective implementation is achieved by adapting policy goals to local contexts through interaction betweeng various stakeholders. Policy implementation and development are viewed as interdependent processes, with local power structures playing a decisive role. A key methodological innovation of this perspective is "backward mapping" (Elmore, 1979), which begins the analysis at the street level and traces it back to policymakers. Hjern and Porter (1981) broadened the focus from individual organisations to complex "implementation structures" involving numerous stakeholders.

The bottom-up approach owes a great deal to research on "street-level bureaucracies" (Lipsky, 1980), which emerged around the same time and focused on policy delivery. Michael Lipsky's work introduced a significant change in perspective, demonstrating that bureaucrats in the lower echelons of state bureaucracies have different incentives and perspectives when it comes toimplementing (or not implementing) policies. His work provided evidence that bureaucratic autonomy and discretion are essential not only for implementation, but also for the provision of adequate services. As "street-level bureaucrats" work directly with those affected by the daily norms, they can exercise the discretion enshrined in these policies to make delivery more efficient and targeted.

An eminent example of this line of research originates from forestry research and predates Lipsky's well-known account. Without being classified as such, Kaufman's "Forest Ranger" describes a bottom-up implementation approach par excellence (Kaufman, 1960). In his account, although overarching policies originate from the centralised US Forest Service, forest rangers themselves are granted significant autonomy and discretion in decision-making at the local level. They often have to navigate complex situations, balancing environmental protection, public interests, and resource management using their local knowledge and judgment. This flexibility in decision-making and the way in which the rangers adapt to the specific conditions of their environments, demonstrates a bottom-up style of implementation, where the frontline workers hold considerable influence over how policies are enacted. More than six decades after its publication, Kaufman's classic continues to receive attention (Koontz, 2007; Tipple & Wellman, 1991). Many other well-known classics in natural resource management, well known to forest researchers, emphasise the participation of local communities in decision-making and policy adaptation, and can also be considered part of this field (Agrawal & Gibson, 2001; Ostrom, 1990). From an EU studies perspective, the bottom-up approach became particularly visible in analyses of environmental policies centred on the so-called "stickiness" of national policy traditions and styles. This was seen as both hindering the downloading of objectives from the European level (Héritier, 1996) and its "uploading" (Börzel, 2002).

4.3.1 *Examples of Forest-related Studies*

The bottom-up perspective is rarely used as a starting point in EU forest-related implementation studies, unlike the top-down perspective. One study that follows a bottom-up approach is Borrass (2014) work, which uses a street-level bureaucracy approach. Borrass compares the implementation of the European Habitats Directive in case studies from Germany and the UK, focusing particularly on the role of local administrative officers in decision-making and the shaping of distinct practices within specific spatial and temporal contexts. His analysis revealed that the strategies employed by individual local officers influence the variations in management and species requirements observed in local forestry practices. Variations in implementation are especially evident when policies at the national and sub-national policies are ambiguous and open to interpretation. Additionally, institutional constraints can broaden or limit the scope and scale of adjustments (ibid.).

Another example of a bottom-up approach is a quantitative study of the implementation of the European Timber Regulation (EUTR) in Germany, which focused on wood-importing operators (Köthke, 2020). Although this study is theoretically grounded in change theories, the author investigated how the different characteristics of operators, such as operator size and sector affiliation, affect awareness, knowledge, compliance, and changes in import activities. The author concludes that the greatest gaps in the operators' practical implementation are "insufficient awareness and knowledge" (Köthke, 2020: 12). Following other studies, such as Leipold (2017), she therefore suggests providing targeted information and capacity building, as well as carrying out compliance checks.

In summary, unlike the top-down approach, the bottom-up approach mainly emphasises the role of street-level bureaucrats, who have a significant impact on how policy implementation is understood as a form of policy delivery. Despite its descriptive and reconstructive focus on actual policy delivery, this more empirical approach is not widely used in implementation studies in EU forest policy research. Therefore, combining this with other theoretical perspectives seems appropriate, as we will demonstrate in the next section.

4.4 Hybrid Perspectives

In recognition of the limitations of both top-down and bottom-up approaches, several leading implementation researchers have proposed a hybrid perspective that combines elements of both (Sabatier & Mazmanian, 1979). This approach aims to strike a balance between the hierarchical control emphasised by top-down methods and the adaptability and local discretion championed by scholars of the bottom-up approach. The hybrid perspective acknowledges that effective implementation requires coordination across multiple levels of governance. While clear policy directives and hierarchical structures are necessary, it posits that local implementers must also have the flexibility to adapt policies to specific contexts. Hybrid models emphasise collaborative governance, in which multiple actors at different levels work together to achieve policy goals. In the European context, this was also acknowledged from the early 2000s onwards, when scholars began to consider more aspects of comparative research, albeit at the expense of being able to identify decisive factors, such as those that contribute to successful policy implementation. Falkner et al. (2007) developed a "world of compliance" typology based on a large-scale yet qualitative approach to comparative policy implementation. This typology highlights country-specific conditions as explanatory factors, which is reminiscent of administrative culture approaches.

4.4.1 Examples of Forest-related Studies

In the context of the implementation of forest-related policies, hybrid approaches can be seen in multi-level governance frameworks, such as those seen in the European Union's Forest Strategy. These frameworks strike a balance between centralised EU directives and adaptations by MS to address local forest conditions and stakeholder needs (Pietarinen et al., 2025). Studies on multi-level governance provide valuable insights into hybrid implementation. For instance, EU forestry policies frequently necessitate a balance between directives and local adaptations by MS. Several studies have emerged that take a hybrid perspective on implementation, particularly with regard to the EU Timber Regulation. Leipold (2017) followed Barrett (2004) by building her analysis on an "interactive and negotiative model of implementation" (Leipold, 2017: 256) when examining stakeholders' perceptions of the EUTR's implementation performance in Germany. This analysis covers both the translation of the EUTR into German law and its compliance assessment. She concludes that the performance assessments vary among the interviewed stakeholders, depending on their understandings of the goal. Therefore, implementation may have different effects even when applied consistently across MS.

Another study of the implementation of the EUTR used Europeanisation theory and the Advocacy Coalition Framework (ACF) to examine the process in Romania (Gavrilut et al., 2021). Building on an argument also made in the context of environmental governance more broadly (Buzogány, 2015), the authors argue that Romania, as a CEE state, Romania may encounter greater challenges than other EU MS in implementing the legislation. This is partly offset by the civil society, environmental NGOs, and "responsible economic operators" (Gavrilut et al., 2021: 42), who use EU structures to facilitate the transition. This finding confirms the outcomes of other studies, which demonstrate that policy implementation can also be bolstered by NGOs and through protests (Davidescu & Buzogány, 2021; Dimitrova & Buzogány, 2014). In contrast, Logmani et al. (2017) show that international and domestic environmental NGOs, as well as the EU Commission, have only limited opportunities to influence political processes. The authors focus on the implementation of Natura 2000 in Poland, specifically the prominent case of the Białowieża forest. They demonstrate that Poland's production-oriented bureaucracy strategically redefined the conservation instrument during implementation. They conclude that, in the implementation phase, EU rules are customised to domestic political realities, and that more focus is therefore needed on bottom-up approaches in implementation studies. Aggestam and Pülzl (2020) have used a hybrid approach to analyse the implementation of the EU Forest Action Plan, adopting a dynamic view of the Europeanisation process whereby up- and downloading of policy positions and preferences are exchanged between the EU and its MS. They show that significant variations in the implementation of a non-legally binding policy instrument become apparent due to the varied strategies of the MS, in contrast to legally binding policies. Their theoretical contribution to the literature on theimplementation of soft policy instruments lies in their call for a "more nuanced and varied approach to the implementation of soft policy instruments" at the national level. They therefore propose adding several new categories to those suggested by Börzel (2002) for studying non-binding policy instruments.

In summary, the hybrid perspective on policy implementation synthesises elements of the top-down and bottom-up approaches in order to address conceptual weaknesses, such as failing to link implementation to policy formulation processes and studying it in segments, by rejecting the classic policy stage model (Jann & Wegrich, 2017). Most forest policy scholars have employed different theoretical approaches to perform these hybrid forms of analysis. Of particular note is the Advocacy Coalition framework, although some have also contributed to the broader policy science literature.

4.5 Ideas versus Interests in Implementation

Ideas influence the agenda-setting process, the formulation of policies and their implementation by determining which representations of problems and potential solutions are recognised by policymakers (Hall, 1993; Surel, 2000). Discourses are understood as the construction of meanings and interpretations of policy problems, operating both strategically and reflectively (Hajer, 1995; Van Hulst et al., 2025). They influence policy implementation through three key aspects: (a) the interconnectedness of frames throughout all stages of the policy process, (b) the influence of informal norms, and (c) the effect of discursive contestations and resistances.

Firstly, discourses define the scope for transformative change in the implementation phase by embedding meanings in policy design. Frames established in earlier stages set the parameters within which implementation occurs (Fischer & Forester, 1993). However, actors involved in implementation can reinterpret policies, sometimes altering their original framing. This highlights that decision-making is an ongoing process of meaning construction, spanning both policy formulation and implementation. Secondly, professional cultures and underlying norms significantly influence policy implementation. Bureaucrats, judicial officials, and frontline professionals interpret policies through the lens of their institutional and professional contexts. These interpretations affect how institutional resources and policy instruments are understood and applied. Consequently, implementation is not merely a technical process; actors' professional norms and values also influence it. Thirdly, policy implementation is inherently contested, with different stakeholders competing to influence policy outcomes. While some stakeholders aim to maintain the original policy intent, while others seek modifications or even obstruct implementation. Discursive contestations continue beyond policy adoption as competing actors negotiate alternative framings. For instance, courts may interpret policies differently depending on legal frameworks and the perspectives of individual judges, which can lead to variations in implementation across institutional levels. Unlike the ideas approach, interest-based approaches of implementation can be approached from a top-down, bottom-up, or hybrid perspective.

4.5.1 Examples of Forest-related Studies

A discursive-institutional approach was applied in a study on the implementation of Natura 2000 in Spain to examine how the adoption of a specific rationality was translated into discourses and institutionalised (Pecurul-Botines et al., 2014, 2019). The authors conclude that certain discourses, such as those concerning participation, gained momentum during the implementation process. However, this cannot be generalised, as in another case in this study, discourses, like those based on rights-based rationalities, have been ignored by hierarchical rationalities. Another study using a discourse approach focuses on the practical implementation of bioeconomy policies studies in Finland (Halonen et al., 2022). The authors show that discourses conflict with the downscaling of bioeconomy policy objectives and aspirations to practical implementation, and they recommend a more realistic identification of the challenges and the use of sensitive stakeholder dialogue across regions.

Interest-driven research on implementation comprises those studies focusing on the preferences and values of forest owners regarding the implementation of specific EU policies, on the assumption that these can either facilitate or impede implementation at the ground level. One such study used a discrete choice experiment to examine the preferences of private forest owners in Slovenia regarding the potential implementation of EU Green Deal goals (Plevnik & Japelj, 2023). This study revealed significant reluctance among older individuals and those residing in rural areas to participate in the implementation of the EU Green Deal. In a comparative study at the national level, Westin et al. (2023) show that the characteristics and value orientation of the forest owners may be less important than the country's context inshaping the willingness to implement.

4.6 Conclusions

This chapter highlights that, although implementation research has long provided valuable insights into European policy processes, its application within the forest policy field remains underdeveloped and fragmented. Few studies explicitly define their approach to implementation or their theoretical basis (Geitzenauer et al., 2016), highlighting a more general gap lack of theory in forest policy research (Arts, 2012). The field comprises a diverse range of theoretical approaches, ranging from ideational perspectives such as discourse analysis to interest-driven frameworks, including the Pathways of Influence model. Some studies apply intermediate frameworks such as the ACF. Empirical studies on EU forest policy suggest that the implementation approaches used tend to be more top-down than bottom-up, although hybrid models are gaining recognition. While the bottom-up perspective is recognised as important, and particularly so at the local level, the effectiveness of implementation often depends on existing political arrangements within the national or regional context. A significant proportion of the literature on forest policy implementation focuses on Natura 2000, with more recent attention also given to the EU Timber Regulation (EUTR). Most studies examine legally binding policies, suggesting that non-legally binding policies may produce even greater variation in implementation outcomes (Aggestam & Pülzl, 2020). Overall, recent research has shifted from viewing implementation failures primarily as non-compliance to examining how MS adapt EU rules to domestic contexts. This reframing of perspectives acknowledges the diversity of administrative traditions, political capacities, and stakeholder constellations, offering forest policy scholars more nuanced insights into themes currently debated in the broader field of (EU) implementation research.

First, a key discussion focuses on the importance of different contexts and of differentiated implementation. Several studies demonstrate that implementation variation does not necessarily indicate failure but rather reflects the interpretation and adaptation of policies to local administrative and political realities (Brendler & And Thomann, 2024). New approaches to environmental policy implementation emphasise that decentralised enforcement, digital monitoring, and networked governance could affect implementation while recognising the diversity of MS (Bondarouk & Mastenbroek, 2018). Allowing flexibility in implementation is essential for achieving environmental objectives in different contexts (Zhelyazkova et al., 2024). However, excessive differentiation could undermine effectiveness. In the context of forest policy implementation research, this implies shifting the focus from measuring compliance to examining how EU objectives such as those in the Regulation on Deforestation-Free Products Regulation (EUDR) or the Nature Restoration Law (NRL), are adapted in practice by forest stakeholders. Therefore, future forest policy research should explore how differentiated implementation via voluntary measures, regional pilot schemes, or adaptive regulatory instruments can foster implementation without compromising ecological integrity.

A second line of research emphasises capacity building and cooperative enforcement. Scholars have demonstrated that effective implementation depends more on supportive frameworks than on sanctions, the latter of which are difficult to monitor in complex and technically demanding policy areas anyway. Therefore, the analysis of forest policy implementation should shift its focus to how capacity building (e.g., through advisory services or knowledge mobilisation) might enhance compliance and stewardship among forest owners and operators. As the bottom-up perspective highljghts, local actors, such as foresters, play a central role in implementation. Additionally, a new trend recognises the roles of business actors, local-level authorities, and civil society in implementation (Buzogány, 2018; Köthke & Sotirov, 2024; Tosun & Schaub, 2023). As forest policy increasingly intersects with climate, biodiversity, and bioeconomy goals, implementation research must also address how diverse target groups engage with, resist, or reframe policy demands.

These developments in EU environmental implementation research provide a valuable opportunity to advance forest policy scholarship. By integrating context-sensitive, actor-centred, and multi-level approaches, research in this field can help to more accurately trace how policies are enacted in practice and the contribution of implementation to sustainable forest governance. As the EU embarks on an era of increasingly ambitious forest-related legislation, understanding implementation as a collaborative, adaptive, and structurally supported process will be crucial to the success of its policy framework. Key topics include implementing the EU climate law (e.g., the LULUCF framework), in which forests play a pivotal role in climate change adaptation. The EUDR, the new NRL, and the Renewable Energy Directive provide further examples of such topics. Studies have already been conducted that examine the impact of the LULUCF legislation and the EUDR on forests (Gulbrandsen, 2024; Johnston et al., 2025) and others have investigated forest-related "Brussels-effects" in neighbouring countries (Bradford, 2020), such as Ukraine (Buzogány, 2016; Buzogány & Varga, 2025). Furthermore, as democracy comes under increasing pressure, examining the impact of different governance regimes is expected to become more important in the future, as the study of forest policy under a populist government by Niedziałkowski et al. (2025) shows for Poland.

It will become increasingly important is to understand how all EU forest-related policy objectives (both legally binding and non-binding) can be implemented simultaneously across the 27 MS and to identify the potential trade-offs and synergies can be generated, given the significant differences in forest types across the Union. This broader understanding will enable us, as forest policy researchers, to provide insights into the success of EU decision-making regarding forests. This is important to know not only because around 30% of EU land is forested and contains important protected habitats, but also because the EU banking system is exposed to nature degradation through corporate lending and forests themselves are under threat due to climate change and biodiversity loss (Ceglar et al., 2025; Mohr et al., 2025). Recent projections suggest that disturbance-induced losses in Europe's forestry sector could more than double under severe climate change scenarios. This highlights the urgent need to bridge the gap between regulatory ambition and implementation reality.

Acknowledgement The authors would like to thank Lisa Fritsch, of the University of Freiburg, for her valuable empirical support in reviewing forest implementation research and for identifying particularly helpful examples that informed and strengthened this study.

References

Aggestam, F., & Pülzl, H. (2020). Downloading Europe: A regional comparison in the uptake of the EU forest action plan. *Sustainability, 12*(10), 3999.

Agrawal, A., & Gibson, C. C. (2001). The role of community in natural resource conservation. In *Communities and the environment: Ethnicity, gender, and the state in community-based conservation* (pp. 1–31).

Arts, B. (2012). Forests policy analysis and theory use: Overview and trends. *Forest Policy and Economics, 16*, 7–13.

Barrett, S. M. (2004). Implementation studies: Time for a revival? Personal reflections on 20 years of implementation studies. *Public Administration, 82*(2), 249–262.

Bondarouk, E., & Mastenbroek, E. (2018). Reconsidering EU compliance: Implementation performance in the field of environmental policy. *Environmental Policy and Governance, 28*(1), 15–27.

Borrass, L. (2014). Varying practices of implementing the habitats directive in German and British forests. *Forest Policy and Economics, 38*, 151–160.

Börzel, T. A. (2002). Pace-setting, foot-dragging, and fence-sitting: Member state responses to Europeanization. *Journal of Common Market Studies, 40*(2), 193–214.

Börzel, T. A. (2003). *Environmental leaders and laggards in Europe. Why there is (not) a "Southern Problem"*. Ashgate.

Börzel, T. A., & Buzogány, A. (2019). Compliance with EU environmental law. The Iceberg is Melting. *Environmental Politics, 28*(2), 315–341.

Bradford, A. (2020). *The Brussels effect: How the European Union rules the world*. Oxford University Press.

Brendler, V., & And Thomann, E. (2024). Does institutional misfit trigger customisation instead of non-compliance? *West European Politics, 47*(3), 515–542.
Buzogány, A. (2015). Building governance on fragile grounds: Lessonsfrom Romania. *Environment and Planning C: Government and Policy, 33*(5), 901–918.
Buzogány, A. (2016). EU-Russia regulatory competition and business interests in post-Soviet countries: The case of forestry and chemical security in Ukraine. *Eurasian Geography and Economics, 57*(1), 138–159.
Buzogány, A. (2018). Neighbourhood countries: Promoting environmental protection close to home. In C. Adelle, K. Biedenkopf, & D. Torney (Eds.), *European Union external environmental policy* (pp. 233–252). PalgraveMacmillan.
Buzogány, A., & Pülzl, H. (2024). Top-down and bottom-up implementation. In F. Sager (Ed.), *Handbook of public policy implementation* (pp. 116–126). Edward Elgar Publishing.
Buzogány, A., & Varga, M. (2025). Breaking the rural underdevelopment trap? Eastern enlargement, agricultural policy and lessons for Ukraine. *Journal of European Public Policy*, 1–22.
Ceglar, A., Marques, A., Boldrini, S., Lelli, C., Toreti, A., Parisi, L., & Heemskerk, I. (2025). European banks face significant vulnerability to ecosystem degradation and climate change. *Communications Earth & Environment, 6*(1), 750.
Davidescu, S., & Buzogány, A. (2021). Cutting deals: Transnational advocacy networks and the European Union timber regulation at the eastern border. *The International Spectator, 56*(3), 105–118.
Dimitrova, A., & Buzogány, A. (2014). Post-accession policy-making in Bulgaria and Romania: Can non-state actors use EU rules to promote better governance? *JCMS: Journal of Common Market Studies, 52*(1), 139–156.
Elmore, R. F. (1979). Backward mapping: Implementation research and policy decisions. *Political Science Quarterly, 94*(4), 601–616.
Falkner, G., Hartlapp, M., & Treib, O. (2007). Worlds of compliance: Why leading approaches to the implementation of EU legislation are only sometimes-true theories. *European Journal of Political Research, 46*(3), 395–416.
Fischer, F., & Forester, J. (1993). *The argumentative turn in policy analysis and planning*. Duke University Press.
Galiana, L., Aguilar, S., & Lázaro, A. (2013). An assessment of the effects of forest-related policies upon wildland fires in the European Union: Applying the subsidiarity principle. *Forest Policy and Economics, 29*, 36–44.
Gavrilut, I., Feiler, L., & Sotirov, M. (2021). Synthetic assessment of the governance of forests and protected areas, related EU policies, and their domestic implementation. In *Europeanization of environmental policies and their limitations: Capacity building* (pp. 29–49). Springer.
Geitzenauer, M., Hogl, K., & Weiss, G. (2016). The implementation of Natura 2000 in Austria—A European policy in a federal system. *Land Use Policy, 52*, 120–135.
Gulbrandsen, L. H. (2024). Implementing the EU LULUCF regulation in Norway: Short-term and long-term policy coherence challenges. *Forest Policy and Economics, 166*, 103270.
Hajer, M. A. (1995). *The politics of environmental discourse: Ecological modernization and the policy process*. Clarendon Press.
Hall, P. A. (1993). Policy paradigms, social learning and the state. In *The case of economic policy-making in Britain* (No. 25, pp. 275–296).
Halonen, M., Näyhä, A., & Kuhmonen, I. (2022). Regional sustainability transition through forest-based bioeconomy? Development actors' perspectives on related policies, power, and justice. *Forest Policy and Economics, 142*, 102775.
Héritier, A. (1996). The accommodation of diversity in european policy-making and its outcomes: Regulatory policy as a patchwork. *Journal of European Public Policy, 3*(2), 149–167.
Hill, M., & Hupe, P. L. (2009). *Implementing public policy: An introduction to the study of operational governance*. Sage Publications Ltd.
Hjern, B., & Porter, D. O. (1981). Implementation structures: A new unit of administrative analysis. *Organization Studies, 2*(3), 211–227.

Hogl, K., Kleinschmit, D., & Rayner, J. (2016). Achieving policy integration across fragmented policy domains: Forests, agriculture, climate and energy. *Environment and Planning C: Government and Policy, 34*(3), 399–414.

Jann, W., & Wegrich, K. (2017). Theories of the policy cycle. In *Handbook of public policy analysis* (pp. 69–88). Routledge.

Johnston, C., Guo, J., & Prestemon, J. P. (2025). The European Union deforestation regulation: Implications for the global forest sector. *Forest Policy and Economics, 173*, 103462.

Kaufman, H. (1960). *The forest ranger*. Johns Hopkins Press.

Koontz, T. M. (2007). Federal and state public forest administration in the new millennium: Revisiting Herbert Kaufman's the forest ranger. *Public Administration Review, 67*(1), 152–164.

Köthke, M. (2020). Implementation of the European timber regulation by German importing operators: An empirical investigation. *Forest Policy and Economics, 111*, 102028.

Köthke, M., & Sotirov, M. (2024). How and why do economic operators comply with EU law? Analysis of firm-level responses to the EU timber regulation in Germany. *JCMS: Journal of Common Market Studies*, *62*(6), 1594–1614.

Křenová, Z., & Kindlmann, P. (2015). Natura 2000—Solution for Eastern Europe or just a good start? The Šumava National Park as a test case. *Biological Conservation, 186*, 268–275.

Leipold, S. (2017). How to move companies to source responsibly? German implementation of the European Timber Regulation between persuasion and coercion. *Forest Policy and Economics, 82*, 41–51.

Lipsky, M. (1980). *Street-level bureaucracy: The dilemmas of individuals in the public service*. Sage.

Logmani, J., Krott, M., Lecyk, M. T., & Giessen, L. (2017). Customizing elements of the international forest regime complex in Poland? Non-implementation of a national forest programme and redefined transposition of NATURA 2000 in Bialowieza Forest. *Forest Policy and Economics, 74*, 81–90.

Mohr, J. S., Bastit, F., Grünig, M., Knoke, T., Rammer, W., Senf, C., Thom, D., & Seidl, R. (2025). Rising cost of disturbances for forestry in Europe under climate change. *Nature Climate Change*, 1–6.

Niedziałkowski, K., Konczal, A., & Mielewczyk, M. (2025). "Hands off our forests!"—The impact of the authoritarian rule on polish forest policy in the context of the European Green Deal. *Forest Policy and Economics, 171*, 103402.

Ostrom, E. (1990). *Governing the commons*. Cambridge University Press.

Overdevest, C., & Zeitlin, J. (2018). Experimentalism in transnational forest governance: Implementing European Union forest law enforcement, governance and trade (FLEGT) voluntary partnership agreements in Indonesia and Ghana. *Regulation & Governance, 12*(1), 64–87.

Partzsch, L. (2021). European Union's proxy accountability for tropical deforestation. *Environmental Politics, 30*(4), 600–621.

Partzsch, L., Müller, L. M., & Sacherer, A.-K. (2023). Can supply chain laws prevent deforestation in the Democratic Republic of the Congo and Indonesia? *Forest Policy and Economics, 148*, 102903.

Pecurul-Botines, M., Di Gregorio, M., & Paavola, J. (2014). Discourses of conflict and collaboration and institutional context in the implementation of forest conservation policies in Soria, Spain. *Biodiversity and Conservation, 23*(14), 3483–3499.

Pecurul-Botines, M., Di Gregorio, M., & Paavola, J. (2019). Multi-level processes and the institutionalization of forest conservation discourses: Insights from Natura 2000. *Forest Policy and Economics, 105*, 136–145.

Pecurul-Botines, M., Giurca, A., Pülzl, H., Niedzialkowski, K., Brockhaus, M., Bouriaud, L., Hajtmarová, Š., Brukas, V., Hoogstra-Klein, M., Konczal, A., Marcinekova, L., Øistad, K,. Pezdevšek Malovrh, S., Pietarinen, N., Secco, L., Roux, J.-L., Wolfslehner, B., & Winkel, G. (2025). Conflicts and policy priorities drive differences in the EU Forest strategy implementation: Insights from 15 European countries. *Forest Policy and Economics, 181*, 103652. https://doi.org/10.1016/j.forpol.2025.103652

Pietarinen, N., Pecurul-Botines, M., & Brockhaus, M. (2025). Politics of delay hinder the implementation of EU forest strategy in Finland. *Ambio, 54*(12), 2154–2169.

Plevnik, K., & Japelj, A. (2023). Uncovering the latent preferences of Slovenia's private forest owners in the context of enhancing forest ecosystem services through a hypothetical scheme. *Forests, 14*(12), 2346.

Pressman, J. L., & Wildavsky, A. (1973). *Implementation: How great expectations in Washington are dashed in Oakland; or, why it is amazing that federal programs work at all.* University of California Press.

Pülzl, H., & Treib, O. (2006). Policy implementation. In *Handbook of public policy analysis: Theory, politics and methods.* CRC Press/Taylor & Francis.

Roux, J.-L., Pülzl, H., Sotirov, M., & Winkel, G. (2025). Understanding EU forest policy governance through a cultural theory lens. *Policy Sciences, 58*(1), 111–144.

Sabatier, P., & Mazmanian, D. (1979). The conditions of effective implementation: A guide to accomplishing policy objectives. In *Policy analysis* (pp. 481–504).

Sarvašová, Z., Ali, T., Đorđević, I., Lukmine, D., Quiroga, S., Suárez, C., Hrib, M., Rondeux, J., Mantzanas, K. T., & Franz, K. (2019). Natura 2000 payments for private forest owners in rural development programmes 2007–2013—A comparative view. *Forest Policy and Economics, 99*, 123–135.

Sotirov, M., Lovric, M., & Winkel, G. (2015a). Environmental governance between Europeanization and domestic politics: Analyzing the implementation of EU biodiversity conservation policy in Bulgaria and Croatia. *Environment and Planning C.*

Sotirov, M., Lovric, M., & Winkel, G. (2015b). Symbolic transformation of environmental governance: Implementation of EU biodiversity policy in Bulgaria and Croatia between Europeanization and domestic politics. *Environment and Planning C: Government and Policy, 33*(5), 986–1004.

Sotirov, M., Winkel, G., & Eckerberg, K. (2021). The coalitional politics of the European Union's environmental forest policy: Biodiversity conservation, timber legality, and climate protection. *Ambio, 50*(12), 2153–2167.

Surel, Y. (2000). The role of cognitive and normative frames in policy-making. *Journal of European Public Policy, 7*(3), 495–512.

Tipple, T. J., & Wellman, J. D. (1991). Herbert Kaufman's forest ranger thirty years later: From simplicity and homogeneity to complexity and diversity. *Public Administration Review*, 421–428.

Tosun, J., & Schaub, S. (2023). Environmental policy implementation. In *Routledge handbook of environmental policy* (pp. 136–150). Routledge.

Van Hulst, M., Metze, T., Dewulf, A., De Vries, J., Van Bommel, S., & Van Ostaijen, M. (2025). Discourse, framing and narrative: Three ways of doing critical, interpretive policy analysis. *Critical Policy Studies, 19*(1), 74–96.

Versluis, E. (2007). Even rules, uneven practices: Opening the 'black box' of EU law in action. *West European Politics, 30*(1), 50–67.

Westin, K., Bolte, A., Haeler, E., Haltia, E., Jandl, R., Juutinen, A., Kuhlmey, K., Lidestav, G., Mäkipää, R., & Rosenkranz, L. (2023). Forest values and application of different management activities among small-scale forest owners in five EU countries. *Forest Policy and Economics, 146*, 102881.

Winter, S., Borrass, L., Geitzenauer, M., Blondet, M., Breibeck, R., Weiss, G., & Winkel, G. (2014). The impact of Natura 2000 on forest management: A socio-ecological analysis in the continental region of the European Union. *Biodiversity and Conservation, 23*(14), 3451–3482.

Zhelyazkova, A., Thomann, E., Ruffing, E., & Princen, S. (2024). Differentiated policy implementation in the European Union. *West European Politics, 47*(3), 439–465.

Chapter 5
Frames in Transition: Prospects for Transformative Change in EU Forest Governance

Filip Aggestam, Johanna Trummer, and Jerylee Wilkes-Allemann

Abstract This chapter explores how the European Union (EU) has framed forests between 1965 and 2025, and the implications of these frames for transformative change. An AI-based frame analysis of 492 EU policy documents identifies seven recurrent frames: “Policy architect”, “Standards and trade”, “Ecological crisis”, “Eco-modernisation”, “Environmental justice”, “Territorial equity and cohesion”, and “Science and systems thinking”. The chapter then maps their distribution across binding and non-binding policy instruments. It also introduces the concept of “transformative hotspots” to describe periods during which shifts in values and institutional changes align to enable a transformation of governance. Three such hotspots emerged: the 1990s, when justice and cohesion entered forest-relevant policy, the 2000s, which were characterised by climate integration and the expansion of due diligence mechanisms, and the 2020s, which have so far been characterised by stronger ecological-crisis and compliance-oriented frames, supported by target-driven and monitoring-focused instruments. It can also be noted that throughout these periods, EU forest governance evolved through layering rather than replacement: conservative, compliance-centred frames dominate legislation while crisis- and justice-oriented frames are concentrated in strategies and programming. When viewed through the lens of transformation, it is clear that the EU has developed enabling capacities (e.g., monitoring and data systems) and steering mechanisms (e.g., cross-sectoral regulations). However, shifts in values and power remain partial. The chapter argues that bridging the soft-hard policy gap by embedding minimum rights-based safeguards in policy, ensuring multi-level implementation capacity, and institutionalising co-production could increase the likelihood of transformative outcomes.

F. Aggestam (✉)
ORBIS Green Policy and Governance, Forest Policy Research Network—European Forest Institute (FPRN-EFI), Vienna, Austria
e-mail: aggestam@orbis-gpg.com

J. Trummer · J. Wilkes-Allemann
School of Agricultural, Forest and Food Sciences, Forest Science, Bern University of Applied Sciences, Zollikofen, Switzerland

F. Aggestam (ed.), *Governing Forests Through Transformative Change*, Advances in Global Change Research 70, https://doi.org/10.1007/978-3-032-19072-7_5

5.1 Forest Governance in Transition

Forests have become an integral part of a much broader global discourse around sustainability and natural resource use, being linked to multiple societal challenges, such as climate change, biodiversity loss, and landscape resilience. In the European Union (EU), forests cover around 39%[1] of the total land area and is at present subject to an increasing number of policy instruments, including regulatory frameworks, financial incentives, strategies, and monitoring systems. However, EU forest governance remains fragmented across sectors, governance levels, and the diverse range of perspectives on what forests are and how they should be used and managed (Aggestam et al., 2025, 2026a, 2026b; Aggestam, 2024a)

The significant increase in forest-relevant policy instruments has also led to conflicting expectations, meaning that forests are now expected to provide multiple services at the same time, ranging from acting as carbon sinks, providing livelihoods, preserving biodiversity, and serving as a source for cultural identity as well as improving well-being. The EU has responded to these environmental and socio-economic developments by launching policy initiatives like the European Green Deal (EC, 2019),[2] the EU Forest Strategy for 2030 (EC, 2021)[3] and the Nature Restoration Regulation (EP & CoE, 2024)[4] with the intent to make existing governance systems more sustainable. However, it remains an open question whether any of these policy changes represent a deeper structural transformation, if this transformation is lasting, or whether these changes primarily reinforce existing governance structures.

5.1.1 From Frames to Transformation

Transforming governance requires more than designing new policy instruments or institutions, it requires a change in how problems are framed, which goals are prioritised, and what knowledge is valued. It also concerns which actors are involved as well as how they interact and how they are empowered. Based on recent academic work on transformative change (Dryzek & Pickering, 2018; Larson et al., 2021; Patterson et al., 2017), this chapter defines transformation as a change in the nature and structure of governance, where values, perspectives, and institutional logics evolve due to changing demands, such as sustainability issues. More details on how transformative change is defined and characterised for this work is provided in Chap. 3.

One way to review these developments and potential transformations is to study policy frames (Aggestam, 2024b; Kahneman & Tversky, 2000; van Hulst & Yanow,

[1] See https://ec.europa.eu/eurostat/databrowser (accessed 6.10.25).

[2] See https://commission.europa.eu/strategy-and-policy/priorities-2019-2024/european-green-deal_en (accessed 6.10.25).

[3] See https://environment.ec.europa.eu/strategy/forest-strategy_en (accessed 6.10.25).

[4] See https://environment.ec.europa.eu/topics/nature-and-biodiversity/nature-restoration-regulation_en (accessed 6.10.25).

2014). Policy frames can be viewed as embedded interpretive structures that represent values and perspectives (e.g., what is at stake and why it matters?) and preferences (e.g., what should be done?) within a specific policy environment. Frame analysis has, in this context, been widely applied to understand how perspectives change over time, how competing visions interact, and how policies are legitimised (Bauer & Knill, 2014; Björnehed & Erikson, 2018; Dewulf et al., 2004). In other words, frames are viewed as an integral component of the governance structure, not merely as rhetorical devices. They shape how institutions function, how conflicts are negotiated, and how transitions may be enabled or prevented.

5.1.2 New Tools for Complex Questions: Analysing EU Policy Documents

Frames are used in this chapter to explain and understand EU forest governance structures and pathways for transformative change. However, this requires analysing policy developments over an extended time-period, which generally involves hundreds of EU policy documents. This represents a methodological challenge for traditional frame analysis approaches, as they struggle to scale up, mainly, due to the substantial financial and human resources needed to analyse large volumes of text as well as the process of identifying patterns and changes over time and policy instruments (Björnehed & Erikson, 2018; Elomina & Pülzl, 2021; Goffman, 1974; Rein & Schön, 1996; Vliegenthart & van Zoonen, 2011). To overcome this issue, this chapter will use an AI-based approach to carry out a frame analysis that will evaluate and identify prevalent perspectives and value orientations within a large number of EU forest-focused and forest-related policy documents (Alshami et al., 2023; Bijker et al., 2024; Fan et al., 2023).

The aim is to apply new analytical tools and explore how AI-based approaches can enable an innovative, large-scale review of how forest-related frames have evolved since the start of EU forest policymaking. Another objective is to examine whether these developments reflect, enable, or constrain transformative change in EU forest governance, whether prospective or already ongoing. The use of AI is intended to deepen and extend the analytical capabilities of policy research (Curry et al., 2024; del Barrio & Gatica-Perez, 2023). However, it should be noted that its application has generated significant pushback from different research communities. Concerns have been raised regarding transparency, reproducibility, and epistemology (Bender et al., 2021; Chubb, 2023). With this in mind, this chapter will reflect on the capabilities and limitations of large language models (LLMs), which are being used increasingly in social science research (Bijker et al., 2024; Curry et al., 2024; Ryazanov et al., 2024). The chapter will as such also contribute to a growing body of work on the use of AI in policy research.

5.1.3 Aim of the Chapter

These methodological and policy-relevant developments raise an important question: Do recent shifts in EU forest governance reflect an adaptation to new pressures or could it be signalling a more significant transformation? One way to address this question is to examine how forests have been framed in EU policy documents, over time, with particular attention to moments that may be linked to broader systemic reorientations. Having this question in mind, this chapter has two main objectives. First, it sets out to define and characterise frames that have affected EU forest policy developments from the 1960s onwards with the support of AI. Second, it will utilise these frames to assess whether the current EU policy environment has been (or is) favourable to transformative change over time.

5.2 Conceptual Framework and Background

5.2.1 Framing Forest Governance

Frames and framing are commonly used in policy research, particularly in environmental and sustainability research (Buijs et al., 2011; Kaufman & Gray, 2003; Shmueli et al., 2006). In the context of forests, governance systems[5] are characterised not only by institutions and policy instruments, but also by how issues are perceived, constructed and communicated. In contrast, "governance" refers to the processes of steering, coordination, and decision-making through which these actors establish and implement goals within the system. At its core, a frame is a structured interpretation of a policy issue which defines the problem, identifies those responsible, outlines the relevant values, and suggests potential policy responses (Benford & Snow, 2000; Goffman, 1974; Schön & Rein, 1994).

Frames do, however, serve both cognitive and strategic functions, mainly as frames help actors (individuals and institutions) make sense of complex (forest/environmental) problems while helping to legitimise decision-making (Entman, 2007; Feindt & Oels, 2005). In policy research, frame analysis is therefore a useful approach and tool when trying to understand how policy agendas evolve, how competing perspectives (or worldviews) and values interact, and how institutions adapt to or resist policy change. Crucially, frames are embedded in the policy-making process and influence the development of forest governance processes and governance systems over time (Daviter, 2007; van Hulst & Yanow, 2014).

In the context of EU forest-related policy, where forests are embedded in many policy domains, such as climate, biodiversity, energy, and trade, a frame analysis

[5] Referring the broader architecture of actors (both state and non-state), formal rules, informal norms, policy instruments and knowledge/monitoring infrastructures operating across levels and sectors.

offers an approach that can reveal how this single natural resource is interpreted and prioritised across sectors and over time. With multiple actors involved, such as EU institutions, EU Member States (MS), industry, NGOs, and scientific bodies, all pushing for different perspectives of what forests are and how they should be used, frames become important when trying to understand EU forest governance. In addition, this chapter adopts this perspective not only to classify the different ways in which forests are being framed, but also to evaluate whether frames reflect, enable, or constrain transformative change.

5.2.2 Moving from Frames to Transformation

Building on the conceptualisation of transformative change introduced in Chap. 3, this chapter considers transformation as the systemic and value-based (or normative) reconfiguration of forest governance, institutions, and actor constellations (Dryzek & Pickering, 2018; Larson et al., 2021; Patterson et al., 2017). Rather than reiterating that framework, this section focuses on how frames can signal, enable, or constrain transformation within such policy systems. More specifically, forest policy frames may:

- *Reveal changes in governance paradigms*, such as the transition from productive to justice-based or ecocentric forest management.
- *Identify emerging actor roles or alliances*, such as the growing importance of science, civil society, or intersectoral coordination.
- *Legitimise institutional or instrumental change*, such as new regulatory tools or cross-sectoral coordination mechanisms.

Frames can thus serve as indicators of changes in policy logics and governance systems.[6] In the context of forest policy, they offer a way to assess whether evolving definitions and narratives around forests reflect shifts in how forests are valued, who is empowered to govern them, and which policy priorities are being prioritised (e.g., timber production versus set-aside). In addition, in this chapter, frames are not only identified and classified but also evaluated in terms of their potential to reflect, support, or prevent transformative change. More specifically, each frame is grouped into one of three categories based on its orientation towards change:

1. **Transformative**: Frames that imply significant reorientation of values, actor roles, and governance approaches.
2. **Transitional or enabling**: Frames that support change within dominant paradigms, often through integration, innovation, or agenda expansion.
3. **Conservative or stabilising**: Frames that reinforce existing governance models, primarily through procedural, legal, or compliance-oriented approaches.

[6] "Governance" refers to processes of steering and coordination, and "governance systems" refer to the broader architecture of actors, rules, instruments, and knowledge infrastructures.

In this chapter, frames are first identified and then evaluated in terms of their potential to effect transformative change. This is achieved by mapping each frame to one of the three categories, conservative/stabilising, transitional/enabling or transformative, and aligning these categories with the three transformation dimensions ("for", "of" and "in"), where applicable. This approach follows the analytical framework introduced in Chap. 3 and is summarised in Table 5.1. Implementation details can be found in the Methods section below.

In addition to identifying and classifying forest-related frames, this chapter reviews changes in perspectives, values, and policy narratives to establish whether these changes cluster around specific periods of strategic or institutional change. Rather than assuming that transformation develops along a linear trajectory, the chapter introduces the concept of "framing shifts", which implies that marginal frames can gain prominence, dominant narratives can be revised, and institutional arrangements can adapt to new pressures. This is consistent with work on frame realignment and transformation in social movement framing (Benford & Snow, 2000; Snow et al., 1986) as well as with discursive shifts in policy discourse (Hajer, 1997; Rein & Schön, 1996).

These developments will be identified through a longitudinal analysis of EU forest-related policy documents (see Methods) and are interpreted in relation to major EU policy milestones. For example, the emergence of climate-related framing is likely to coincide with the adoption of the Kyoto Protocol and the expansion of Land Use, Land-Use Change, and Forestry (LULUCF) accounting in EU legislation (EP & CoE, 2013, 2018). More recently, the European Green Deal and the Nature Restoration Regulation represent major policy milestones that changed how forests are conceptualised and framed at the EU level.

As will be shown in the analysis, these framing shifts do not always lead to structural transformation. However, they provide insight into how EU forest-related

Table 5.1 Linking frames to types of transformative change

Transformative type	What changes?	Frame indicators	Examples
Governance *for* transformation *(capacities and infrastructure)*	Data, knowledge, monitoring, advisory capacity, and coordination mechanisms	Enabling narratives, systems thinking, and capacity-building frames	Frames emphasising data, science-policy integration, or spatial monitoring
Governance *of* transformation *(steering and policy mixes)*	Policy instruments, targets, and cross-sectoral alignment	Coordinating instruments, regulatory innovation, and cross-sectoral narratives	Eco-modernisation or Green Deal alignment
Transformation *in* governance *(perspectives, rights and power)*	Values, decision rights, and distributional rules	Paradigm-challenging, justice—or crisis-based reframing of problems and values	Forest justice, planetary boundaries, systemic risk discourse

Source Adapted from Patterson et al. (2017)

policy discourses have evolved, whether through adaptation, incorporation of new values, or resistance to more disruptive change. Moreover, to assess the potential for transformative change, the concept of a "transformative hotspot" is introduced in this chapter. This is an integrative term for a moment in time where changes in framing, institutional design, and value orientations occur together in ways that may signal transformation. It combines the concept of a "policy window," which occurs when the streams of problems, policies, and politics meet (Kingdon, 1984) with the idea of path dependencies and "critical junctures", representing brief periods that shape future policy pathways (Capoccia & Kelemen, 2007). Along the same line of argument, it also draws on the notion of "punctuated equilibrium," where periods of stability are interrupted by short bursts of change in agendas and institutions (Baumgartner & Jones, 2009). Finally, it also reflects work on "leverage points for transformation" in terms of policy interventions that can increase the likelihood of systemic change (Abson et al., 2017; Meadows, 1999). The analysis will focus on three such hotspots between the 1990s and the 2020s (see Methods). Each of these hotspots represents a particular phase when EU forest governance has evolved (over the last 30 years).

5.3 Methods and Data

This chapter builds on an earlier publication: *Sixty years of EU forest policy discourse: An AI-assisted frame analysis of policy documents* (Aggestam et al., 2026a). It draws from the same dataset and approach but extends the analysis to focus on the transformative potential of forest-related frames and associated implications for EU forest governance. Further details on the methodological approach, including database construction, prompt design, and AI platform comparisons, can be found in the original publication, particularly its supplementary materials, and are only summarised here (Aggestam et al., 2026b).

5.3.1 Document Screening and Sampling

The analysis is based on a sample of 492 EU policy documents that refer to forests, covering the period from 1965 to 2025. The documents were found through a two-step search and screening process using the EUR-Lex platform. First, an initial sample of 15,545 entries was retrieved using "forest" as a keyword filter on 21 January 2025. Secondly, documents were screened to ensure that forests were addressed substantively rather than merely mentioned in passing.

The sample includes different policy documents, such as regulations, directives, communications, decisions, strategies and working documents. The number of documents per decade was capped at 100 (except for earlier decades where fewer documents were available) to ensure temporal balance. Policy proposals were excluded unless they were significant for historical comparison, such as a proposed EU forest regulation that was never adopted (e.g., the European Commissions proposal for a Council Directive concerning forestry measures in 1974).

5.3.2 Data Structuring and Preparation

Each policy document was converted into a structured Word file containing the full text and a metadata section. The metadata included the document's title, reference, type, decade, year of publication, source URL and categorisations such as legal status (hard versus soft policy instrument) and thematic relevance (forest-focused, forest-related or supportive). To enable scalable querying and integration, all files were then imported into a customised SQLite database. This made it possible to manage content, carry out batch analysis, and cross-reference by decade, document type, and category.

5.3.3 Frame Identification Using AI

The analysis was carried out using ChatGPT (version 4.0 'o' model) and Julius.ai to identify frames across the dataset. Using two different AI models enabled cross-platform triangulation and comparison to reduce platform-specific bias. The frame analysis followed an iterative process using theory-informed prompts (Benford & Snow, 2000; Goffman, 1974; Schön & Rein, 1994). More information on the prompts that were used can be found in Aggestam et al. (2026b).

In a nutshell, ChatGPT and Julius were asked to:

- Identify recurring narratives and problem definitions.
- Characterise dominant value orientations (e.g., anthropocentric or ecocentric).
- Group related discursive elements into frame typologies.

The same queries were run on both platforms using the structured SQLite database to improve reliability. The results were then compared and synthesised, and adjustments were made based on manual review and expert feedback. A total of seven principal frames and two nested frames were identified.

These initial frame outputs were validated through internal consistency checks and by comparing them across platform runs. To assess robustness, the same analysis was rerun on ChatGPT using a separate account to examine model variability. While the principal frames showed consistency, variations were observed across the nested frames. Following this triangulation process, a synthesised frame typology was developed.

5.3.4 Analysing the Potential for Transformative Change

The second part of the analysis focused on evaluating the transformative potential of each frame. Building on the classification introduced in the conceptual framework (see Table 5.1), this step applied it systematically across the frame typology.

Although the dataset consists only of EU policy documents and therefore does not directly capture implementation dynamics, these texts nevertheless provide insights into how forest governance change is framed, the types of institutional solutions that are legitimised, and the values or actors that are prioritised. Each frame was for this reason reviewed to determine whether it implied:

1. **Paradigm orientation**: The frame signals a shift in underlying governance values, such as moving from productivist to ecocentric, justice-based, or integrative paradigms.
2. **Actor reconfiguration**: The frame increases the visibility of new actor groups, such as civil society, scientific institutions, marginalised communities, or argues for the redistribution of agency across governance levels. This had to be inferred from the associated policy documents.
3. **Instrumental or institutional innovation**: The frame is associated with novel governance tools, coordination mechanisms, or implementation strategies, as set out in the associated policy documents.

Based on these dimensions, the frames were grouped according to the three categories noted above: conservative/stabilising, transitional/enabling or transformative. This classification allowed for a grounded review of how forest policy frames may support, enable, or prevent transformative change within EU forest governance.

5.3.5 Transformative Hotspots

Three "transformative hotspots" were identified as important moments in EU forest policy, characterised by shifts in principal frames alongside relevant institutional or policy developments. The hotspots were selected based on the following criteria:

- Emergence or resurgence of previously marginalised or new frames.
- Temporal alignment with major policy moments, such as the Rio Earth Summit, the Kyoto Protocol, and the European Green Deal.
- Signals of evolving value orientations or shifts in the visibility of different actor coalitions within policy narratives.

Each hotspot was analysed using ChatGPT and the available policy database to carry out a targeted review of policy documents from the relevant period. The analysis assessed whether the observed changes corresponded to governance that was for, of, or in transformation, as characterised by Patterson et al. (2017), and whether these developments reflected incremental adaptation, transitional experimentation, or signs of structural reconfiguration in terms of EU forest governance.

5.4 Interpreting Forest Policy Frames

This section examines how EU forest governance has been framed over time, considering the implications for institutional arrangements, shifting priorities, and the potential for transformative change. At the heart of the analysis is a typology of seven forest-related frames, which are introduced below, reflecting how forests are defined, why they matter, and how they are governed at the EU level.

By tracing the emergence, evolution and persistence of these forest-relevant frames, this section sets out to assess how EU forest governance has responded to changing pressures or crisis over time. First, the results identify and characterise the principal frames and their implications for governance. Then, it examines the temporal evolution and layering of these frames. Finally, it evaluates the frames' transformative potential for EU forest governance.

5.4.1 The Database

As noted above, the frame analysis is based on a database that contains 492 EU policy documents that explicitly reference forests. The sample includes a variety of legally binding instruments, such as regulations, directives and decisions, as well as non-binding instruments, such as communications, strategies and working documents.

The content of the policy database reflects the evolving structure of EU forest governance over six decades. For example, prior to the 1990s, the number of forest-relevant policy documents was relatively limited. From the 1990s onwards, however, there was a significant increase, particularly in the form of non-binding (soft) policy instruments and strategies. The most recent decade (2020–2025) encompasses key policy initiatives related to the European Green Deal, the Nature Restoration Regulation, and the Deforestation-Free Products Regulation. Further details on the search criteria, database development, and structure can be found in Aggestam et al. (2026b, 2026c).

5.4.2 Forest Frames and Governance Rationales

The analysis of EU forest-related policy documents resulted in the identification of seven frames, five principal frames and two nested frames (Table 5.2). These frames represent how the EU has conceptualised and governed forests over the last 60 years. The frames differ in their problem definitions, value orientations, and governance rationales, reflecting different institutional perspectives, assumptions about forests, challenges, and forest governance priorities.

Each frame operates across three dimensions of the policy discourse: (1) diagnostic framing: how forest-related challenges are defined, (2) prognostic framing:

Table 5.2 Principal and nested frames in EU forest policy

Frame		Narrative	Governance rationale	Transformative type
1. Policy architect		Frames forest-related challenges as primarily institutional and regulatory. Emphasises legal clarity, procedural coherence, and the role of monitoring and enforcement mechanisms across EU Member States (MS)	Rule-based governance, institutional stability, and legal compliance	Conservative/ stabilising
Nested frame	2 Standards and trade	Emphasises the regulation and standardisation of forest products, including timber, biomass, and reproductive material. Highlights certification schemes, product traceability, and international trade alignment	Market regulation, trade harmonisation, soft enforcement	Conservative/ stabilising
3. Ecological crisis		Depicts forests as ecological systems under threat from climate change, biodiversity loss, and degradation. Calls for urgent conservation and restoration interventions	Urgency-driven governance, precautionary principle, ecological restoration	Transformative
4. Eco-modernisation		Integrates environmental protection with economic growth through innovation, green technologies, and sustainable bioeconomy strategies	Market-based solutions, green growth, public–private partnerships	Transitional/ enabling
5. Environmental justice		Highlights issues of fairness, procedural inclusion, and the rights of communities affected by forest-related decisions. Links environmental governance with distributive justice and participation	Rights-based governance, participatory legitimacy, inclusion	Transformative

(continued)

Table 5.2 (continued)

Frame		Narrative	Governance rationale	Transformative type
Nested frame	6. Territorial equity and cohesion	Frames forests as integral to regional development, rural resilience, and social cohesion in structurally disadvantaged areas	Regional policy integration, spatial equity, rural livelihoods	Transitional/ enabling
7. Science and systems thinking		Emphasises evidence-based policymaking, spatial monitoring, and integration of scientific knowledge. Highlights the role of data, modelling, and international reporting systems	Coordination, adaptive management, cross-scale legitimacy	Transitional/ enabling

Note Actor categories, frequencies, and the distribution of policy instruments are provided in Aggestam et al. (2026b)

types of solutions or instruments promoted and (3) motivational framing: what justifies or legitimises action. For example, the "Policy architect" frame defines forest issues as being caused by fragmented governance and institutional misalignment (*diagnostic*), proposes legal harmonisation and policy coherence as solutions (*prognostic*), and justifies intervention by appealing to legal certainty and institutional stability (*motivational*). In the same way, the "Standards and trade" nested frame focuses on risks related to market credibility and traceability. It promotes certification and product regulation as tools and draws its legitimacy from the need to harmonise trade and consumer protection.

In contrast, frames such as "Ecological crisis" emphasise systemic environmental degradation and biodiversity loss, advocate for restoration and conservation strategies, and use ecological urgency and planetary boundaries as justification. The "Environmental justice" frame defines challenges in terms of exclusion and inequitable impacts. It proposes participatory and rights-based governance as a response, basing its legitimacy on fairness and procedural inclusion. The other frames, such as "Eco-modernisation" and "Science and systems thinking", promote integration, innovation, and coordination. Eco-modernisation links environmental goals with green growth and technological advancement, while science and systems thinking frame the policy problem as a need for better evidence, monitoring, and cross-sectoral learning.

Taken together, these frames reflect competing governance rationales, ranging from rule-based and technocratic approaches to participatory and value-based approaches. They also reflect how EU forest governance has evolved into a multidimensional and layered policy system, covering various perspectives on how forests are understood as well as how forest policy should be formulated and implemented.

It can also be noted that the institutional anchoring of these frames varies across policy instruments. As summarised in Table 5.2, the "Policy architect" frame frequently appears in legally binding instruments (e.g., regulations and directives), while "Science and systems thinking" is more visible where monitoring, reporting, traceability, and coordination are central to implementation. In contrast, frames such as "Eco-modernisation" and "Environmental justice" tend to be found in soft policy instruments, including strategic documents and communications, where agenda-setting and value-based signalling are prioritised over implementation. The sub-frames, such as the "Standards and trade" and "Territorial equity and cohesion" furthermore highlight some of the thematic differences across sectors, particularly between trade and cohesion policy.

As part of a bigger picture, these patterns are significant. They demonstrate that a single paradigm does not drive EU forest governance, but rather multiple, often competing, frames that co-exist alongside one another. In other words, some frames stabilise and reinforce existing governance arrangements through procedural and legal instruments. In contrast, others introduce new problem definitions or value orientations that challenge existing priorities and institutional structures. Understanding these variations and the possible conflicts between these different approaches is important when considering prospects for transformative change.

5.4.3 Moving Towards Layered Forest Governance

As outlined earlier, the frames identified in Table 5.2 demonstrate the complexity behind EU forest governance. Rather than being dominated by a single narrative, EU forest-related policy reflects the co-existence and interaction of multiple, sometimes competing, governance perspectives. This process, often referred to as discursive layering (Schmidt & Radaelli, 2004), helps to explain why stabilising and transformative elements co-exist within the EU's forest policy landscape. Rather than replacing old narratives, new frames tend to build upon or co-exist with older ones, resulting in governance structures that are not aligned.

This layering is evident in certain policy instruments. For example, the Deforestation-Free Products Regulation (EP & CoE, 2023) reflects the "Policy architect" frame in that it emphasises traceability, compliance, and procedural enforcement. It also incorporates elements of the "Science and systems thinking" frame through data-driven monitoring and reporting mechanisms. Likewise, the EU Forest Strategy for 2030 (EC, 2021) covers the "Ecological crisis" and "Environmental justice" frames by calling for urgent restoration, community involvement, and equitable access to ecosystem services, however, these remain within non-binding policy documents. Meanwhile, instruments such as the EU Bioeconomy Strategy (EC, 2018) represent the "Eco-modernisation" frame by promoting innovation, biomass markets. and green growth, although with a broad reference to sustainability. These examples demonstrate that frames do not replace one another, but rather co-exist across policy instruments, each with varying degrees of importance and institutional influence.

Moreover, this layering has consequences. It influences institutional priorities, the distribution of responsibilities across actors and sectors, as well as the conceptualisation and operationalisation of policy change. For example, the frequent appearance of the "Policy architect" frame in legally binding instruments highlights the EU's continued reliance on compliance-based governance. Arguably, the minimal presence of justice-oriented frames in such policy instruments reveals the limitations of redistributive or participatory approaches within legally binding frameworks. Put differently, the distribution of frames across different types of policy instruments reflects the underlying power dynamics of EU institutions and the structure of EU policymaking. Strategic and non-binding documents, such as the EU Forest Strategy, allow room for new ideas, alternative values, and emerging concerns, such as forest rights, rural cohesion and ecosystem restoration. However, regulations and directives often only codify selected elements of these agendas into enforceable norms. This highlights that EU forest governance remains dominated by procedural, legal, and technical frames, while ecological, justice-based, and systems-oriented narratives are secondary.

These findings have important implications for EU policymaking. The prevalence of conservative frames in legally binding policy instruments would suggest that institutional gatekeeping restricts the adoption of transformative or justice-oriented approaches in compliance-based policies (Pülzl et al., 2014; Winkel & Sotirov, 2016). Although strategic and non-binding policy documents permit alternative narratives and values, these are often not translated into binding regulatory frameworks (Aggestam & Giurca, 2021). This separation risks further entrenching the fragmented and layered EU forest governance architecture in which compliance-based approaches dominate formal legislation, while participatory and value-based approaches remain confined to non-binding policy instruments (Howlett & Rayner, 2013; Howlett et al., 2010; Schmidt & Radaelli, 2004). In the EU forest-based sector, for instance, this fragmentation has long been recognised as an obstacle to developing coherent policies, with forest-related competencies spread across sectors and levels (Winkel, 2017; Wolfslehner et al., 2020; Aggestam, 2025).

5.4.4 Temporal Evolution of Forest Policy Frames

Traditionally, identifying temporal changes in policy frames has faced significant methodological challenges, mainly due to the size and heterogeneity of the datasets required to capture developments over extended time periods. AI-based tools, as applied in this chapter, can facilitate the identification of patterns and frame trajectories, which would otherwise be too resource-intensive to achieve using manual methods. As such, AI provides a possible approach to analysing complex policy documents over time. However, it is also worth noting that applying AI to policy research raises important concerns, including concerns about transparency, interpretive validity, and epistemological assumptions (Curry et al., 2024). While these issues are only briefly addressed here, Aggestam et al. (2026a) include a detailed reflection on the use of AI in policy research.

5.4.4.1 Longitudinal Trends

The longitudinal analysis reveals a structured, yet non-linear, evolution in forest framing in EU policy between 1965 and 2025. Over time, these changes reflect a growing complexity of forest governance rationales (Table 5.2) as well as changes in the underlying values, priorities, and instruments associated with forest-related policymaking. Rather than evolving gradually, these changes have happened during punctuated phases (Baumgartner & Mahoney, 2008; True et al., 2019), often during periods of institutional restructuring, new or shifting global commitments, or changing socio-economic expectations. For example, the EU enlargements of 2004 and 2007 reframed cohesion and rural development priorities (e.g., bringing forest-rich regions into focus). The recession of 2008–2010 and subsequent growth agendas increased narratives linked to the bioeconomy and biomass, focusing on jobs and regional competitiveness. The 2022 energy security shock increased discussion around wood-based bioenergy and supply resilience, while rising consumer concern shaped due diligence expectations in supply chains. These inflexion points, identified in this chapter as "transformative hotspots", represent moments where changes in framing coincide with broader policy or governance developments.

1960s to 1970s	The discourse surrounding forests was primarily technocratic and regulatory in nature. The "*Policy architect*" and "*Standards and trade*" frames dominated, depicting forests primarily as an administrative domain for standardisation and trade regulation. For example, early legislation concerning the marketing of forest reproductive material emphasised procedural conformity and quality assurance, making no reference to biodiversity or sustainability (EP & CoE, 1966). Forests were not considered ecological or social entities, but rather regulated commodities within internal market frameworks. Environmental concerns were essentially absent during this period
1980s	The emergence of alternative perspectives. Although the legal and regulatory framework remained dominant during this period, elements of the "*Eco-modernisation*" framework began to emerge, particularly in cohesion policy documents promoting rural development, innovation and structural adjustments. For example, the adoption of the Integrated Mediterranean Programme (EP & CoE, 1985), alongside the increasing allocation of forestry-related funding under the European Agricultural Guidance and Guarantee Fund, indicates a shift in EU forest policy towards broader narratives focusing on economic modernisation and territorial efficiency. Although justice-oriented themes, such as "balanced territorial developments", emerged at the rhetorical level, they were mainly presented in procedural terms and remained subordinate to economic objectives

(continued)

(continued)

1990s	A turning point was reached in the 1990s. The "*Environmental justice*" and "*Territorial equity and cohesion*" frames entered strategies and funding instruments, shifting the focus away from predominantly productivist perspectives. For example, despite its non-binding nature, the first EU Forest Strategy (EC, 1998) presented forests as a multifunctional socio-ecological system, emphasising cross-sector coordination and stakeholder involvement. At the same time, rural development reforms incorporated forestry initiatives, including afforestation, restoration, and non-productive investments, into Pillar II of the Common Agricultural Policy (CAP), thus operationalising the goals of spatial cohesion and rural livelihoods (EP & CoE, 1999). Programmes such as LEADER and pre-accession instruments further embedded place based approaches and participation at a regional level. Global and pan-European processes, such as the Rio Forest Principles (1992), Agenda 21, and the MCPFE/Helsinki-Lisbon resolutions, reinforced this shift towards participatory, sustainable forest management (SFM) and distributional concerns. Moreover, language relating to justice and equity increased in soft policy instruments (e.g., strategies, funding regulations and guidance). At the same time, the implementation of the Habitats Directive (e.g., via Natura 2000) provided an ecological reframing of forests (EP & CoE, 1992)
2000s	Discursive complexity reached its peak in the late 2000s and early 2010s. The "*Ecological crisis*" frame gained importance with the adoption of the Kyoto Protocol (in force in 2005), the launch of the EU Emissions Trading Scheme (ETS) in 2005, and the 2009 climate-energy package. These policy developments elevated the importance of forests within the EU. For example, this pathway was formalised when forest carbon sinks were incorporated into EU emissions accounting (EP & CoE, 2013). At the same time, the "Eco-modernisation" frame expanded, presenting trees as a renewable material within the EU Bioeconomy Strategy (EC, 2012), linking forests to innovation- and growth-oriented agendas, including bioenergy under the Renewable Energy Directive—RED I (EP & CoE, 2009). The co-existence of crisis, market, and systems-oriented perspectives during this period indicates a high degree of discursive pluralism, particularly in non-binding policy instruments and strategies
2010s	The focus shifted towards governance integration, ecological urgency, and implementation-oriented coordination in the 2010s. The growing prominence of the "*Ecological crisis*" frame, alongside a stronger "*Policy architect*" orientation, reflected an increasing emphasis on biodiversity loss, restoration needs, monitoring and more data-driven forms of governance. At the same time, the "*Science and systems thinking*" frame became more visible through knowledge infrastructures and coordination tools, most significantly the Forest Information System for Europe (FISE), which was developed under the second EU Forest Strategy (EC, 2013), and the first EU Biodiversity Strategy (EC, 2011). These initiatives expanded the role of monitoring, modelling, and performance-based governance, including the use of harmonised forest data (e.g., JRC-led platforms), remote sensing and reporting interfaces (e.g., Copernicus services) and indicator frameworks aligned with ecosystem accounts. The EU's broader push for evidence-based policymaking (institutionalised via impact assessments, Fitness Checks and the REFIT programme) further emphasised this approach. At the same time, climate instruments such as the LULUCF Regulation reinforced the need for measurement and reporting on forest carbon (EP & CoE, 2018). Nevertheless, participatory and justice-oriented approaches remained largely confined to non-binding policy documents

(continued)

(continued)

2020s	Although not an entire decade, the period from 2020 to 2025 suggests a time of discursive narrowing. The "*Policy architect*" and "*Ecological crisis*" frames have become particularly prominent, especially in legally binding legislation such as the Deforestation-Free Products Regulation and the Nature Restoration Regulation (EP & CoE, 2023, 2024). These instruments prioritise enforceability, monitoring/verification and procedural compliance. At the same time, "Science and systems thinking" has become more visible as a supporting frame in texts centred on implementation, particularly where monitoring, verification, indicators, geospatial data, and evidence-based coordination are central. Considered as a whole, this pattern suggests a partial re-centralisation of procedural and compliance-based frames under conditions of heightened ecological urgency.

EU forest policy has evolved from an initial phase of regulatory consolidation in the 1960s and 1970s to expansion and diversification in the 1990s and 2000s, and more recently, to a period of selective integration and procedural tightening in the 2010s and 2020s. The co-existence of multiple frames, rather than a single dominant paradigm, highlights the complex and contested nature of EU forest governance. Although strategic instruments (e.g., EU Forest Strategies and Bioeconomy Strategies) set out ambitious or transformative objectives, legally binding legislation (e.g., the Deforestation-Free Products Regulation and the Nature Restoration Regulation) generally tends to stabilise existing procedural and institutional arrangements.

5.5 Transformative Hotspots: The 1990s, 2000s and 2020s

5.5.1 Defining the Hotspots

A transformative hotspot is a period during which changes in principal frames, value orientations, and institutional developments (e.g., instrument design, coordination arrangements and legal codification) occur concurrently, potentially reorienting EU forest governance. The volume of new activities does not define hotspots, but rather the degree of alignment, such as when value-based and institutional changes reinforce one another.

Using the frame typology (Table 5.2) and longitudinal analysis (see Sect. 4.4), hotspots can be identified along four dimensions:

1. **Value reframing**: Evidence that problem definitions and policy logics move beyond productivist/administrative baselines towards ecocentric, justice-oriented or systems perspectives.
2. **Actor visibility and roles**: Changes in the knowledge and claims that are legitimised (e.g., increased salience of civil society, regional actors, and scientific/monitoring bodies, as well as the epistemic communities that connect them).

3. **Instrumental/coordination change**: The introduction or strengthening of tools that restructure practice (e.g., due diligence regimes, restoration targets, and cohesion mechanisms).
4. **Institutional anchoring**: The translation of these shifts into hard law or durable arrangements (beyond soft policy instruments).

When interpreted using the "for/of/in transformation" framework (Table 5.1) set out by Patterson et al. (2017), hotspots may indicate:

- *Governance **for** transformation*: Enabling capacities and integrative infrastructures, such as monitoring systems and coordination platforms.
- *Governance **of** transformation*: Deliberate steering via new regulatory approaches or cross-sector alignment.
- *Governance **in** transformation*: Reconfiguration of underlying logics, such as justice or ecological limits frames becoming legally binding.

Hotspots are heuristic and diagnostic. More specifically, they represent credible pathways to reconfigure policy goals, power relationships, knowledge and institutional structures, without assuming definitive outcomes. Using these criteria, the following sections will examine three periods: the 1990s, the 2000s, and the 2020s.

5.5.2 Hotspot 1: The 1990s and the Rise of Justice and Cohesion Narratives

Why is this a hotspot?

The 1990s marked the first period in which equity, participation, and territorial cohesion were consistently embedded in the EU's forest-related discourse. A significant proportion of the 101 policy documents from this period are categorised under the "Environmental justice" and "Territorial equity and cohesion" frames, indicating a shift away from earlier technocratic and trade-centred approaches. This shift is evident in documents such as the (i) first EU Forest Strategy, which emphasised multifunctionality and stakeholder involvement, (ii) the LEADER programme, which promoted bottom-up rural partnerships, and (iii) the Habitats Directive, which facilitated consultations on forest habitats. Together, these examples demonstrate a transition from technocratic and trade-focused perspectives to justice- and cohesion-oriented ones.

The changing discourse around forests coincided with the 1992 Rio Earth Summit and the Forest Principles/Agenda 21, the first EU Forest Strategy (which formalised the multifunctionality of forests at the EU level), and the expansion of instruments focusing on cohesion and rural development. These instruments addressed structural disparities in marginal and forest-dependent regions. Examples include the EAGGF guidance measures and LEADER II programming, which took place between 1994 and 1999. These changes are evident in the increased use of value-laden language regarding access, participation, regional balance, and community well-being, as compared to earlier decades, and in the growing visibility of expert networks framing participation and cohesion as a measure of governance quality.

Principal frames and transformative potential

- **Environmental justice**: Policy documents are beginning to view forests as socio-political spaces rather than merely regulated goods or environmental assets. Features include: (i) calls for access to information and public participation in environmental decision-making; (ii) references to transparency and accountability in the allocation and targeting of rural development funds; and (iii) acknowledgement of the social impacts of ecological decline (e.g., on employment and cultural landscapes). In transformative terms, this represents a normative shift (from resource productivity to rights and legitimacy considerations) and actor reconfiguration (increasing the importance of civil society, local communities and regional authorities). However, incorporation into legally binding instruments remains limited.
- **Territorial equity and cohesion (nested frame)**: Forests are linked to place-based development, rural resilience and balanced territorial outcomes. This nested frame is transitional and enabling as it broadens the scope of forest governance and supports multi-level coordination, for example, between MS and regional programmes, without replacing productivist or administrative approaches. Examples include the emphasis on cross-sectoral coordination in the first EU Forest Strategy and the integration of forestry measures into rural development policy.

Governance changes and institutional reflections

EU forest governance underwent an initial value-based reframing during the 1990s, with forests becoming associated not only with ecological functions, but also with identity, well-being, and rights. This change was accompanied by a more significant participation of non-state and regional actors in strategies and programming documents, most notably civil society, rural stakeholders, and regional authorities, reflecting a broader recognition of stakeholder views during policy design. In practice, the 1990s saw the introduction of soft (non-binding) coordination through the first EU Forest Strategy (EC, 1998), as well as stronger programming links between the environment, cohesion, and rural development, such as via the EAGGF Guidance and LEADER programmes. Most institutional innovations remained non-binding (e.g., strategies and funding regulations/guidance), while the binding instruments, including the continued implementation of the Habitats Directive (EP & CoE, 1992), reframed forests ecologically while keeping a mainly procedural legal position. This implies that justice and participation were embedded in strategic and funding instruments but were not widely translated into binding legislation.

Assessment

A transformation in forest governance occurred at the discursive level in the 1990s, with limited institutional impacts. The level of change was moderate. For instance, although justice and cohesion considerations became mainstream in soft instruments such as strategies, programming, regulations and guidance, their institutionalisation in hard instruments remained limited. Furthermore, the stability of these changes appears to be fragile, as later periods demonstrate a trend towards procedural re-centring in binding instruments. In transformative terms, the decade scored highly in terms of discourse, moderately in terms of programme design and poorly in terms of empowering actors as set out by key binding instruments. Arguably, the 1990s represent a threshold moment. Justice- and territory-oriented perspectives broadened the policy meaning of forests beyond productivity and market regulation, creating credible pathways (e.g., via strategies and funding regimes) for more inclusive discussion on forests. Although these perspectives did not change the binding instruments at the time, they probably established the discursive and programmatic basis for the evolving climate and biodiversity frames found in the 2000s.

5.5.3 Hotspot 2: The 2000s and Climate Integration and Bioeconomy Expansion

Why is this a hotspot?

The 2000s marked a second turning point in EU forest governance. This decade shows the highest co-occurrence of the "Ecological crisis" and "Eco-modernisation" frames in the database, reinforced by the "Standards and trade" nested frame. This expansion (in terms of values and perspectives) occurred alongside significant developments in climate and market governance, including the Kyoto Protocol (effective from 2005), the EU climate-energy package (EP & CoE, 2009), the first Renewable Energy Directive (RED I; EP & CoE, 2009), FLEGT (CoE, 2005; EC, 2008), and the first EU Timber Regulation (EP & CoE, 2010). Together, these policy instruments incorporated forests into climate accounting, linked them to bioenergy targets, and formalised supply-chain due diligence. This implies that forest issues became embedded within a broader regulatory framework covering climate, energy, and trade policy. This was made possible by epistemic communities that standardised monitoring, reporting and verification (MRV) and forest-relevant metrics, such as, through the IPCC's good-practice guidance for LULUCF, the JRC's work on methodological inventories and land-cover change, and the FOREST EUROPE criteria and indicators. These developments provided a shared technical basis for forest-relevant instruments.

Principal frames and transformative potential

- **Ecological crisis**: Forests are increasingly recognised as important for climate mitigation and biodiversity, with more references to carbon sinks, ecosystem resilience, and restoration. The 2009 climate-energy package (EP & CoE, 2009) increased the status of forests within EU climate governance, setting the stage for the integration of land use and carbon accounting (EP & CoE, 2013). There is strong transformative potential when considering problem definitions, such as ecological limits, systemic risks, as well as instrumental innovations, such as MRV for land use. However, justice-oriented value claims remain secondary in binding instruments.
- **Eco-modernisation**: In parallel, forests are being positioned as renewable assets that underpin bioenergy and the bioeconomy, as seen in RED I and the initial discussions around the Bioeconomy Strategy (adopted in 2012). Policy texts emphasise innovation, resource efficiency, and green growth, often through public–private partnerships and cohesion funding. This frame is transitional and enabling in that it integrates environmental objectives into growth-oriented regimes without displacing productivist perspectives.
- **Standards and trade (nested frame)**: Market governance shifts from a reliance on voluntary certification to mandatory due diligence, as set out in the FLEGT regulation and EUTR, which introduced the requirement for legality verification and risk assessment across supply chains (including Voluntary Partnership Agreements (VPAs) and customs interfaces). While this constitutes institutional innovation with high practical significance for the forest-based sector, supporting compliance-oriented governance and creating new linkages between trade, customs and forest policy, it remains conservative, emphasising enforcement over value-based reframing.

Governance changes and institutional reflections

EU forest governance underwent a dual reorientation in the 2000s. First, the crises framing brought ecological limits and systemic risks to the centre of policy discussions around forests, second, eco-modernisation recast forests as a strategic resource for low-carbon growth. Institutionally, this change was reflected in a reconfiguration of roles for different stakeholders. For example, scientific bodies increased in importance through IPCC-aligned metrics and JRC/EEA data infrastructures, NGOs translated anti-illegal logging campaigns into the FLEGT/EUTR. At the same time, industry actors became more important for bioenergy supply chains and related standard-setting processes. These shifts were operationalised through instrumental and coordination changes, such as the expansion of MRV for land use and forestry (e.g., LULUCF), the introduction of due diligence schemes (e.g., FLEGT/EUTR) and energy-climate integration (RED I), all of which were supported by cross-DG coordination (e.g., DG ENV, CLIMA, ENER, TRADE and AGRI). Compared to the 1990s, many of these innovations were anchored in legally binding instruments, representing a shift away from strategic signalling. Nevertheless, justice and participation remained secondary, with involvement mainly addressed through consultations and stakeholder forums (e.g., VPAs under FLEGT), rather than through rights-based provisions in hard instruments. Arguably, this has been reinforced by expert-led (epistemic) communities that prioritise measurement, verification, and risk management.

Assessment

In this decade, change is best characterised as the governance of transformation. This is driven by new regulatory combinations and cross-sectoral interlinkages (climate-energy-trade), as well as elements of governance *for* transformation, such as the development of MRV capacities and supply-chain infrastructures (e.g., FLEGT/EUTR). While the shift is significant in terms of coordination and compliance mechanisms, it is moderate in terms of perspectives on forests, as eco-modernisation internalises environmental objectives while retaining growth and productivist perspectives. This configuration appears relatively stable because key innovations are based on binding instruments, such as the climate-energy package, FLEGT, EUTR and LULUCF. In transformative terms, substantial progress was made on tools and integration, but only partial progress was made on values and power. For example, crisis and market frames co-dominate, while the justice frame remains peripheral. Overall, forests became increasingly embedded within EU climate policy, the energy transition, and due diligence regimes, resulting in a clear regulatory and infrastructural shift. This hotspot consolidated enabling capacities (MRV and supply-chain control) and steering mechanisms (energy and climate targets) but did not fully change the underlying governance structures. Accordingly, this period is best interpreted as a transitional period.

5.5.4 Hotspot 3: The 2020s and Target-Driven Integration and Procedural Recentralisation

Why is this a hotspot?

The 2020s mark a third turning point in EU forest governance, following the European Green Deal's shift from setting the agenda to implementation and enforcement. The database consists of binding legislation such as the Deforestation-Free Products Regulation (EP & CoE, 2023), the Nature Restoration Regulation (EP & CoE, 2024) and the revised LULUCF framework (EP & CoE, 2018), which establish a forest regime that is driven by targets and MRV. In terms of narratives, this period is characterised by the prominence of the "Policy architect" and "Ecological crisis" frames within hard policy instruments, alongside a more visible implementation-oriented role for "Science and systems thinking". The combination of quantified targets, expanded due diligence obligations, along supply chains, and geospatial/data traceability infrastructures suggests a coordinated consolidation of forest-relevant policy focused on compliance and monitoring.

Principal frames and transformative potential

- **Policy architect**: Binding legislation that focuses on enforceability, conformity, and risk control. For example, the EUDR builds on the EUTR and introduces deforestation-free due diligence, geolocation of plots, and risk mitigation measures. In addition, although not at the MS level, the Nature Restoration Regulation introduces restoration obligations, including those for forest ecosystems, with accompanying monitoring and reporting requirements. Meanwhile, the revised LULUCF framework tightens removals accounting towards the 2030 targets. The transformative potential resides in institutional capacity building (e.g., competent authorities and data systems) and regulatory coherence across sectors. However, the framework remains procedural and compliance-oriented.
- **Science and systems thinking**: The 2020s see the consolidation of integrated monitoring and performance governance, as reflected in the increased use of geospatial data, remote sensing/Copernicus, supply-chain traceability and EU-level information systems (including FISE interfaces). This framework provides governance ***for*** transformation by strengthening MRV and cross-sectoral coordination (e.g., between DG ENV-CLIMA-ENER-TRADE-AGRI). However, it does not change distributive arrangements or actor power, and decision authority largely remains with existing institutional actors.
- **Ecological crisis and environmental justice**: Crisis and justice rationales underpin strategic narratives, such as the Green Deal and the Biodiversity Strategy, and stakeholder positions. However, in legislation, their value-based content is largely proceduralised and translated into targets, eligibility criteria, and compliance checks. Justice remains minimal in binding instruments, with participation organised mainly via consultations and impact assessments. Territorial equity concerns remain mainly through funding and programming instruments.

Governance changes and institutional reflections

Although only the first half of the decade has passed, the 2020s have seen a shift towards target-driven integration from a variety of perspectives. Forests are becoming increasingly embedded in restoration obligations, carbon-removal accounting, and deforestation-free supply chains, primarily through legally binding instruments. This reorientation is accompanied by a reconfiguration of roles for the various stakeholders involved, such as competent authorities, customs/market surveillance bodies, and data providers (e.g., the JRC, EEA, and Copernicus) are becoming increasingly important, NGOs are retaining their watchdog and implementation functions, industry is focusing on compliance and traceability, and MS authorities remain central to enforcement. Operationally, the period expands the existing MRV infrastructure, including geolocation requirements, remote sensing and risk-scoring systems, and strengthens implementation toolkits through guidance, delegated acts, and compliance-based frameworks. Unlike the 1990s, and more extensively than in the 2000s, these changes are anchored in hard policy instruments, thus consolidating a compliance-centred focus. Justice-oriented language remains limited to legislative provisions, such as for consultations and stakeholder forums.

Assessment

The 2020s are best characterised as a period of governance of transformation, with policymaking focused on targets, due diligence regimes, and MRV requirements. This is supported by elements of transformative governance, such as expanded data infrastructure and inter-service coordination. As in the 2000s, change is significant in instrumental terms (as suggested by the development of compliance-based infrastructure, traceability systems, and restoration targets). However, it is only moderately paradigmatic, as crisis- and justice-oriented perspectives are largely proceduralised in legislation rather than being used to redefine underlying policy processes. This configuration appears relatively stable, nevertheless, outcomes remain dependent on the capacity of MS and cross-sector implementation. In terms of transformation, the period delivers substantial advances in tools and integration, but only partial progress in terms of values and power. The policy architect and ecological crisis frames dominate, while science and systems thinking is more visible as an implementation-oriented supporting frame and the justice frame remains secondary. In effect, the 2020s represent a consolidation phase in which forests are systematically and increasingly integrated into restoration, carbon accounting, and deforestation-free trade through stricter regulatory coordination and monitoring. However, this consolidation operates more as a technocratic approach that leaves distributive and participatory issues unresolved. This may be driven in part by epistemic communities that prefer technical standards over value-based changes.

5.5.5 Synthesis—Patterns Across Hotspots

The three hotspots highlight a shift from a period of new frames and perspectives, such as justice and territorial cohesion, to a period of consolidating policy instruments relevant to forests (Table 5.3). In the 1990s, the scope of forest-related policies expanded with the introduction of justice and territorial cohesion framings. However, most of these changes took the form of non-binding instruments (e.g., the EU Forest Strategy). Then, in the 2000s, forests were integrated into climate, energy and trade instruments, demonstrating regulatory uptake without a change in values (for example, eco-modernisation internalised environmental goals, but the productivist perspective remained). The 2020s saw the consolidation of compliance-focused approaches, with targets, traceability and MRV being enshrined in legislation. Policy architect and systems frames remained dominant, while justice and crisis frames continued to be primarily incorporated into soft instruments. These developments reflect layering rather than replacement. Innovations first appeared in non-binding strategies and programming in the 1990s. They were then instrumentalised through climate/market frameworks in the 2000s, before being increasingly turned into target-driven, enforcement-oriented approaches in the 2020s. The result is a governance framework with increased capacity and coordination, but no significant value-based changes. Commitments to distribution and participation remain weakly embedded in non-binding instruments, meaning that verification and risk control procedures have a greater influence on forests.

The analysis suggests that normative shifts are more likely to change forest governance when the following four enabling conditions[7] occur at the same time:

1. **Normative change embedded in binding legislation**. For example, only enforceable safeguards, such as justiciable rights and reviewability, can change incentives and behaviour. Procedural participation alone can be ignored or minimised during implementation.
2. **MRV coupled with justice and outcome indicators**. What is measured (and funded) is what is managed. Without distributional and territorial metrics, systems will optimise for carbon/area targets while neglecting equity and local impacts.
3. **Multi-level implementation capacity**. Rules travel through municipalities and competent authorities. Without adequate staffing, funding, data access, and mandates, ambitious legislation stalls or is unevenly implemented.
4. **Institutional gateways for deliberation**. Structured co-production and dispute resolution keep local knowledge and conflict management within the core of the policy, reducing resistance, improving fit-to-place and sustaining compliance over time.

These enabling conditions interact with several recurring trade-offs. For instance, there is a trade-off between capacity and conflict resolution. More specifically, MRV and due diligence systems can strengthen governance for/of transformation, but they risk depoliticising conflicts when justice becomes proceduralised rather than rights-based. Another example is the distinction between coordination and closure, whereby cross-DG alignment can improve implementation. However, it may also recentralise policy around technical compliance, thereby narrowing the scope for alternative value orientations and perspectives. Finally, there is the issue of scalability versus inclusion, while supply-chain and data infrastructures expand oversight, actor empowerment remains weak (e.g., legislation fails to embed participation and benefit-sharing).

One outstanding issue is the policy implications of these findings. The results emphasise the importance of explicitly linking soft and hard policy instruments. For instance, when justice frames are only present in strategies, it is essential to incorporate minimum rights-based safeguards into relevant binding instruments (e.g., participation with reviewable outcomes). MRV should also be applicable to alternative approaches, for example, by pairing restoration and carbon indicators with territorial and social inclusion metrics and connecting performance to funding conditions. Effective implementation also depends on the capacity of MS, notably sustained

[7] These four conditions cover the minimum complementary functions that a governance system needs for a shift to be effective, ranging from having enforceable norms to delivery capacity to institutionalised deliberation/co-production.

Table 5.3 Comparing transformative hotspots over time

Dimension	1990s	2000s	2020s	Transformative potential
Values	Value broadening toward justice and territorial inclusion. Multifunctionality formalised in strategy	Dual logics in terms of ecological limits and low-carbon growth. Justice remains secondary	Procedural narrowing where crisis and justice is proceduralised. Compliance and systems logics dominate	From broadening values to procedural narrowing
Principal frames	Environmental justice, Territorial equity & cohesion and Policy architect	Ecological crisis, Eco-modernisation, Standards & trade, Policy architect	Policy architect, Ecological crisis, and a more visible Science & systems thinking strand	
Actors	Visibility of civil society and regions in non-binding instruments	Increasing importance of scientific bodies, NGOs (illegality), and industry (bioenergy and standards)	Competent authorities, customs/market surveillance, data providers, NGOs watchdog role	From inclusion in strategies to technocratic enforcement and compliance
Instruments & coordination	Soft coordination (forest strategy). Cohesion & rural programming links	MRV. FLEGT/ EUTR. RED I. Cross-DG coordination	EUDR/NRR/ LULUCF tightening, geolocation, remote sensing, risk scoring; delegated acts/penalties	Increase in forest-relevant tools and coordination
Institutional anchoring	Mainly non-binding instruments	Mixed, with some binding instruments	Mainly compliance-centred and binding instruments	Progressive hardening over time
Transformative type	Transformation in governance (discursive). Partial institutional response	Governance of transformation (steering via new policy mixes). Enabling capacities	Governance of transformation, while limited transformation in governance	Discursive to instrumental to procedural consolidation
Transformative potential	Threshold moment; opening pathways	Transitional period. Infrastructure consolidation	Consolidation and delivery phase. Value narrowing	Directional change without changing values

investments in competent authorities, interoperable data systems, and local delivery so that enforcement can extend beyond Brussels (or the capital). In parallel, co-production (scientific and societal) must be embedded within relevant policy instruments in order to become institutionalised rather than relying mainly on technical consultations. Finally, there must be systematic coherence checks across policy domains, which should also include compliance and safeguards to ensure proper implementation.

5.6 Discussion and Conclusions

This chapter demonstrates that EU forest governance has evolved through discursive layering rather than paradigm replacement. Technocratic frames, such as the "Policy architect" frame and the nested frame "Standards and trade", have persisted since the 1960s and frequently define the procedural elements in binding instruments. Alongside these frames, transformative narratives (e.g., "Ecological crisis" and "Environmental justice") and enabling frames (e.g., "Eco-modernisation" and "Science and systems thinking") have expanded the range of policy instruments affecting EU forest governance, although with uneven institutional changes. The result is a hybrid governance system in which multiple frames co-exist, interact, and, at times, compete, with their institutional setups varying systematically depending on the type of instrument (e.g., hard versus soft) and period.

The three transformative hotspots further outline some of the temporal mechanics behind this layering. For example, in the 1990s, justice and territorial cohesion were incorporated into strategies and funding regulations, broadening the meaning of forests beyond productivist and administrative perspectives, however, with weak embedding in binding instruments. In the 2000s, crisis and eco-modernisation frames co-dominated as forests were integrated into climate-, energy- and trade-related instruments (e.g., MRV and due diligence). This resulted in a significant instrumental and infrastructural shift without a corresponding change in values. In the 2020s, the forest-relevant policy instruments increasingly coalesced around target- and MRV-driven implementation (e.g., EUDR and LULUCF), re-centring forest governance around compliance mechanisms and data infrastructures. Taken together, the hotspots trace a trajectory from a discursive opening in the 1990s through to regulatory consolidation in the 2000s to compliance-centred integration in the 2020s.

5.6.1 Implications for Transformative Change

From a transformative lens, substantial governance *for* transformation (such as national capacities, MRV, and knowledge infrastructure) and governance *of* transformation (such as forest governance being characterised by cross-sectoral policy mixes) have evolved over the last couple of decades. However, persistent changes in terms of values, power configurations, and rights remain minimal or partial in binding instruments.

Two structural features help to explain this:

- **Instrumental asymmetry**. Crisis and justice frames are most visible in non-binding strategies and programming documents, whereas policy architect and systems frames dominate binding instruments. This soft-hard gap effectively decouples value-based ambitions from enforceable safeguards, which may help to explain why distributive and participatory policy objectives rarely influence regulatory design or enforcement criteria.
- **Epistemic and administrative centrality**. The significant increase in monitoring, modelling, and risk management practices improves implementation capacity but also implies a technocratic approach. This narrows the space for rights-based or place-based policy objectives unless they are explicitly codified into binding instruments.

In other words, the conditions of policy design that make value-based changes more likely to reconfigure forest governance include normative anchoring in hard instruments, equity-sensitive MRV, as well as multi-level implementation capacity and institutional pathways that allow for deliberation. These conditions align with the hotspot synthesis and suggest ways to bridge the gap between soft and hard policy instruments while avoiding a compliance-centred approach.

5.6.2 *Methodological Reflections*

Using AI tools made it possible to conduct a large-scale longitudinal frame analysis of a diverse set of EU forest-related policy documents. The approach made it possible to identify nested relationships (e.g., the "Standards and trade" frame which was nested within the broader "Policy architect" frame, and "Territorial equity" as a spatialised and justice nested frame) as well as period-specific co-occurrence patterns that would have been challenging to detect through manual coding. However, AI also introduces traceability and replicability constraints, including possible prompt-induced biases, token-window limits, and occasional inconsistencies when parsing structured inputs. Given these limitations, LLM outputs should complement, rather than substitute, human analysis. This should be paired with transparent prompt disclosure, validation subsets, and protocol versioning. A more extensive methodological discussion around opportunities and risks of AI-supported policy research is provided in Aggestam et al. (2026a).

5.6.3 Conclusions

This chapter characterises EU forest governance as a layered and hybrid policy system in which seven recurring frames co-exist and interact over time. The regulatory-technocratic frames emphasise that compliance-based approaches underlie the majority of binding and forest-related instruments. In contrast, the crisis, justice, eco-modernisation and systems frames expand agendas primarily through strategies, programming documents and knowledge infrastructure. The three hotspots illustrate the progression from discursive opening to regulatory consolidation and, most recently, compliance-centred integration. The net effect is greater capacity and coordination, notably via MRV and due diligence, without a shift in the underlying value orientations of binding policy instruments.

The analysis implies that transformation depends on closing the soft-hard policy gap by:

- Embedding minimum safeguards (e.g., justice/equity) in legislation.
- Making MRV important for equity (e.g., pairing restoration/carbon indicators with territorial and social metrics and linking results to funding).
- Investing in multi-level enforcement capacity (e.g., competent authorities, interoperable data, sub-national delivery).
- Institutionalising co-production beyond technical consultations.

Without these steps, the range of values that shape EU forest governance and associated binding instruments is likely to narrow further as a result of continued procedural tightening.

Finally, methodologically, the AI-based frame analysis extended the scope and qualitative aspects of the research process (e.g., enabling longitudinal pattern detection). However, its epistemic limitations, such as black-box inference, prompt sensitivity and input-parsing constraints, require transparent protocols, validation and theory-led interpretation. AI should be a complement, not a substitute, offering a practical and resource-saving approach to track policy frames in EU forest governance.

References

Abson, D. J., Fischer, J., Leventon, J., Newig, J., Schomerus, T., Vilsmaier, U., Von Wehrden, H., Abernethy, P., Ives, C. D., Jager, N. W., & Lang, D. J. (2017). Leverage points for sustainability transformation. *Ambio, 46*(1), 30–39. https://doi.org/10.1007/s13280-016-0800-y

Aggestam, F. (2024a). *European forest policy and governance an integrative analytical framework (1st edition)*. Routledge.

Aggestam, F. (2024b). Through smoke to policy: Framing the EU forest fire policy landscape. *Land, 13*(9), 1450. https://doi.org/10.3390/land13091450

Aggestam, F. (2025). *European forest policy and governance: An integrative analytical framework* (1st ed.). Routledge. https://doi.org/10.4324/9781003198987

Aggestam, F., & Giurca, A. (2021). The art of the "green" deal: Policy pathways for the EU forest strategy. *Forest Policy and Economics, 128*, Article 102456. https://doi.org/10.1016/j.forpol.2021.102456

Aggestam, F., Trummer, J., & Wilkes-Allemann, J. (2026a). Sixty years of EU forest policy discourse: An AI-assisted frame analysis of policy documents. *Forest Policy and Economics, In press.*

Aggestam, F., Trummer, J., & Wilkes-Allemann, J. (2026b). Supplementary materials—Sixty years of EU forest policy discourse: An AI-assisted frame analysis of policy documents. *Forest Policy and Economics, In press.*

Aggestam, F., Trummer, J., Wilkes-Allemann, J. (2026c). EU forest policy database (1960–2025). Mendeley Data, V1. https://doi.org/10.17632/7mdts45mmp.1

Aggestam, F., Weiss, G., Elomina, J., & Pülzl, H. (2025). *Taking a bird's eye view of EU wood-based policy. Untangling policy, institutional, and actor frameworks affecting the wood based sectors* (Knowledge to Action) [Knowledge to Action]. European Forest Institute. https://doi.org/10.36333/k2a07

Alshami, A., Elsayed, M., Ali, E., Eltoukhy, A. E. E., & Zayed, T. (2023). Harnessing the power of ChatGPT for automating systematic review process: Methodology, case study, limitations, and future directions. *Systems, 11*(7), 351. https://doi.org/10.3390/systems11070351

del Barrio, D., & Gatica-Perez, D. (2023). Framing the news: From human perception to large language model inferences. *Proceedings of the 2023 ACM International Conference on Multimedia Retrieval*, pp. 627–635. https://doi.org/10.1145/3591106.3592278

Bauer, M. W., & Knill, C. (2014). A conceptual framework for the comparative analysis of policy change: Measurement, explanation and strategies of policy dismantling. *Journal of Comparative Policy Analysis: Research and Practice, 16*(1), 28–44. https://doi.org/10.1080/13876988.2014.885186

Baumgartner, F. R., & Jones, B. D. (2009). *Agendas and instability in American politics* (2. ed). The University of Chicago Press.

Baumgartner, F. R., & Mahoney, C. (2008). The two faces of framing: Individual-level framing and collective issue definition in the European Union. *European Union Politics, 9*, 435–449. https://doi.org/10.1177/1465116508093492

Bender, E. M., Gebru, T., McMillan-Major, A., & Shmitchell, S. (2021). On the dangers of stochastic parrots: Can language models be too big? In *Proceedings of the 2021 ACM Conference on Fairness, Accountability, and Transparency*, 610–623. https://doi.org/10.1145/3442188.3445922

Benford, R. D., & Snow, D. A. (2000). Framing processes and social movements: An overview and assessment. *Annual Review of Sociology, 26*, 611–639.

Bijker, R., Merkouris, S. S., Dowling, N. A., & Rodda, S. N. (2024). ChatGPT for automated qualitative research: Content analysis. *Journal of Medical Internet Research, 26*, Article e59050. https://doi.org/10.2196/59050

Björnehed, E., & Erikson, J. (2018). Making the most of the frame: Developing the analytical potential of frame analysis. *Policy Studies, 39*(2), 109–126. https://doi.org/10.1080/01442872.2018.1434874

Buijs, A. E., Arts, J. M. B., Elands, H. M. B., & Lengkeek, J. (2011). Beyond environmental frames: The social representation and cultural resonance of nature in conflicts over a Dutch woodland. *Geoforum, 42*, 329–341. https://doi.org/10.1016/j.geoforum.2010.12.008

Capoccia, G., & Kelemen, R. D. (2007). The study of critical junctures: Theory, narrative, and counterfactuals in historical institutionalism. *World Politics, 59*(3), 341–369. JSTOR.

Chubb, L. A. (2023). Me and the machines: Possibilities and pitfalls of using artificial intelligence for qualitative data analysis. *International Journal of Qualitative Methods, 22*, 16094069231193592. https://doi.org/10.1177/16094069231193593

CoE. (2005). *Council Regulation (EC) No 2173/2005 of 20 December 2005 on the establishment of a FLEGT licensing scheme for imports of timber into the European Community*. Official Journal of the European Union.

Curry, N., Baker, P., & Brookes, G. (2024). Generative AI for corpus approaches to discourse studies: A critical evaluation of ChatGPT. *Applied Corpus Linguistics, 4*(1), Article 100082. https://doi.org/10.1016/j.acorp.2023.100082
Daviter, F. (2007). Policy framing in the European Union. *Journal of European Public Policy, 14*(4), 654–666. https://doi.org/10.1080/13501760701314474
Dewulf, A., Craps, M., & Dercon, G. (2004). How issues get framed and reframed when different communities meet: A multi-level analysis of a collaborative soil conservation initiative in the Ecuadorian Andes. *Journal of Community and Applied Social Psychology, 14*, 177–192.
Dryzek, J. S., & Pickering, J. (2018). *The politics of the anthropocene* (1st ed.). Oxford University Press Oxford. https://doi.org/10.1093/oso/9780198809616.001.0001
EC. (1998). *COM 1998/649 final: A forestry strategy for the European Union*. European Commission.
EC. (2008). *Commission Regulation (EC) No 1024/2008 of 17 October 2008 laying down detailed measures for the implementation of Council Regulation (EC) No 2173/2005 on the establishment of a FLEGT licensing scheme for imports of timber into the European Community*. Official Journal of the European Union.
EC. (2011). Our life insurance, our natural capital: An EU biodiversity strategy to 2020. In *COM(2011) 244 final*. European Commission.
EC. (2012). *COM(2012) 60 final: Innovating for Sustainable Growth: A Bioeconomy for Europe*. European Commission.
EC. (2013). *COM(2013) 659 final: A new EU Forest Strategy: For forests and the forest-based sector*. European Commission. http://ec.europa.eu/agriculture/forest/strategy/communication_en.pdf
EC. (2018). *COM(2018) 673 final: A sustainable Bioeconomy for Europe: Strengthening the connection between economy, society and the environment*. European Commission.
EC. (2019). *COM(2019) 640 final: The European Green Deal*. European Commission. https://eur-lex.europa.eu/legal-content/EN/TXT/?uri=uriserv:OJ.C_.2020.104.01.0003.01.ENG. Retrieved 8 Apr 2025
EC. (2021). *COM(2021) 572 final: New EU Forest Strategy for 2030*. European Commission.
Elomina, J., & Pülzl, H. (2021). How are forests framed? An analysis of EU forest policy. *Forest Policy and Economics, 127*, Article 102448. https://doi.org/10.1016/j.forpol.2021.102448
Entman, R. M. (2007). Framing bias: Media in the distribution of power. *Journal of Communication, 57*(1), 163–173. https://doi.org/10.1111/j.1460-2466.2006.00336.x
EP & CoE. (1966). *Council Directive 66/404/EEC of 14 June 1966 on the marketing of forest reproductive material*. Official Journal of the European Communities.
EP & CoE. (1985). *Council Regulation (EEC) No 2088/85 of 23 July 1985 concerning the integrated Mediterranean programmes*. Official Journal of the European Communities.
EP & CoE. (1992). *Directive 92/43/EEC: The conservation of natural habitats and wild fauna and flora*. Official Journal of the European Communities. https://eur-lex.europa.eu/legal-content/EN/TXT/?uri=CELEX%3A31992L0043
EP & CoE. (1999). *Council Regulation (EC) No 1257/1999 of 17 May 1999 on support for rural development from the European Agricultural Guidance and Guarantee Fund (EAGGF) and amending and repealing certain Regulations*. Official Journal of the European Communities.
EP & CoE. (2009). *Directive 2009/28/EC of the European Parliament and of the Council of 23 April 2009 on the promotion of the use of energy from renewable sources and amending and subsequently repealing Directives 2001/77/EC and 2003/30/EC*. Official Journal of the European Union.
EP & CoE. (2010). *Regulation 995/2010: Laying down the obligations of operators who place timber and timber products on the market*. Official Journal of the European Union.
EP & CoE. (2013). *Decision No 529/2013/EU of the European Parliament and of the Council of 21 May 2013 on accounting rules on greenhouse gas emissions and removals resulting from activities relating to land use, land-use change and forestry and on information concerning actions relating to those activities*. Official Journal of the European Union.

EP & CoE. (2018). *Regulation (EU) 2018/841 of the European Parliament and of the Council of 30 May 2018 on the inclusion of greenhouse gas emissions and removals from land use, land use change and forestry in the 2030 climate and energy framework, and amending Regulation (EU) No 525/2013 and Decision No 529/2013/EU*. Official Journal of the European Union.

EP & CoE. (2023). *Regulation 2023/1115: Making available on the Union market and the export from the Union of certain commodities and products associated with deforestation and forest degradation and repealing Regulation (EU) No 995/2010*. Official Journal of the European Union.

EP & CoE. (2024). *Regulation (EU) 2024/1991: Nature restoration and amending Regulation (EU) 2022/869*. Official Journal of the European Union. https://eur-lex.europa.eu/eli/reg/2024/1991/oj

Fan, Y., Agaru, P., Li, P., & Li, H. (2023). *Uncovering the potential of ChatGPT for discourse analysis in dialogue: An empirical study* (version 2). arXiv. https://doi.org/10.48550/ARXIV.2305.08391

Feindt, P. H., & Oels, A. (2005). Does discourse matter? Discourse analysis in environmental policy making. *Journal of Environmental Policy & Planning, 7*(3), 161–173. https://doi.org/10.1080/15239080500339638

Goffman, E. (1974). *Frame analysis—an essay on the organisation of experience*. Harper & Row.

Hajer, M. (1997). *The politics of environmental discourse: Ecological modernization and the policy process*. Oxford University Press.

Howlett, M., Rayner, J., Goehler, D., Heidbreder, E., Perron-Welch, F., Rukundo, O., Verkooijen, P., & C., W. (2010). Overcoming the challenges to integration: Embracing complexity in forest policy design through multi-level governance. In J. Rayner, A. Buck, & P. Katila (Eds.), *Embracing complexity—meeting the challenges of international forest governance* (pp. 93–110). International Union of Forest Research Organisations (IUFRO).

Howlett, M., & Rayner, J. (2013). Patching vs packaging in policy formulation: Assessing policy portfolio design. *Politics and Governance, 1*(2), 170–182.

van Hulst, M., & Yanow, D. (2014). From policy 'frames' to 'framing': Theorizing a more dynamic, political approach. *The American Review of Public Administration*. https://doi.org/10.1177/0275074014533142

Kahneman, D., & Tversky, A. (2000). *Choices*. Cambridge University Press.

Kaufman, S., & Gray, B. (2003). Using retrospective and prospective frame elicitation to evaluate environmental disputes. In L. Bingham & R. O'Leary (Eds.), *The Promise and Performance of Environmental Conflict Resolution*. Resources for the Future.

Kingdon, J. W. (1984). *Agendas*. Alternatives and Public Policies.

Larson, A. M., Mausch, K., Bourne, M., Luttrell, C., Schoneveld, G., Cronkleton, P., Locatelli, B., Catacutan, D., Cerutti, P., Chomba, S., Djoudi, H., Ihalainen, M., Lawry, S., Minang, P., Monterroso, I., Myers, R., Naito, D., Pham, T. T., Reed, J., Stoian, D. (2021). Hot topics in governance for forests and trees: Towards a (just) transformative research agenda. *Forest Policy and Economics, 131*, 102567. https://doi.org/10.1016/j.forpol.2021.102567

Meadows, D. (1999). *Leverage points: Places to intervene in a system*. The Sustainability Institute. https://donellameadows.org/archives/leverage-points-places-to-intervene-in-a-system/

Patterson, J., Schulz, K., Vervoort, J., Van Der Hel, S., Widerberg, O., Adler, C., Hurlbert, M., Anderton, K., Sethi, M., & Barau, A. (2017). Exploring the governance and politics of transformations towards sustainability. *Environmental Innovation and Societal Transitions, 24*, 1–16. https://doi.org/10.1016/j.eist.2016.09.001

Pülzl, H., Kleinschmit, D., & Arts, B. (2014). Bioeconomy—an emerging meta-discourse affecting forest discourses? *Scandinavian Journal of Forest Research, 29*(4), 386–393. https://doi.org/10.1080/02827581.2014.920044

Rein, M., & Schön, D. A. (1996). Frame-critical policy analysis and frame-reflective policy practice. *Knowledge and Policy, 9*(1), 85–104.

Ryazanov, I., Öhman, C., & Björklund, J. (2024). How ChatGPT changed the media's narratives on AI: A semi-automated narrative analysis through frame semantics. *Minds and Machines, 35*(1). https://doi.org/10.1007/s11023-024-09705-w

Schmidt, V. A., & Radaelli, C. M. (2004). Policy change and discourse in Europe: Conceptual and methodological issues. *West European Politics, 27*(2), 183–210. https://doi.org/10.1080/0140238042000214874

Schön, D. A., & Rein, M. (1994). *Frame reflection: Toward the resolution of intractable policy controversies*. Basic Books.

Shmueli, D., Elliott, M., & Kaufman, S. (2006). Frame changes and the management of intractable conflicts. *Conflict Resolution Quarterly, 24*(2), 207–218. https://doi.org/10.1002/crq.169

Snow, D. A., Rochford, E. B., Worden, S. K., & Benford, R. D. (1986). Frame alignment processes, micromobilization, and movement participation. *American Sociological Review, 51*, 464–481.

True, J. L., Jones, B. D., & Baumgartner, F. R. (2019). *Punctuated-equilibrium theory: Explaining stability and change in public policymaking*. Routledge.

Vliegenthart, R., & van Zoonen, L. (2011). Power to the frame: Bringing sociology back to frame analysis. *European Journal of Communication, 26*(2), 101–115. https://doi.org/10.1177/0267323111404838

Winkel, G., & Sotirov, M. (2016). Whose integration is this? European forest policy between the gospel of coordination, institutional competition, and a new spirit of integration. *Environment and Planning C: Politics and Space, 34*(3), 496–514, https://doi.org/10.1068/c1356j

Winkel, G. (2017). *Towards a sustainable European forest-based bioeconomy-assessment and the way forward. What Science Can Tell Us 8*. European Forest Institute. https://www.efi.int/publications-bank.

Wolfslehner, B., Pülzl, H., Kleinschmit, D., Aggestam, F., Winkel, G., Candel, J., Eckerberg, K., Feindt, P., McDermott, C., Secco, L., Sotirov, M., Lackner, M., & Lazya Roux, J. (2020). *European forest governance post-2020* (From Science to Policy 10). European Forest Institute. https://doi.org/10.36333/fs10.

Part III
Cross-Cutting Dimensions of Forest Governance

Chapter 6
Roots of Ownership: Relations Between Forest Tenure and Forest Governance

Gerhard Weiss and Ivana Zivojinovic

Abstract The type of ownership (or tenure) is an integral part of the institutional framework for forestry, influencing both forest management and policy implementation. Forest ownership and governance are interlinked with different forest ownership types requiring corresponding administrative structures and policy instruments. Patterns of forest ownership also influence the policy discourse and the policymaking processes. This chapter aims to present the latest knowledge on the distribution, trends and impacts of forest ownership in Europe. It provides an overview of the historical development of forest ownership, forest management, and forest governance co-developed in Europe, as well as a theoretical section on property rights. These contributions help to improve our understanding of current forest-related issues and possible solutions. The main finding is that the scientific, timber-production-oriented concept of the "normal forest" still dominates forestry thinking and correlates with the assumption of a rational, commercially oriented "normal forest owner" in forest policy. In reality, however, the decisions of forest owners' are guided by many other goals and values. Furthermore, ownership types beyond public and private are highly diverse and increasing due to current societal changes. In designing policy instruments, governance implications include the need for policymakers to address the variety of real types of owners, rather than an "ideal type" or "normal" forest owner. Moreover, the growing number of "non-traditional" owners with various objectives and preferences could be considered as an asset rather than a threat to forest management, especially in promoting climate-smart forestry and transitioning to a sustainable society.

G. Weiss (✉) · I. Zivojinovic
University of Natural Resources and Life Sciences, Vienna, Austria
e-mail: gerhard.weiss@boku.ac.at

F. Aggestam (ed.), *Governing Forests Through Transformative Change*, Advances in Global Change Research 70, https://doi.org/10.1007/978-3-032-19072-7_6

6.1 The Relevance of Forest Ownership for Forest Governance

Who owns this forest? What is the ownership pattern of forests in your country? These are the key questions to consider when discussing any aspect of forest governance. The type of ownership (or tenure) is a fundamental aspect of the institutional framework for any forest-related issues, influencing the management of forests and the implementation of policies. It also indirectly sets the framework for any forest-related discourse in a country and the entire policymaking process. This chapter aims to discuss these questions in more detail.

But how much is actually known about the relationship between forest ownership and governance? There has recently been growing interest in this question in Europe. However, it is interesting that there is not a really large body of knowledge about it. Ownership is a sensitive political issue that is usually not questioned. Governments rarely alter the distribution of tenure, except during major institutional changes, such as the nationalisation of property in socialist Eastern Europe and its subsequent restitution following the fall of communism. Similarly, any regulation of forest management alters property rights, and such changes are therefore among the most difficult to implement due to opposition from forest landowners. The sensitivity of the topic is also evident in the State of Europe's Forests reports, which do not extensively address this issue (Forest Europe, 2020). They present the basic distribution of public and private ownership but do not analyse forest management within the different ownership categories further. Similarly, scientific research on the impact of different forms of ownership on forest management and the provision of ecosystem services is similarly sensitive, and such research topics are rarely financed.

When discussing the relationship between forest ownership, forest management, and governance, it is important to first clarify the type of questions this involves. The most straightforward question would concern the difference between public and private ownership. Another question might be which is better, but the answer depends on what one wants from the forests. The answer will inherently depend on value judgements. However, before addressing this, it is necessary to identify the impact of ownership—a question that is not easy to answer in itself. This is for two reasons: first, there is a lack of comparative studies, second, it is difficult to isolate the effects of ownership since various factors interfere with the plain form of tenure. Besides tenure, forest governance comprises other institutional and policy factors, which make it difficult to determine the influence of ownership alone. While it is reasonable to assume that private owners would pursue different objectives to public owners, management orientations vary within both types pf forest.

Another question relates to the implications of different types or patterns of forest ownership mean for the governance system, e. g., the necessary administrative structure and policy instruments for implementing policy goals. A further question concerns its implications for policymaking, given that landowners are important actors in the policymaking system. The aim of this chapter is to present the current knowledge on these three issues with a focus on Europe and to discuss their implications for transformative societal change.

Fundamentally, this chapter demonstrates that the simple distinction between private and public ownership is too simplistic to capture the diverse forms of ownership within and between these two fundamental types. From an institutional perspective, common ownership is more complex than has been suggested thus far, encompassing a range of examples, including old and new forms of common or joint ownership. The chapter also draws attention to the significant differences between individual smallholders and large private companies, and between state forests and those under municipal or communal ownership. It is also important to distinguish between the formal ownership of land and the ownership or use rights for trees and forest products. Additionally, the size of the property can significantly impact the owner's ability to manage the forest or achieve their management goals. Gender is another aspect with a range of potential implications for forest governance that has largely been neglected thus far. For instance, men and women may have different goals and approaches to forest management, or different support needs—questions that have hardly been addressed in research and practice yet. Furthermore, women are underrepresented in the institutional structures of the both public and private forestry sectors.

The primary message of this chapter is that more effective forest policies require a more thorough consideration of the diversity of ownership types. This is particularly pertinent in the context of the global biodiversity and climate crises, which present significant challenges for forest management. A new vision for sustainable forest management (SFM) should prioritise biodiversity and climate-smart practices. Implementing this policy goal requires an understanding of the diverse goals and motivations of different types of owners.

This chapter aims to go even further. It analyses ownership as part of the broader institutional framework of current forest management, including the concept of sustainability. The concept of SFM emerged in the context of the industrialisation. Transforming our economy requires shifting the focus to include the forest sector, necessitating a different approach to forest management.

Against this backdrop, the chapter begins with an overview of the historical development of current ownership patterns and of the emergence of modern forest management. It then presents and discusses fundamental concepts related to ownership and tenure within the institutional framework of forest governance. The central sections present our current knowledge of the impact of forest ownership on forest management and its outcomes across Europe, as well as the relationship between ownership and policy. Finally, the chapter concludes by outlining the main implications of these findings for forest governance and the transition to a sustainable society.[1]

---

[1] This chapter draws on the results of the European research initiative COST Action FACESMAP. This initiative produced a variety of publications, including a thorough study of forest ownership in the ECE region (UNECE/FAO, 2020). To provide a broader perspective, the chapter also incorporates relevant studies on forest history, management, the governance of ecosystem services, and innovation.

6.2 Historical Roots of Ownership, Forms of Forest Management and Forest Governance in Europe

Examining the historical development of forest ownership in Europe not only clarifies the current distribution of ownership forms but also highlights the impact of tenure or stewardship on forest management practices. This history is closely linked to scientific forestry concepts, the idea of SFM and modern forest governance. Understanding this history if fundamental to grasping current forest management practices and how these can be adapted to new challenges.

Before examining the influence of forest ownership on contemporary governance, this chapter provides an overview of the discourse on SFM. Foresters often claim that modern forest laws and the scientific management approaches have prevented the depletion of European forests. This is connected to the idea that SFM emerged alongside modern forest management in the form of "rational forestry" (Plochmann, 1985). Some even claim to have discovered the concept of "sustainability" when Hans Carl von Carlowitz first used the term "*Nachhaltigkeit*" in his book "Sylvicultura oeconomica" (Carlowitz, 1713). Of course, forests and other resources have been managed both sustainably and unsustainably throughout history. The analysis should first identify the various historical forms of forest management and connect them to their institutional contexts, including specific actors, interests, and power dynamics. Such a social-ecological analysis approach reveals the larger social, economic and institutional transformations visible that occurred behind the changing forest management systems. It will also highlight that the modern form of forest management was not primarily an instrument for sustainability or ecological conservation, but rather a means of promoting industrial forest use over against traditional, farm-related forest use by local populations.

6.2.1 Historical Development of Forest Governance and Ownership in Europe

Distinct periods of governance are revealed when forest history is reviewed. Forests have been managed by local communities, absolute monarchies, market economy, and state regulations (Weiss, 2001). The socialist period in Eastern Europe also warrants mention in recent history, as it had a significant impact on forests and their management (Kuemmerle et al., 2009; Plieninger & Schaich, 2014).

In ancient or prehistoric times, before the establishment of sovereignty (until around the fourteenth century), forests were either freely accessible or, in areas close to settlements, were typically used as a common resource (Jeanrenaud, 2001). Although knowledge sources are scarce, it can be assumed that forests were often managed communally by the local population (Mölder et al., 2021). After the Middle Ages, monarchs or sovereigns claimed unused forests as state land, and landlords began to acquire the power to regulate the local commons or to appropriate all rights of use, as ownership (Weiss, 2001). While the old German law was based on the concept of rights of use, the new Roman law introduced the institution of property. It is important to recognise that the literature on the commons and the historical roots of German forestry may be influenced by the authors' values (Sieferle, 1982 [2001]; Weiss, 2001). Given the historical context and limited data, interpretations should be made with caution. With the rise of scientific forestry in the eighteenth and nineteenth centuries, the commons were often abolished and the land divided among commoners and right holders into smaller private parcels, resulting in the fragmentation (or parcelisation) of forests which was often exacerbated by inheritance processes (Mölder et al., 2021). This process of privatisation and the promotion of "efficient" forest management practices led to to the abandonment of historical management systems associated with traditional farming, such as coppicing or wood pasturing, and the conversion of forests into high forests (Mölder et al., 2021). In many regions of Europe, the common or community ownership of forests has been maintained in some form of joint private ownership to this day (Lawrence et al., 2021).

Following the establishment of territorial sovereignty, the sovereigns enacted forest laws, which were often justified by alleged timber shortages. Initially, historians studied this as a societal crisis driven by an ecological or energy problem, but it was later revealed that it was primarily a pretext **to assert control over natural resources**, particularly against local landowners and populations (Jeanrenaud, 2001; Radkau, 1986; Sieferle, 1982). The so-called "wood famine" (in German, "*Holznot*") was widely used to allocate resources to strategic industries, generate revenue through manufacturing, and expand state power via forest legislation and administration (Weiss, 2001).

6.2.2 Sustained Yield of Timber and the Emergence of Modern Forest Management

Although it is often claimed that forest laws and the rational, scientific form of forest management saved the threatened forest resources, the evidence shows that, rather than protecting them, sovereigns strongly supported the exploitation of forest resources, initially for fuelwood to support the salt and mining industries, and later for providing timber.

Carlowitz's "Sylvicultura oeconomica" is a perfect illustration of how forest management responded to the societal context of the time, characterised by an increasing need for wood to support emerging industries. It was one of the first comprehensive technical books dedicated to forestry (Huss & Gadow, 2012). It was written by an Enlightenment-era chief miner and forest enthusiast. Forestry was essential for mining, providing support structures in the mines and fueling further processing stages, before mineral coal became available (Sieferle, 1982). The book describes the many values of (managed) forests and the many uses for wood in great detail. It also highlights the numerous pressures on forest use and the apparent shortage of wood (in German, "*Holzmangel*"), which is referenced in the book's lengthy title. The book describes the often depleted state of the forests and laments their careless use, while primarily dealing with reforestation and tending to them. This perfectly illustrates the various needs of the population and forest-related industries at that time, as well as the customary practices and existing knowledge concerning forest management. Finally, it presents a proposed form of forest management to ensure a continuous wood supply. Carlowitz's aimed for the "continuous, steady and sustained use" ("*continuirliche beständige und nachhaltende Nutzung*", Carlowitz, 1713) of the forests for an important wood supply. Although he did not coin the term "sustainability" ("*Nachhaltigkeit*"), he used it alongside several synonyms to express a core principle of the forest management he advocated. "Sustained" was just one such term, and rather than presenting it as new, he framed it as a long-standing, natural approach to forestry (Huss & Gadow, 2012). While promoting the systematic and sustained management of forests, he criticised the negligence ("*Nachlässigkeit*") in contemporary practice (Huss & Gadow, 2012), indicating that although good examples existed, they were not widely applied.

"Sylvicultura oeconomica" was written to meet the needs of the mining industry. It is important to note is that modern forest management is closely tied to timber production. The new demand for timber, compared to other uses of wood (e.g., fuelling the salt or iron works), can be assumed to have been a significant factor in shaping the way forest management is practised today, and thus the appearance of most forests (Weiss, 2001). Today's forest management system is oriented towards the optimal production of timber, i.e. sawn wood. This implies several aspects that are specific to this product, such as the need for straight logs and a long enough rotation period to reach the required log. It also involves the exclusion of other uses such as grazing, needle litter collection and other agricultural uses, which could not only might reduce the amount of wood, but also the quality of sawn wood.

Conflicts among forest users involved both the quantity and quality of forest management (Weiss, 2001). A commonly held view of how 'modern' forest management has succeeded in practice is that coal has replaced wood as a fuel source. Nevertheless, demand for forest resources has remained high despite this shift. Forest laws promoting modern forest management, which often excluded local people and livestock in order to meet the rising timber demand driven by the central role of timber in construction and economic growth (Sieferle, 1982). The technical core of today's forest management system aims **to produce high-quality timber efficiently** (Mitscherlich, 1963). Against this background, it becomes clear that the scientific approach to forest management was closely tied to the interests of powerful industries and states, rather than to preserving forests for their ecological value or for the everyday needs of the population (Scott, 1998; Weiss, 2001). The concept of implied sustainability is known as the "sustained yield" concept, and the scientific concept at its core is the "normal forest" concept—a forest with multiple even-aged stands representing each age class, which effectively provides continuous timber production (Leslie, 1966). The development of this concept is inextricably linked to the economic and political circumstances of the time, and has repercussions for the institutional frameworks for forest tenure and management.

The introduction and enforcement of scientific forestry are closely linked to changes in ownership and tenure. In many European regions, state authorities campaigned for the abolition of the commons and the privatisation of communal woodlands. Agricultural and forest commons were often divided among the commoners and those with usage rights (Mölder et al., 2021). Where the interests of different groups overlap in the same forest area, common ownership regimes may be more appropriate than privatised ownership. Additionally, parcellation makes managing small properties very difficult, and many small parcels are abandoned in European regions. A prototypical example of the regulatory and management problems created by private ownership is the protection of mountain forests that safeguard settlements or traffic lines from rockfall or avalanches (Glück, 2002; Weiss, 2001).

6.2.3 How Forest Ownership, Governance and Management are Closely Intertwined

Overall, the analysis suggests that the growing interest in forests has resulted in access shifting from local populations to states or landlords, and shifted the concept of stewardship shifting from common use to private (or public) ownership. **Modern forest sciences** have played a crucial role in exerting control and power over forest resources and people (Winkel, 2012). The concepts underlying scientific forestry are connected to an approach to forest management that prioritises resource extraction and is embedded in an autocratic governance system. This approach is oriented towards maximising a steady supply of resources for large industries and, eventually, fiscal income for the state (Scott, 1998). Simplifying forests into a "one-commodity machine" has significantly impacted biodiversity and increased risk, leading to the need for restoration forestry, and neglecting all other forest uses and social values. More recent debates about SFM, sometimes referred to as multiple-use forestry, highlight the social values of forests and forestry. This can be seen as a critique of the sustained yield approach to SFM, which is often associated with industrially oriented forest management (Glück, 1987; Koch & Kennedy, 1991). This new, more inclusive approach to forests and forest management is reflected in the definition of the Forest Principles at the United Nations Conference on Environment and Development (UNCED) in 1992:

> Forest resources and forest lands should be sustainably managed to meet the social, economic, ecological, cultural, and spiritual needs of present and future generations (UN, 1992)

This was also included in the definition of SFM by the Ministerial Conference on the Protection of Forests in Europe in 1993 (Resolution H1, Forest Europe, 1993). These rising inclusive and integrated definitions do not necessarily replace conventional thinking in forest policy. In fact, scholars have found that neoliberal, commercially and economically oriented discourses still dominate global forestry discourse and practice. This results in the structural exclusion or suppression of national or local views, groups or communities (Leipold, 2014), including female images of forest professionals (Lidestav & Egan Sjölander, 2007).

The scientific forestry model had far-reaching effects, resulting in a widespread shift from common ownership to private or state ownership. It promoted a standardised management system designed for large-scale, industrial forests. One could question whether even-aged spruce monocultures were ever adequate for forest farming. State authorities strongly promoted them after World War II. However, this system requires regular maintenance, which suits large, regularly managed forest holdings better than small-scale farm forests. These monocultures are also more susceptible to biotic and abiotic damage, a risk that is exacerbated by climate change. This highlights the need to adapt forest management systems to different ownership regimes, such as those of farmers or urban owners. At the same time, ownership regimes can be adapted to societal needs. For example, local commons can be created or recreated when the public interest overrides private interests. In Northern Italy, for example, new legislation permits the common management of abandoned land if the owners do not assert their interests.

The European COST Action FACESMAP, which focused on forest ownership, revealed that **different ownership types** have distinct management objectives, attitudes, skills and capacities. Given these differences in objectives, preconditions and contexts, different management approaches, information and advice are required. As shown above, this applies both to large forest holdings and small-scale farm owners, as well as for emerging new, non-traditional and urban forest owners. Specific forest management approaches for these different owner types have yet to been developed. These new management methods do not necessarily entail entirely novel silvicultural techniques, but rather the appropriate application to specific situations and owners' preferences. In this respect, analysis should not be limited to silvicultural prescriptions. It should also consider the socio-economic context, the wider policy framework, and emerging forms of management or new business models (Weiss et al., 2017). For instance, forest owners who simply aim to keep the forest stable rather than expecting an income, would want a model where management interventions are minimised and only cover the costs. Conversly, forest enthusiasts without a background in forestry but with an entrepreneurial attitude may be interested in exploring new opportunities with their forest land. They could develop new forest structures for recreational or aesthetic purposes, as well as new products or services for their own use or for sale. There are interesting examples of such fresh and innovative market products that have been developed in these circumstances (Weiss et al., 2023). New organisational models and ownership forms have emerged, including charities, social enterprises, joint ownership and community stewardship models (Ludvig et al., 2019, 2020; Weiss et al., in progress).

6.3 Basic Concepts of Forest Ownership

Several terms are used to describe forest ownership and tenure. While some of these terms appear straightforward, others are less clearly defined. Are property, ownership and tenure the same thing? In fact, even the basic term of ownership is understood differently in different contexts and countries. To achieve a clear understanding, these terms require explanation and definition. This not only brings clarity to any discussion about forest ownership and helps us to gain a deeper understanding of the concepts and their implications.

6.3.1 Towards a Comprehensive Definition of Forest Ownership

For the purposes of technical discussions and international data collection, the FAO defines forest ownership as follows:

> [Forest ownership] generally refers to the legal right to freely and exclusively use, control, transfer, or otherwise benefit from a forest. Ownership can be acquired through transfers such as sales, donations, and inheritance (FAO, 2018, p. 16).

The following important explanatory note further clarifies this definition:

> Forest ownership refers to the ownership of the trees growing on land classified as forest, regardless of whether or not the ownership of these trees coincides with the ownership of the land itself (FAO, 2018, p. 16).

While the initial definition indicates that ownership should include comprehensive and exclusive **rights** to rule and reign over one's property, the subsequent clarification begins to relativise this notion. Not only may the ownership of the land differ from the ownership of the trees, but the entire use of the forest land is also often regulated. Animals, for instance, may fall under various separate regulations. For example, farm animals may or may not be permitted to graze in the forest, and this right may be held by the landowner, a neighbour, or another rights holder. Game or wildlife management typically falls under a different jurisdiction, with hunting rights are held by the landowner in some countries and by the state in others. Rare or endangered species may again be administered under a different policy, meaning their use or management may be specifically regulated or fully restricted. Rights over different elements of a forest resource may be distributed among several individuals or groups, which complicates the identification of rights holders. In many cases, these rights are restricted by law. The definition of tenure arrangements is illustrated by the various regulations governing property rights, as shown below.

Taking these considerations into account, the UNECE/FAO study on forest ownership in the ECE region (UNECE/FAO, 2020) provides a more comprehensive definition, stating that:

forest ownership must be understood as a "multi-layered system of relations between the legally entitled holder of the resource and the rights and duties involved in relation to the forest resource" (Weiss & Nichiforel, 2020).

According to this definition, **forest ownership appears is a complex system of features** comprising the general institutional setting and the allocation of property rights, the nature of ownership, the character of the owning entity, and the regulation and organisation of forest management or stewardship (Weiss & Nichiforel, 2020). All of these aspects may have a stronger or weaker public or private character. Tenure arrangements are various combinations of property rights being attributed, formally or informally, to the legal holder of the resource or to other resource users (Weiss & Nichiforel, 2020).

6.3.2 *Forest Ownership as Bundles of Rights*

According to the legal regulations governing different forest products or uses, ownership rights are often divided into various use rights over specific features of the forest resource. These features may include particular products or ecosystem services. These are typically defined as timber, firewood, game, wild forest fruits or mushrooms, grazing or other agricultural use, water (for drinking or power production), and access for recreational purposes. Protection against natural hazards may also be regulated; for instance, in Austria, where this ecosystem service is granted to the beneficiary. Biodiversity conservation and landscape amenities are often reserved for the public, requiring land or forest owners to maintain outstanding landscape features or preserve protected species and habitats. In this way, the rights associated with a piece of forest can, in this view, be seen as rights over a bundle of uses or ecosystem services that are granted to landowners, other users, and/or the public (Fig. 6.1).

Another, more general and more analytical approach to categorising the bundle of rights associated with ownership is the broader concept of **property rights** (Schlager & Ostrom, 1992). In this framework, the complex property rights system is broken up into five distinct categories of property rights (related to forest resources in Fig. 6.1 and in brackets below, (Weiss & Nichiforel, 2020)):

(1) Access rights (the right to enter a forest land);
(2) Withdrawal rights (the right to harvest timber, firewood, game or other forest products);
(3) Management rights (the right to plan forest activities and transform the forest);
(4) Exclusion rights (the right to prevent others from accessing the land or harvesting its resources);
(5) Alienation rights (the right to sell forest land or forest products, or to lease management and exclusion rights).

Following this framework, not all rights are allocated to the landowners; some are assigned to specific user groups or rights holders, and some are reserved for

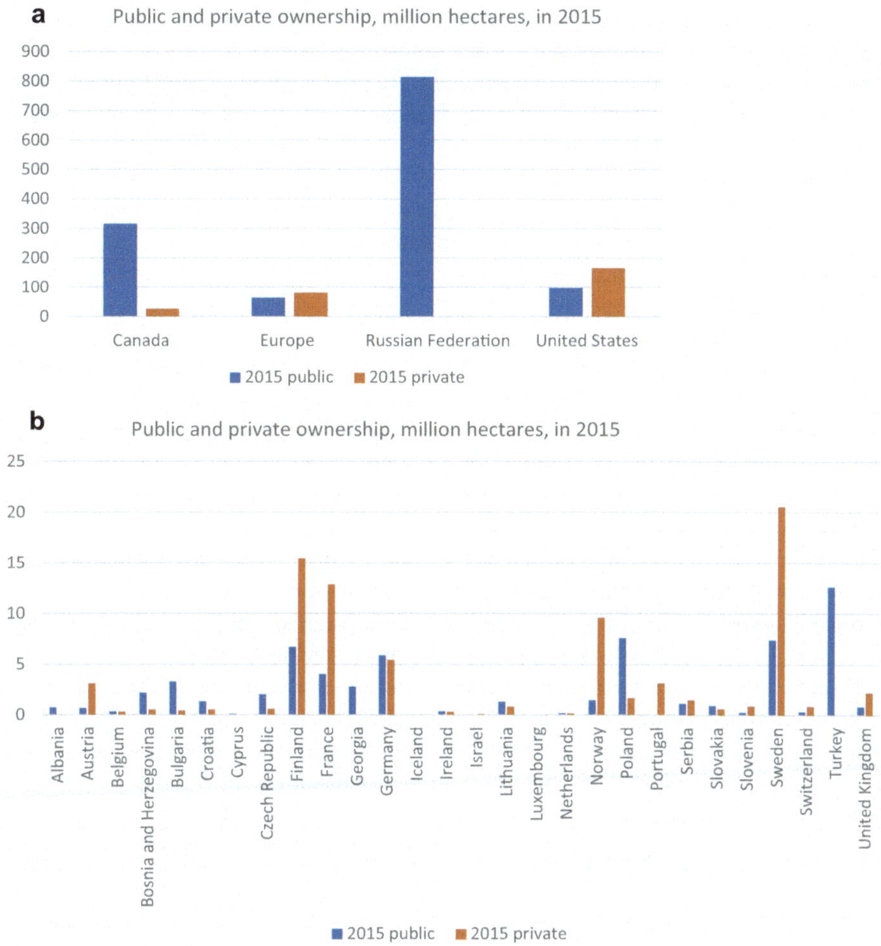

Fig. 6.1 Bundles of ownership rights over various ecosystem services (left), or over property rights categories (right) (source: own figure)

public authorities. Specific right holders are often neighbouring farmers or beneficiaries of the protective functions of mountain forests. User groups may include recreationists or collectors of wild forest products. The general public benefits from protected species or habitats for biodiversity conservation. The so-called "everyman's right" that exists in some European countries, particularly in Northern Europe, may include the right to access forests for recreational purposes and/or to collect non-wood products.

One implication of this division of rights for different groups is that the full potential of some goods is not realised. For example, this occurs when the same party does not hold the use and management rights to the forest. When public access is permitted for the collection of mushrooms or berries (e. g., when the withdrawal rights apply to everyone), forest management is most likely oriented towards timber production (the landowner has the withdrawal right), rather than towards the collection of mushrooms or berries. Ideally, the forest would be managed to produce both, which might require a lighter canopy cover. However, the owner's incentive is to maximise timber production. Granting the owner the exclusive rights over mushrooms and berries would enable them the possibility to harvest and market all forest products, and the forest management would balance the production of all products according to their profitability. In practice, however, the collection costs could make the market price of the forest fruits too high for the landowner to make a profit. In this case, public access would be more appropriate for forest visitors who view the collection of mushrooms and berries as a leisure activity.

Landowners or rights holders themselves may grant the use of the forest (in its entirety or for specific services) or the utilisation of their rights to others through private contracts, e.g., to leaseholders. Landowners are granted alienation rights, allowing them to sell forest products or lease the management or hunting rights. Specific beneficiary groups are often granted the use rights for their own use only, such as grazing right holders or recreational users. Alienation would also eventually include the right to destroy a product or resource. Most forest laws primarily aim to maintain all forest resources and limit the conversion or clearing of forest land to specific purposes (usually when in the public interest) and according to strict procedures (with decision-making powers given to state authorities).

Tenure is the generic term used to describe the allocation of property rights, which are categoried according to the level of control held by the rights holder. According to the five categories of property rights, the following types of right holders can be distinguished (Schlager & Ostrom, 1992; further explanations in Weiss & Nichiforel, 2020):

- "Owner", the legal holder of a resource who is granted the full set of property rights;
- "Proprietor", who differs from the owner in that they do not hold the right of alienation;
- "Claimant", who has no right of exclusion or alienation but has the authority to decide on management and withdrawal;
- "Authorised user", who is allowed to use certain defined resource benefits, as granted by the owner or proprietor;
- "Authorised entrant", who is granted the right to enter the forest land and enjoy non-subtractive benefits, but not to harvest any wood or non-wood products.

In this framework, the owner is specifically connected with the full set of rights. Therefore, in certain context, it may be more appropriate to use the more general term "right holder" when referring to one of the other tenure categories. Another term used in such situations is "stewardship", which may concern owners or other forms of tenure. "Forest stewardship" may refer to various forms of tenure arrangements, particularly those without full ownership rights, which are often found in community forestry or when ownership is not the primary focus, as in the case of traditional commons. In such cases, there was often no concept of ownership per se, but rather a joint use of the resource (Jeanrenaud, 2001).

6.3.3 *Formal and Informal Rights and the Quest for Implementation*

The extent to which owners can effectively decide and implement forest management objectives effectively will depend on the extent to which various property rights are granted to them (Nichiforel et al., 2018). Throughout Europe, national forest laws outline certain basic provisions regarding forest management, including restrictions on converting forests for other forms of land uses and an obligation to regenerate forests after final felling. However, there are disparities exist across countries in the extent to which forest laws and authorities prescribe forest management. In some countries, management rights are withheld from the owners, enabling public authorities to decide whether Forest Management Plans (FMPs) are needed and when the right to their develop them lies with the authorities (Nichiforel et al., 2018).

The distribution of property rights in European private forestry was analysed using a systematic comparison of forest legislation across 31 European countries (at the national or subnational levels) (Fig. 6.2, Nichiforel et al., 2018). The study developed a standardised method for assessing the level of property rights of private forest owners according to legal provisions. This method used 37 indicators to characterise five categories of property rights: access, withdrawal, management, exclusion, and alienation. The result was a standardised Property Rights Index in Forestry (PRIF).

Consequently, only two countries—Finland and the Netherlands—grant forest owners full freedom to decide how much timber to harvest (Nichiforel et al., 2018). In most countries, general silvicultural restrictions apply, such as maximum harvesting amounts, minimum harvesting ages and the need for felling licences. In Eastern and Southeastern European (ESEE) countries, owners typically have little or no influence over the amount or timing of timber harvests. While forest owners in Europe have the right to collect non-timber products, such as mushrooms and berries; however, in the ESEE region, they require approval to sell these products on the market. In most countries, the forest owners have limited capacity to prevent the general public from harvesting non-timber forest products. Although owners mostly have the legal right to restrict public harvesting and picking of non-wood forest products for commercial use, it is difficult to control in practice. Hunting rights and game ownership are much more strictly regulated. In most countries, game either belongs to nobody (res nullius) or to the state. In a few cases, hunting associations own the game. Often, the owners cannot influence the hunting quota. In some countries, the owners always have the right to hunt, but this often requires a certain size of property. In many countries, owners do not have this right and are often not compensated. The major factor differentiating the regulatory frameworks is the freedom of owners to decide and implement forest management objectives, as mentioned. While in more "liberal"

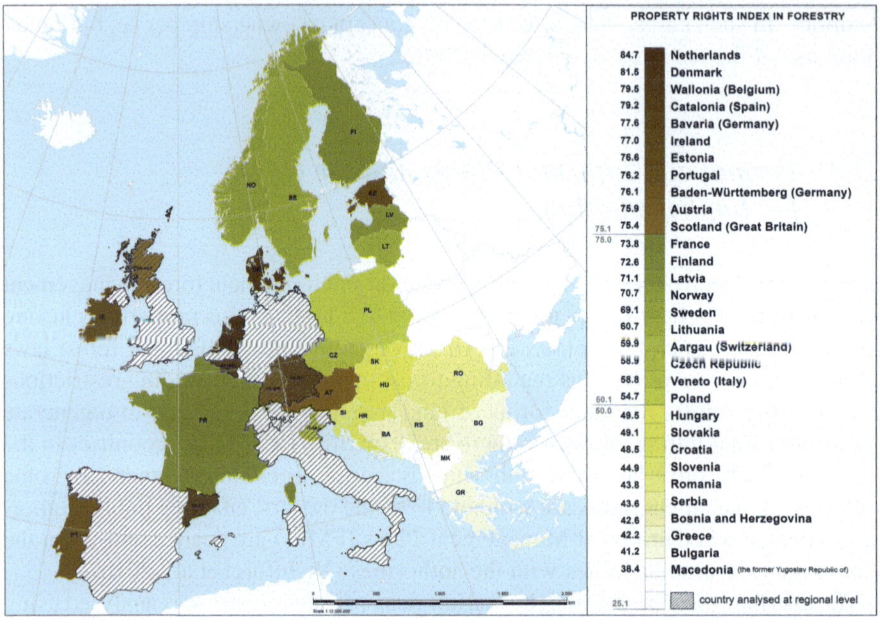

Fig. 6.2 Distribution of the property rights index in forestry (PRIF) across Europe. *Source* Nichiforel et al. (2018)

countries do not require FMPs are only required for specific purposes (e.g., for large-scale clear-cuts or when requesting subsidies), in countries with stricter frameworks, they are mandatory and may even be prescribed by public authorities. Overall, the most regulated regimes with the lowest PRIF values are typically found in the former socialist countries in ESEE (except the Baltics) (Nichiforel et al., 2018).

In a subsequent step, Nichiforel et al. (2020) assessed trends in the allocation of property rights in forestry. They identified a trend towards deregulation from the 1980s to the 1990s in most Western European countries. Since then, forest management policies have been liberalised to some extent, while new restrictions have been introduced to address environmental concerns, such as biodiversity and climate change policies. By contrast, former socialist countries underwent significant changes to management, exclusion and withdrawal following the 1990s. As shown above, most former socialist countries still maintain a strongly state-centred approach to private forest management (Nichiforel et al., 2020).

It is important to recognise that property rights are not necessarily fully regulated and may be based on traditional or **customary rights** (Weiss & Nichiforel, 2020). Although traditional, customary tenure arrangements may have been translated into law, they often remain informal, based on locally recognised rights that are not formally recognised. In practice, situations arise in which de jure rights are handled differently in everyday life (de facto). This may be due to the coexistence of parallel systems of old traditions and new legal provisions that have not been effectively streamlined. For example, the collection of non-wood forest products is often based on local customs, even when these are not formally recognised, or when they differ from legal provisions. Many de facto rights also result from a lack of enforcement of formally defined rights. In some cases, this may reflect the dominance of tradition over more recent regulation. In other situations, the continuous toleration of the theft of forest products establishes quasi-rights that conflict with the de jure rights. De facto rights may, conversely, constitute the owner of the resource illegally using it (e.g., felling without formal approval or exceeding the allowable quantities).

Property rights are defined within the **legal frameworks** of national states, usually within different legislative jurisdictions and potentially at various administrative levels. Although forests are complex ecosystems, their administration is usually divided into different sectors or policy fields. This creates coordination problems when implementing regulations across different administrative units. Regulations may contradict each other or may be difficult to implement when divided between different public administrations. Contradictions in forest management regulations may arise because different policies are formulated for different goals and based on different knowledge systems. For example, forest policies aim to optimise timber production within a SFM framework, whereas hunting policies aim to establish healthy game populations within a habitat management framework. Forest policies and silvicultural concepts view cultivated trees as valuable and consider browsing by game to be a threat and a disturbance. In contrast, hunting policies and wildlife management concepts aim to increase game populations and view the forest as a habitat that provides food and shelter for wildlife.

Contradictions often arise during the implementation phase, too. Inevitably, the policies serve different interests, and the formulation and implementation of the laws depend on the power of the various interest groups. The distribution of power across interest groups, as well as among the relevant authorities, determines which policy goals can be more successfully enforced than others. A lack of power often results in non-enforcement and implementation gaps. This may affect public interests that are not being enforced against landowners, or the interests of certain groups that are not being enforced against others.

The question has been raised as to what type of ownership would be better suited to securing the conservation of forests and their sustainable management. In short, studies have revealed both positive and negative examples in both categories (Mölder et al., 2021). Some have claimed that only those with full ownership rights can efficiently manage resources. However, research into common property regimes indicates that proprietors' rights can be sufficient to encourage long-term investment and harvesting of resources (Ostrom et al., 1994). Participants in common property regimes are mostly proprietors.

Most studies agree that security and stability of ownership are key to SFM. While other factors also matter, **tenure security** is generally considered to be more important than the specific form of tenure (Robinson et al., 2014). Security in this context refers to the owners' confidence that their rights will be respected by others, including the state, and that these rights will not be arbitrarily withdrawn. Regarding what constitutes secure rights, it is worth noting that tenure security is highly contextual and can be provided by either customary or statutory institutions (Katila et al., 2020). In the European context, issues related to sustainability and tenure security have been significant in light of recent societal changes and current institutional conditions in the ESEE. A study of forest disturbance (e.g., clear-cutting) across different land ownership types in Poland, during and after the socialist era revealed that disturbance peaked during the immediate transition period for all ownership types (Kuemmerle et al., 2009). In a comprehensive study of the causes of illegal logging in the ESEE region, Bouriaud (2005) identified two key drivers: rural poverty and weak public enforcement capacity. In Romania, insecure property rights following forest restitution resulted in unauthorised logging, as owners lacked the means to protect their forests. Fearing timber theft, many people harvested more timber than was permitted in order to avoid losing it to others. Underlying reasons for illegal logging can be seen in inappropriate institutional and policy frameworks, where policies overlook the population's need for firewood, for instance.

6.3.4 Types of Forest Ownership—Beyond Public and Private

Returning to the distinction between public and private, several aspects require clarification. These adjectives may be connected to goods, land property, and the entity that owns a piece of land. In institutional economics, private goods, private ownership, and private owners are all distinct concepts. Furthermore, there are always more than just these two categories, public and private.

Before defining property regimes, it is useful to discuss the concept of **public and private goods** from an institutional economic perspective. This concept characterises the nature of goods or services and links them to possible or suitable property regimes. It is important to distinguish between the private or public nature of the goods and the character of the property regime. In terms of their economic nature, goods can be categorised into four ideal types of goods (Glück, 2000; Mantau, 2001; McKean, 1998; Ostrom et al., 1994; Table 6.1). These are characterised by two continuous dimensions: rivalry and excludability. The first dimension is rivalry over the good (or service), which refers to the fact that it can only be consumed once, either by one consumer or by multiple consumers (also characterised as their subtractability). The second dimension is how easy or difficult it is to exclude others from consumption. Private goods, by nature, can only be consumed once and it is easy to exclude others from consumption. Such goods are consequently easy to market. Timber is the best example from forestry. Public goods can be enjoyed by many and are difficult to exclude from. Examples include services such as biodiversity conservation, clean air, scenic views and carbon sequestration, all of which help to mitigate climate change. However, they are difficult to market because people may enjoy these goods or services without paying, so-called "free riders".

Club/toll goods are partly non-rival, but exclusion can be practised. Examples include recreational facilities, national parks, and forest roads, which can be used by many people without affecting other users, at least when they are not used frequently.

Table 6.1 Classification of goods according to physical characteristics

		Exclusion	
		Difficult or costly	Easy
Consumer rivalry	High	*Common-pool resources* (e.g., forest, pasture, groundwater aquifers)	*Private goods* (e.g., timber, Christmas trees)
	Low	*Public goods* (e.g., biological diversity, protection against avalanches)	*Club or toll goods* (e.g., forest road, recreational facility, national park)

Sources Ostrom et al. (1994), McKean (1998), cited from: Glück (2000)

Club or entrance fees, as well as tolls, can be charged to generate revenue for maintenance and to prevent overcrowding. Common pool resources are such that they are subtractable, but it is difficult to exclude users. Consequently, they are best managed and consumed jointly by the beneficiaries or interested users. This applies to pastures, rangeland, as well as hunting, fishing, and groundwater aquifers. In a certain sense, forests as a whole can be considered a common-pool resource, and joint management in a common-property regime is often the most suitable approach. Mountain forests, for example, have been described in this way as they provide multiple benefits, particularly in the highly populated mountain valleys such as those in Austria and Switzerland (Glück, 2002; Kissling-Näf et al., 2002). These benefits include public goods, such as protection against natural hazards, amenity values, and biodiversity, as well as private goods such as timber and game. However, there are competing relationships and trade-offs in production.

The nature of goods is linked to **ideal property regimes**. In general, private goods are suited to private ownership and exchange on markets. Public bodies should manage public goods, which often lack private ownership and are available for public use or open access. Club or toll goods require the formation of joint ownership or management bodies, such as clubs, associations, or societies. For common-pool resources, establishing common property regimes is essential. It is important to emphasise the distinction between common property regimes and open-access situations. While common property regimes have management structures, open access is not regulated at all. The so-called "Tragedy of the Commons" as described by Hardin (1968) is in fact an open-access situation, not a characteristic of commons (Feeny et al., 1990). Hardin's example is pastureland that anyone can use to graze their animals, which inevitably leads to the overuse and destruction of the resource. Similar situations occur, for instance, in the open oceans, where overfishing is a severe problem. However, common or collective property use and stewardship has been an important property regime, including for forests (Ostrom et al., 1994).

Unfortunately, these definitions are not used coherently in the literature. Furthermore, to be precise, it is necessary to distinguish between (i) the alienable nature of the ownership, and (ii) the nature of the entity that owns the forest. According to the first definition, public ownership is assigned to all citizens, meaning that the resource cannot be sold (Bouriaud & Schmithüsen, 2005). Thus, under a "public" regime, the benefits of the resource should be maintained for future generations. According to this interpretation, if the law allows forests in state or municipal ownership to be sold, it should considered "private". This understanding of public ownership is found in some European countries with regards to their state-owned forests, however, this characteristic is often not easily observable without a detailed analysis of the underlying legal provisions. Therefore, in practice, the latter interpretation is usually applied to forest ownership categories (Weiss & Nichiforel, 2020). This is evident in the Global Forest Resource Assessment (FAO, 2018) and the recent forest ownership study (UNECE/FAO, 2020). In terms of the entity owning the forest, property rights theory distinguishes five legal ownership categories or "resource regimes" (Weiss & Nichiforel, 2020):

- *state property (res publicae)* is owned in the name of citizens and managed by an agent of the government;
- *municipal or communal property (res communalis)* is held by local governmental bodies such as communes, towns, municipalities or other administrative entities;
- *common property (res communis)* is private property held by a group of co-owners with a governance structure which assigns rights and duties;
- *private property (res privatae)* is owned by a specific individual or private legal entity that holds the rights to control its use;
- *open access property (res nullius)* is owned by no one, and consequently, anyone has access and can use it.

The last category is not often found in Europe. However, situations where ownership appears to be unclear, disputed, or unknown, are found in some countries. For instance, this occurs in some ESEE countries when restitution processes have not been finalised.

The above categories are **not consistently used** or understood in research and practice, as they vary depending on the specific situations and the traditions of different disciplines or countries. For example, public ownership is sometimes used solely for state forests, and sometimes for municipal forests too. In Bulgaria, the Czech Republic, Latvia, and Slovenia, municipal forests are assigned to the private ownership category, whereas in Estonia, Poland and Romania, they are assigned to the public ownership category. Common or community property, which is still found across Europe, is seen and officially categorised as either private or public in different countries. In Switzerland, not only forests owned by municipalities and those belonging to citizen communities (*Burgergemeinden*; formerly known as "common" resources) are both considered public. By contrast, Austria, Norway, and the UK classify municipal/local government forests as public, as well as community/common forest land; private land is also classified as such. Some countries, however, classify their communal land (Portugal) or municipal land (Finland and France) as neither public nor private, but as "other types" in the FRA. Community or common ownership often accompanies specific organisational rules that are often protected by law. These include tribal or indigenous forests, as well as common property regimes that date back to historical commons, but which are not necessarily referred to as such (Weiss & Nichiforel, 2020).

In addition to community forests, there are a variety of other forms of joint, philanthropic, or **charitable ownership**. The primary goal of these is to deliver social or environmental benefits rather than to maximise timber production or financial returns. While these may be officially recognised as registered charities and granted tax exemptions, as well as access to charitable funding, they have restricted ownership rights regarding the use of profits or disposal of assets. Church forests are regarded as a private form of ownership in some European countries (e.g., Austria, the Czech Republic, Serbia, Greece, Bulgaria and Estonia), as a public form in others (e.g., Belgium), or as a separate category (e.g., filed under charitable organisations in the UK) (Weiss & Nichiforel, 2020). In conclusion, establishing an intermediate, semi-public category alongside public and private forests may be beneficial (Weiss et al., 2019).

6.3.5 Conflicts and Current Issues

Around ownership, a range of topics has been discussed in policy and research. An older theme is small-scale forest management, but in relation to this, a range of overlapping and interrelated terms has been used recently, including non-industrial private forest owners, fragmented forest ownership, absentee or urban owners, and others. Furthermore, customary and indigenous ownership rights, as well as gender issues, have been discussed.

Small-scale forest management is an established research topic with a specific dedicated journal (Small-scale Forestry). Within private forestry, small-scale family-owned farms are often contrasted with industrial or institutional ownership of forests. The former group is therefore also addressed as non-industrial private forest owners (NIPFOs). A sometimes observed increase in parcelization of forest properties through market exchange or hereditary processes is often problematised under the label of fragmentation. Originally, fragmented forest ownership refers to a property structure within a single forest holding that is divided into parcels located at a distance from each other. In current policy discourse, however, it refers to a growing number of owners, with a trend of decreasing property sizes among average forest owners (Weiss & Nichiforel, 2020).

Research on forest ownership types has applied different methods and approaches (Ficko et al., 2019) and used varied criteria for their classification, including the legal form of ownership, structural attributes, and social characteristics of forest owners and/or forest holdings, and owners' management goals, attitudes and behaviour (Weiss et al., 2019). From the policy concern of a decreasing supply of the industries with raw material, studies have identified **"new"** forest ownership types that are often referred to as "absentee" or "non-resident", "urban" or "non-farm/non-agricultural" forest owners. In this respect, the notion of new owners in terms of how long they have owned their forest must be distinguished from new forest owner types in terms of non-traditional characteristics and goals. New forest ownership types are often characterised in a continuum from more traditional or rural to more non-traditional or urban types of owners. "Urban" can again be understood in different ways in this context, either referring to the owners' place of residence (i.e., in a rural or urban area) or to their social identities or lifestyles, with connected attitudes and behaviours (Weiss & Nichiforel, 2020; Weiss et al., 2019).

Understanding ownership rights requires awareness that, in addition to **formal rights**, customary or other **informally** established land-use rights may exist. While this is inevitable and normal, pronounced differences in understanding between different groups may cause conflicts. The use of wild forest products can conflict with traditional uses by local populations, potentially clashing with the legally defined property rights of landowners (Górriz-Mifsud et al., 2017). In Austria, for example, there have been cases where the intent of large private landowners to exercise their right to exclude mushroom pickers from their forest has caused outrage among resident populations. At the same time, there are cases where landowners issue picking licenses.

Customary tenure often refers to the traditional rights and land ownership of **indigenous communities**. The legal system recognises those rights; however, there may be continuous conflicts over their interpretation or enforcement. Such conflicts often stem from overlapping land claims, limited participatory rights, and unequal power dynamics between Indigenous groups and state or private actors engaged in forestry, mining, and other land-intensive sectors.

In northern Sweden, for example, the Sámi Indigenous Peoples practice reindeer husbandry based on long-standing usufructuary rights. However, these rights often come into conflict with forestry operations that prioritise timber extraction or land conversion. Although the Sámi communities are recognised as having cultural and traditional ties to the land, they frequently face obstacles in influencing land-use decisions that impact their grazing routes and winter-feeding grounds. Forestry operations, such as clear-cutting and soil scarification, significantly reduce the availability of lichen-rich habitats, a critical winter food for reindeer, resulting in ecological degradation and loss of livelihood opportunities for herders (Elomina et al., 2025). While forestry and mining sectors are officially seen as being vital for local economic development, they are increasingly seen by Indigenous communities as threats to traditional land uses and cultural survival. Indigenous stakeholders express concerns about having limited decision-making power in land-use planning, despite being directly affected by industrial expansion. The situation is further complicated by the fact that most forest lands are either state- or corporately owned. In contrast, Sámi reindeer herding communities hold only non-ownership (usufruct) rights, placing them at a systemic disadvantage when negotiating land use (Elomina et al., 2025). The conflicts are not just about resource access, but also about competing worldviews: economic growth versus cultural continuity and ecological integrity. Applied land use strategies often fail to align with the biophysical realities and cultural needs of Indigenous groups. Industrial development is usually assessed through macroeconomic indicators, neglecting the socio-cultural costs borne by local communities (Živojinović et al., 2024).

Gender aspects in forestry and forest ownership have gained increasing attention, revealing persistent structural inequalities. Historically, forestry has been a male-dominated field, with forest ownership and management roles often passed down through patriarchal inheritance systems or associated with physically demanding work perceived as being more suitable for men (FAO, 2006; Lidestav et al., 2020). Today, women account for approximately 31% of private forest owners in Europe; however, many are co-owners listed alongside male spouses, which continues to obscure women's roles in decision-making (Lidestav et al., 2020). A review study and critical assessment, applying three analytical lenses (gender as an empirical variable, a relational and structuralisation category and a meaning category) and using empirical evidence from 16 European countries, concludes that gender matters in forest ownership, management, operations and their understandings (Follo et al., 2017). The study demonstrates that gender structures affect, for example, the actual behaviour of female and male forest owners and the self-evaluation of their forestry competence.

Even when women do enter the sector, they remain underrepresented in leadership and management positions—estimated at only 5–8% in Austria (Ludvig et al., 2024). Despite policy commitments and growing enrolment in forestry-related education, women's full participation remains limited due to gendered stereotypes, unequal access to networks, and work-life balance challenges. Even well-qualified women in leadership roles are seen as exceptions rather than equals, highlighting a deep-rooted cultural bias in the sector. Gender inequality also intersects with innovation and resilience in the forestry sector. Research shows a statistically significant correlation between the perceived importance of gender balance in management and higher organisational innovativeness (Ludvig et al., 2024). The broad perception that forestry is a male-dominated profession, as well as a lack of awareness, interest, and understanding among researchers of the concept of gender and its societal relevance, also constrain the integration of gender aspects into research (Mai et al., 2011). Among others, gender mainstreaming would enhance the quality of understanding between forestry organisations and society (Hehn et al., 2006). Changing the limited awareness of the role of genders in society and the forest sector, as well as the need for improved gender equality, also needs to be reflected in education. On this topic, an IFSA Open Letter on Gender and Forest Education says:

> Gender must be addressed more openly and forestry curricula revised, in order to challenge the norms, counteract structural oppressions and transform the forestry sector to a more equitable one. (IFSA, 2022)

Across Europe, "old" **forestry commons**—some with statutes dating back eight centuries—coexist with "new" commons that have emerged only in the past two or three decades (Lawrence et al., 2021). Ancient institutions, such as the Italian *Regole* or Swedish forest commons, have repeatedly adapted to wars, state appropriations, and changing markets, proving that collective rules, inherited shares, and place-based identities can remain resilient for a millennium. By contrast, "new" commons have been catalysed by post-socialist restitution, land-reform legislation and urban social innovation, for example Slovenian agrarian commons revived after restitution,

Scottish community buyouts created since 2003, and shareholder-based forests established for city dwellers in Germany, demonstrating how novel legal tools and shifting notions of "community of interest" enable collective ownership even where no mediaeval precedent exists. Lawrence et al. (2021) argue that the continuum from very old to very new commons provides a "natural experiment": long-standing models reveal how communities can steward forests across centuries, while recent creations test fresh combinations of rights, business models and multi-level links with the state, all of which enrich Europe's repertoire for adaptive, socially just forest governance today.

6.3.6 *Confusions and Conclusions*

This suggests that simple terms, such as ownership, public, and private, are not clear. Indeed, scrutiny of these terms reveals inconsistent usage in the literature and in practice, with meanings that vary according to context. This variability is typical of scientific concepts, but a precise understanding remains essential. As indicated early on, forest ownership is a complex system comprising various interrelated features and a network of relations between different holders of rights on the forest resource, embedded within wider, geographically distinct institutional and political systems.

Applying property-rights theory provides a clearer understanding of the bundle of rights and the various forms of tenure arrangements. When using the terms private and public, care should be taken not to conflate (i) the economic nature of the goods or services, (ii) the property regime, (iii) the nature of the ownership and (iv) the character of the owning entity. It should also be recognised that ownership categories extend far beyond a simple public–private dichotomy, both within and beyond those categories.

This section has shown that basic concepts related to forest ownership and tenure must be applied with great care to avoid ambiguity or confusion. There are additional sources of confusion when entering various current debates in forest governance. For instance, in how far the interplay between formal (statutory) and informal (customary) forest tenure or use rights may be a problem, such as concerning traditional forest uses by local populations or Indigenous groups. Further important insights emerge when specifying the questions at hand: who owns which types of forests, under what conditions, for what purposes, and whether the gender of forest owners matters.

The concept of ownership and related terms may seem vague, unclear or ambiguous. However, it rather points to the fact that reality around ownership is **highly complex** and important to know. A clear understanding of these concepts is also a prerequisite for comprehending the facts and trends presented in the following section.

6.4 Distribution, Trends and Impacts of Forest Ownership in Europe

The datasets compiled by COST Action FACESMAP and UNECE/FAO (2020), based on surveys from 35 ECE countries, provide updated and publicly available information on forest ownership patterns, trends, and related governance structures (https://unece.org/forests/forest-ownership-unece-region). Although systematic comparison across countries is limited due to varying definitions and methods, the datasets still offer valuable insights and help address key policy questions and assumptions.

6.4.1 Ownership Patterns in Europe

Overall, according to the UNECE/FAO study, approximately 56% of European forests are **privately owned**, while 44% **are public** (excluding Russia) (Lawrence, 2020). The State of Europe's Forests report also presents this basic information, reporting that 53.5% of Europe's forests (including Russia) are publicly owned, and 46.5% are in private ownership (Forest Europe, 2020). There are, however, great differences across countries (Fig. 6.3), as in some countries practically 100% of the forests are in public ownership (e.g., Turkey or Georgia), in some still dominating public (e.g., Poland, Albania, Bosnia and Herzegovina, Czech Republic or Bulgaria), in some the shares are more evenly split (e.g., Ireland or Germany). In other countries, private ownership predominates (the United Kingdom, Norway, Sweden, Finland, France, Austria or Portugal).

From the data, the countries' ownership patterns may be clustered into the following:

- Dominating private ownership by individuals/families: e.g., Norway (although with some indigenous ownership), Finland, France, Austria, Portugal;
- Dominating private ownership with mixed types: Sweden (with share of private businesses), Switzerland (individual and common), Slovakia (individual and common);
- Mixed ownership structure: e.g., Germany (private and pronounced shares of sub-national state and municipal forests), Netherlands (national and local public forests and a mix of individuals and private institutions);
- Dominating state ownership (national): e.g., Georgia, Poland, Croatia, Czech Republic, Bulgaria, Slovakia);
- Dominating municipal ownership: Albania.

When examining the geographical distribution of private and public forest ownership across Europe, certain patterns emerge (Fig. 6.4). Private ownership predominates in Western Europe. In contrast, the opposite is true in Central, Eastern, and Southeastern Europe. While in the North, around 70% of the forests are privately

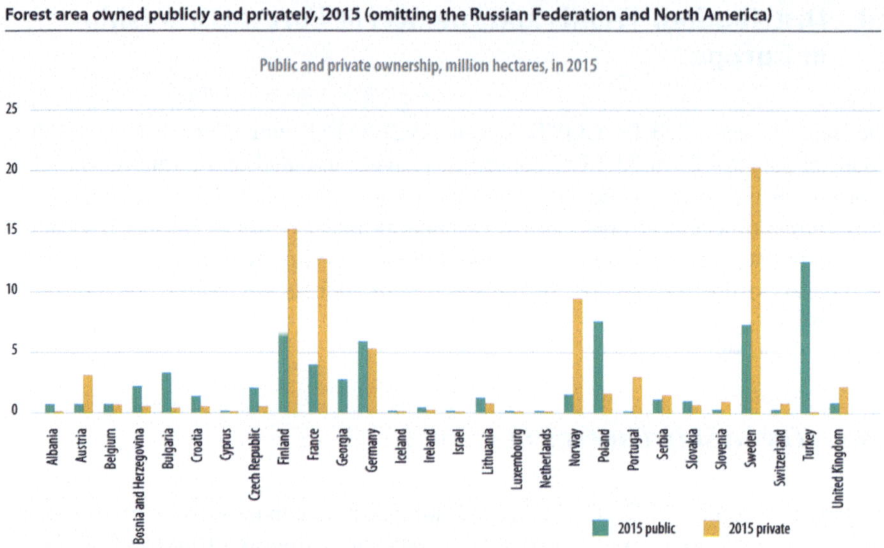

Fig. 6.3 Public and private forest area in European countries, year 2015, in Mio ha. *Source* UNECE/FAO (2020)

owned, in Southeastern Europe, around 90% are public (see also Forest Europe, 2020).

As discussed in Sect. 3.4, important forms of ownership exist beyond the public–private dichotomy. The UNECE/FAO Forest Ownership Study (UNECE/FAO, 2020) attempted to capture these different forms of ownership, categorising them under public (national, sub-national, and local government) and private (specific types). The majority of public forests are state-owned, i.e. on the national level (78%), 9% are owned by subnational governments, and local governments hold 13% (data for 25 countries from the ECE region, Aggarwal et al.,2020). Municipal (or communal) ownership appears to be the dominant form of public ownership in Switzerland, Luxembourg, Albania, Portugal, and France. Sub-national ownership is prevalent in Germany, the UK, and Canada. In other countries, it is mostly characterised by state forests. Within the private category, the majority is owned by individuals and families (77%), private enterprises own 17%, other institutions own around 1.5%, indigenous communities own around 2%, and some 2.5% is other private common property. Indigenous forest land is mostly reported by Norway and the U.S. (Canada is not included in the dataset). Common property is mostly attributed to Switzerland and Slovakia, as well as to Poland, Serbia, Sweden, Finland, and the U.S. (Aggarwal et al., 2020). Note that churches are not specifically categorised here, and that any categories besides more "regular" state or private ownership may be allocated inconsistently across countries' official reporting systems.

Other studies that have tried to map "other" forms of ownership beyond public and private (Wong et al., 2020) find that over 5% of forests in Germany are owned

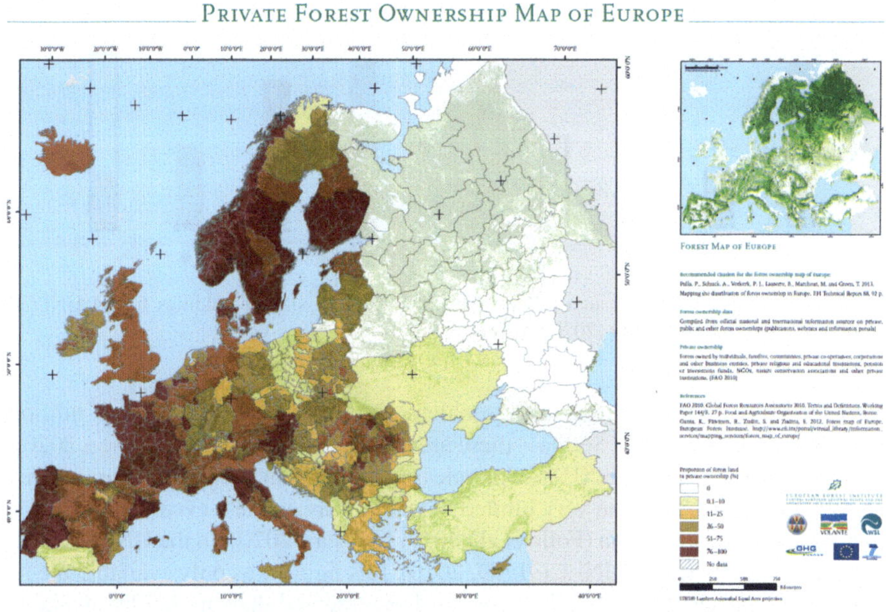

Fig. 6.4 Distribution of private and public ownership across Europe (red—strong private, yellow—strong public). *Source* Pulla et al. (2013)

by some environmental NGO, and more than 1% in the UK (e.g., National Trust). Several countries are reported to have more than 5% of their forests owned in some form of common ownership, which may be officially categorised as private, public, indigenous, joint, or other types of ownership: Norway, Poland, Slovakia, Austria, Slovenia, Romania, Italy, and Portugal. Churches or other religious institutions own at least 1% of forest land in Sweden, Slovakia, Slovenia, Serbia, Romania and Greece (Wong et al., 2020).

The **average size** of forest properties across Europe is difficult to determine due to the limited and inconsistent data. The State of Europe's Forests report from 2003 provided an average size of private holdings at 13 ha and an average size of public holdings of approximately 1,300 ha (data for Europe excluding Russia, Forest Europe, 2003). The current report does not give numbers since there is no complete dataset across countries. Most of the public holdings in Europe are between 11 and 500 ha; the majority of private forest properties (almost two million holdings reported by 18 countries) fall in the size class up to 10 ha (Forest Europe, 2020). The average size of properties does not reveal much—the data instead provides a picture of significant differences in property sizes, both within public and private ownership. Small-scale holdings prevail in private forests in Europe, with 88% of all forest holdings being smaller than 10 ha (Aggarwal et al., 2020; Wong et al., 2020; Fig. 6.5).

Average private holdings in the countries range from around 1 ha (e.g., Bulgaria and Croatia) to over 100 (Slovakia), from where such data have been reported.

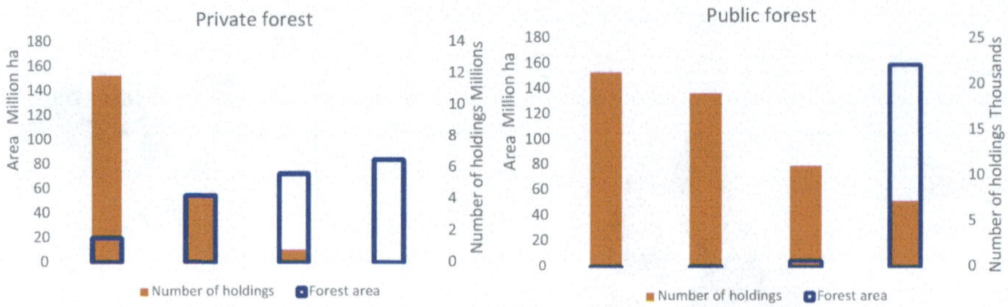

Fig. 6.5 Distribution of forest area and number of owners by size of holdings for 24 UNECE countries. *Source* UNECE/FAO (2020)

Average public holdings range from less than 100 ha (in the UK) to over 1 million hectares (ha) (Bulgaria and Serbia—these are single-state forests) (Lawrence, 2020).

As reported above, around 31% of private forest owners in Europe are **female**. This is a rough estimate based on very limited data that is available—even when scrutinising national statistical data (Follo et al., 2017; Lidestav, 2020). In the UNECE/FAO Forest Ownership Study (UNECE/FAO, 2020), this figure was only available for 13 European countries. More detailed, gender-disaggregated data are hardly available in any country. The authors claim that "numbers matter" as they may increase the visibility of women, and this knowledge would indicate any gendered differences between men and women as forest owners.

The reason for the lack of that data may lie in the fact that, in most European countries, private forest ownership has traditionally been regarded as being held by families, usually in connection with farms (Lidestav, 2020). According to patriarchal norms, the head of the household was a man. With this understanding that the "traditional forest owner" is typically a family/household headed by a man, there is no need to identify the gender of the individuals legally owning the land (Lidestav, 2020). As these traditional norms become increasingly outdated, the individual woman and man must be identified and should be reported as the owners.

6.4.2 *Recent and Ongoing Changes in Forest Ownership*

Across Europe, forest area has increased by about 9% over the past 30 years. This expansion is largely due to afforestation programmes, natural regeneration, and the abandonment of agricultural land. These changes are evident in both public and private forest lands. They are particularly visible in countries such as Spain, Portugal, and parts of Eastern Europe, where former agricultural lands have been converted into forest cover. This trend contributes positively to climate mitigation and biodiversity objectives but also brings new challenges in terms of ownership and governance structures (Forest Europe, 2020; Lidestav et al., 2020; Živojinović et al., 2015).

6.4.2.1 Shifts in Ownership Structure: From Public to Private

One of the most pronounced changes in recent decades has been the **shift from public to private ownership** in many European countries, especially in **Eastern and Southeastern Europe (ESEE)**. These shifts are predominantly a consequence of the **post-socialist transformation** in the 1990s, which led to the restitution and privatisation of forest land.

Restitution refers to the legal process of returning assets that were expropriated—typically during socialist or communist regimes in the twentieth century—to their original private or communal owners, or to their heirs. This process has been one of the most transformative changes in forest ownership in Central, Eastern and Southeastern Europe since the 1990s, profoundly reshaping ownership structures, forest management institutions, and policy frameworks (Dobšinska et al., 2020; Lidestav et al., 2020; Živojinović et al., 2015).

The scale and outcomes of restitution varied widely across countries, depending on national legislation, historical land records, institutional capacity, and political will (Fig. 6.6):

- The Baltic States completed restitution relatively quickly in the early 1990s. Broader forest governance reforms, including additional privatisation of forest

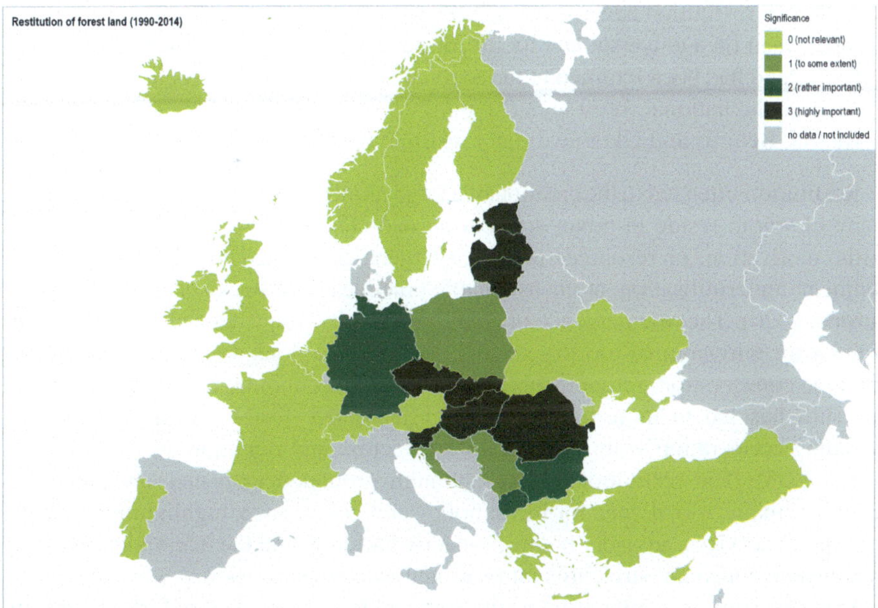

Fig. 6.6 Restitution of forest land in Europe, adapted from Lidestav et al. (2020). *Source* FACESMAP/UNECE/FAO Enquiry

lands, accompanied these processes. Estonia and Latvia moved from nearly 100% public ownership in 1990 to around 50% by 2015 (Lidestav et al., 2020).

- Romania implemented one of the most extensive restitution programs. By 2015, more than 45% of the forest area had been restituted to private individuals, municipalities, and churches. However, the process was often hindered by legal disputes, corruption, overlapping claims, and inadequate cadastral records, which delayed forest governance reform and management effectiveness (Lidestav et al., 2020; Nichiforel et al., 2018). Slovakia, Hungary and Croatia initiated similar processes, albeit more slowly. In Slovakia, boundary disputes and complex claims hampered progress. In the Czech Republic, church restitution was delayed until 2012, when new legislation finally enabled religious institutions to reclaim property that had been confiscated during the communist era.
- In the former German Democratic Republic, restitution was a complex process that involved leasing or selling forest and agricultural properties to new owners, cooperatives, or the state.
- In some ESEE countries, such as those in the former Yugoslavia and Poland, not all forest land had been nationalised during the socialist period. In Poland, the state kept most of its forests under public management. In Bosnia and Herzegovina, the process is hampered by ongoing political conflicts. In North Macedonia, the restitution process was hindered by a lack of documentation and political support, and effectively halted around 2008 (Dobšinska et al., 2020; Stojanovski, 2022).
- In Serbia, restitution began only in 2006, focusing solely on religious institutions, rather than on a universal law for all former owners. By 2022, over 32,000 ha of forest land had been returned to the Serbian Orthodox Church, representing about 1.4% of the national forest area (Živojinović et al., 2025).
- Russia, Belarus and Ukraine did not restitute forest land.

Restitution often led to the creation of a large number of small-scale forest owners, many of whom reside in urban areas or abroad. These new owners often lack the skills, motivation, or resources to manage their forests actively, leading to abandonment, underutilisation, or unsustainable practices (Dobšinska et al., 2020; Forest Europe, 2020). The re-emergence of church forests has reintroduced a major institutional actor into national forestry sectors. These institutions often pursue goals that combine conservation, religious heritage, and socio-economic functions. In Serbia, this shift has led to more adaptive and collaborative forest governance models, including cooperation with local forest contractors and communities (Živojinović et al., 2025). The restitution process, though grounded in justice and historical redress, remains a challenge for forest policy and governance, highlighting the need for support services, advisory bodies, capacity building, and the integration of newly established ownership structures into national forestry strategies.

Privatisation, as distinct from restitution, involves the transfer of forest land from public to private ownership, not necessarily to former owners, but often through sales, leases, or allocation schemes. It has been a major instrument of economic restructuring in post-socialist countries, aimed at reducing state control, increasing efficiency, and stimulating private investment in the forestry sector. Privatisation was implemented in diverse ways across the region:

- In Estonia, privatisation supplemented the restitution process by allowing private individuals and entities to purchase state forest lands, contributing to the rapid emergence of a market-oriented private forest sector.
- In Hungary and Bulgaria, privatisation complemented restitution. However, it often led to a fragmented forest ownership mosaic, where newly established private owners faced obstacles in accessing markets, managing forests sustainably, or investing in forest infrastructure.
- In Albania, a unique approach was taken by distributing forest land to local communities without a clear definition of property rights, creating a system of communal de facto privatisation that has proven difficult to govern (Lidestav et al., 2020; Živojinović et al., 2015).

Besides this major shift in ESEE, privatisation was important only in Sweden and the UK. Sweden decided in 2001 to sell up to 10% of their forest land to strengthen private farming and forestry in rural areas. The UK's plan from 1980 to sell public land to strengthen the private forestry sector faced strong public resistance. It was subsequently revoked, followed by a devolution policy in 2003 (see below) and the introduction of new public policies to allocate land to local communities (Lidestav et al., 2020; Živojinović et al., 2015).

Another trend in state forests is a commercialisation process through the introduction of more private management forms, either the creation of state-owned enterprises (Bulgaria, Czech Republic, Finland and Germany) or leasing to private enterprises (Lithuania, Slovakia, Poland and Ukraine). Leasing public forests to private operators, rather than a full transfer of ownership, aimed to balance state oversight with private efficiency, but raised issues related to transparency, accountability, and long-term forest planning.

Privatisation contributed to the diversification of the private forest owner population, with varying motivations, including commercial forestry, nature conservation, hunting, and personal or family heritage. This increasing heterogeneity challenges traditional policy models based on large-scale, production-oriented forest management (Lidestav et al., 2020).

6.4.2.2 Decentralisation

In parallel with restitution and privatisation, a significant trend shaping forest ownership in Europe is **decentralisation**—the transfer of authority, responsibility, or ownership of forest lands from national governments to sub-national entities, municipalities, communities, or private stakeholders.

Decentralisation has taken various forms across Europe and is particularly pronounced in countries with long-standing traditions of **local self-governance** or **multi-level governance frameworks**:

- In the **United Kingdom**, particularly in **Scotland** and **Wales**, decentralisation has included not only devolution of forest governance from national to regional levels but also **community empowerment and land reform legislation**. In Scotland, the Community Empowerment (Scotland) Act 2015 and the Land Reform (Scotland) Act have enabled local communities to acquire and manage former state forests through mechanisms such as **community buyouts**. These forests are often managed through **community trusts or associations**, aiming to provide local benefits, enhance biodiversity, and support rural development (Lidestav et al., 2020; Živojinović et al., 2015).
- In **Germany**, forest ownership is traditionally decentralised, with forestry jurisdictions and forest ownership shares on the state level, and significant shares of forest land held by **municipalities (communal forests)**. These local governments have autonomy over forest management decisions, often guided by state-level frameworks and regulations. Communal forests play a vital role in public service provision, including recreation, protection, and local income generation (Aggestam et al., 2020a).
- In **Switzerland**, forest governance is highly decentralised, with municipal forest owners and **forest cooperatives (Burgergemeinden),** forms of community ownership dating back centuries, remaining important actors in managing forest resources at the local level, including in mountainous regions.

Decentralisation is often driven by broader political agendas such as regional autonomy, subsidiarity, rural development, and public sector efficiency. However, it also stems from the recognition that **local actors are well-positioned to manage forest resources sustainably**, provided they have the appropriate capacity and support.

6.4.2.3 Emergence of New Joint Ownership Forms

Decentralisation has often been accompanied by the **emergence of new and hybrid forest ownership models** that blend public and private rights, or that redefine the role of local actors in forest management. They do not fit easily into conventional public/private categories. These include:

- **Community and cooperative ownership**, where collective bodies (e.g. residents, associations) jointly own and manage forest land. These forms often emphasise **social and ecological goals**, such as education, biodiversity conservation, and climate action (various commons in Italy, Scotland, Slovenia and Sweden, Lawrence et al., 2021).
- **Public–private partnerships**, where public authorities maintain ownership but delegate certain management rights or responsibilities to private actors, NGOs, or local associations (e.g. York Community Woodland in England; Community Agreements on Scotland's National Forest and Land in Scotland).
- **Charitable ownership**, particularly in the UK, where **non-profit trusts and foundations** own and manage forest land for conservation or public benefit, is a notable

example (e.g., Dundreggan Estate—Trees for Life, Woodland Trust, **John Muir Trust**).

These joint ownership models represent a shift from a **state-centred model of forest governance** to a **networked governance system** with multiple actors involved in decision-making and benefit-sharing. They facilitate flexible, locally oriented **tailored management**, but also present challenges, including (i) c**apacity gaps**, particularly in terms of technical knowledge, financial resources, and administrative skills; (ii) l**egal complexity**, as hybrid ownership structures may fall outside standard regulatory frameworks; and (iii) e**quity concerns**, such as ensuring broad-based participation and avoiding elite capture of communal resources (Lähdesmäki et al., 2023; Lawrence et al., 2021; Louda et al., 2023). In some countries, **support instruments exist**, such as specific a**dvisory services** or **training programs,** f**unding schemes** through rural development funds or ecosystem service payments, or i**nstitutional platforms** for knowledge exchange and joint decision-making.

6.4.2.4 Diversification of Forest Owners and the Rise of "New" Forest Owners

The diversification of forest ownership also happens through broader **societal transformations**—including urbanisation, shifting lifestyles, generational change, and the growing importance of non-timber values. The term "**new forest owners**" refers to individuals or entities who have recently acquired forest land, often through **inheritance, purchase, restitution, or policy-driven allocation**, but who **differ significantly from traditional forest owners** in terms of their background, motivations, and relationship to forest management (Weiss & Nichiforel, 2020; Weiss et al., 2019). Key characteristics of this group include:

- **Urban residence**: Many new forest owners live distant from their forest land, often in cities, and may not visit it regularly.
- **Lack of forestry background**: Unlike traditional owners (e.g. farmers, foresters), new owners often have little or no technical knowledge or experience in managing forests.
- **Diverse motivations**: Their interests may include nature conservation, recreation, aesthetics, heritage, lifestyle, or financial investment—rather than commercial timber production.
- **Small-scale holdings/fragmentation**: Many owners have small and often fragmented plots (smaller than 5 ha, often even smaller than 1 ha), which limits economic viability and makes coordinated management more difficult (Lidestav et al., 2020).

The emergence of new forest owners has profound implications for forest management and the implementation of policy. Policy makers mostly see the problem of abandonment, **passive ownership** or m**anagement inactivity**, as many new owners lack the interest or ability to manage their forests actively. There is also an i**ncreased**

heterogeneity of goals, where, unlike traditional timber-focused management, these new owners may prioritise **non-market values**, such as biodiversity, cultural heritage, or leisure, which challenge conventional policy and support instruments. Forestry actors often express their difficulties in "how to reach" those owners, implying several dimensions: First, owners are often not known through land registries, and they are not in contact lists of forest authorities or advisors. Second, they do not consult sectoral information channels such as traditional advisory services and magazines. Furthermore, due to their different lifestyles and values, they are not effectively reached with traditional forestry language, arguments, and incentives (Wilkes-Allemann et al., 2021).

To address these challenges and harness the potential of new ownership patterns, several countries have tried to develop **targeted policy measures and institutional responses such as**: (i)**advisory and extension services** for new owners such as beginner training or property planning tools (e.g., in **Austria**, **Finland**, **Slovenia**, Lawrence et al., 2020); (ii) f**orest owner associations and cooperatives** to organise small and new owners, promote collaboration, and enhance economies of scale (e.g., **family forest owner movement** in the Nordic countries); (iii) d**igital engagement** through **online platforms and mobile tools** to engage absentee owners and simplify administrative procedures; and (iv) to recognise **multi-functionality** and accommodate **non-commercial uses**, such as conservation, carbon sequestration, and recreation.

6.4.3 Outcomes—What the Type of Ownership Means for Forest Management

How do different types of ownership manage their forests, and how does that impact the condition of forests and the provision of ecosystem services? Such questions are, in fact, difficult to answer and do not necessarily follow a public/private divide.

6.4.3.1 A Public/Private Divide?

A literature review study on the links between land tenure and biodiversity suggests that publicly owned forests tend to have more old-growth or ancient woodland structures (Mölder et al., 2021). Large nature reserves, national parks, and other conservation areas are mostly associated with long-term public ownership, a fact that may be related to (a) that the historical legacy of natural or semi-natural forests often found in long-term public ownership makes them suitable for protected area designations, and (b) the establishment of protected areas tends to be easier in publicly owned forests (Mölder et al., 2021).

A study on forest harvesting patterns across Europe, using data from National Forest Inventories from 13 regions, found a huge **diversity in harvest regimes**, primarily depending on the productivity of the region and the species, but with important deviations related to local forest conditions, subsidies, importance of other forest services, and ownership of the forest (Schelhaas et al., 2018). In the studied regions, privately owned forests had a lower harvest probability than publicly owned forests, attributed to the fact that many private owners have only small properties, and little knowledge of forest management, or do not live on or are not close to their forest property. There is a positive relationship between harvesting intensity and property size, as well as the existence of strong forest owners' associations or other forms of common management. When public forests are managed less intensively, it is usually due to a strong focus on nature conservation, often through the ownership of National and Natural Parks (Schelhaas et al., 2018). Still, those observations are based only on a few case studies and are not generalisable across countries. They do not necessarily characterise the type of forest management in the majority of forests. Let us carefully examine management objectives and outcomes across different types of ownership.

6.4.3.2 Diversity of Goals and Influence Factors

Numerous studies indicate that owners' objectives are typically multi-faceted and, overall, exhibit rather conservative orientations. They typically follow multiple objectives, which combine production, household economy, and intangible benefits (Lawrence, 2020), and understand forest management as a combination of forest maintenance, ecosystem stewardship, and economic activity (Feliciano et al., 2017). While larger forest holdings are more oriented toward profitable forest management than smaller ones, for all of them, the preservation of the capital leads their economic objectives (Rametsteiner et al., 2005). Besides the size of the owned forests, their ecological characteristics also influence forest management decisions (i.e., the perceived ecological, economic and sentimental property values, Tiebel et al., 2022). Furthermore, besides the skills and capacities of the owners for managing their forests, various demographic characteristics influence management orientations towards economic goals (e.g., older age, rural orientation) or conservation (e.g., female gender, higher levels of education, Tiebel et al., 2022). Altogether, the literature clearly shows that forest owners are guided by multiple goals, with **profit not being the primary objective**. This aligns with studies of private enterprises outside the forest sector, which suggest that the strategic decisions of company owners are guided by private and household considerations rather than commercially or profit-oriented ones (Miller & Le Breton-Miller, 2006).

Furthermore, social norms, institutional factors and policy programmes have an impact (Deuffic et al., 2018). Forest management depends on what is allowed or prescribed, as well as on how far public policies support it. Traditionally, public forest policies aim to secure the supply of industries with timber by regulating the maintenance of forest land and its productivity. As shown above, ESEE countries apply rather restrictive approaches, while Western European countries are more liberal.

An interesting case is Finland, where the former prescription of even-aged management was liberalised in 2011 to allow more diverse forms of management and goals connected with forests (Wong et al., 2020).

Various obstacles to innovative and SFM in Europe have been reported across Europe (Wong et al., 2020; Živojinović et al., 2015), many of which relate to the institutional forestry frameworks:

- Fragmentation of forest land and insufficient revenues from forest management;
- Restrictive forest policy frameworks;
- Lack of incentives and/or financial support, lack of a cadastral survey or land registry;
- Illegal logging.

6.4.3.3 Evidence on Forest Management Impacts

What can be said about the impact of these factors on forest management across Europe? The amount and management approach in different ownership categories can be inferred from national forest inventories; however, the systems are only comparable to a limited degree. One basic information available across countries is the area of **forests available for wood supply (FAWS)**. In some countries, the share of FAWS is evenly distributed between public and private ownership; in others, particularly in Scandinavia, the proportion is smaller in public forests (Fig. 6.7, Lawrence, 2020). It is usually the state forests with lower shares of FAWS, as well as municipal forests (Wong et al., 2020). The share of FAWS can be seen as an indication of challenges and restricted possibilities for income generation. The reasons why forests are not available may be related to the natural conditions (e.g., bare land) or to legal restrictions (e.g., protected areas). In Austria, for instance, the state forests own large shares of mountainous terrain above the tree line, which are often protected areas as well.

Growing stock may be used as an indicator of forest conditions. Although this must be used cautiously as an indication of management approach or the delivered ecosystem services, it may be insightful to compare public and private forests across time and/or within countries. On the country level, overall, growing stock per hectare (m^3/ha) in FAWS in public and private forests is comparable and increasing in both (Lawrence, 2020).

The **utilisation rate**, shown as volume harvested as a proportion of net annual increment, is considerably below 100% in most countries, in both public and private forests. For the whole ECE region (including the USA, excluding Russia), the rate is higher in private forests (69%) than in public (61%), but with significant variations (Wong et al., 2020). In many countries, the utilisation rate increased in the last decades, often in both public and private forests. Overall, utilisation rates differ considerably across European countries, including in terms of their accentuation in public and private forests, making it practically impossible to generalise or discern patterns. Due to specific local ecological, climatic, cultural, economic and political contexts, countries differ strongly in their results, even within European regions (Aggestam, 2024).

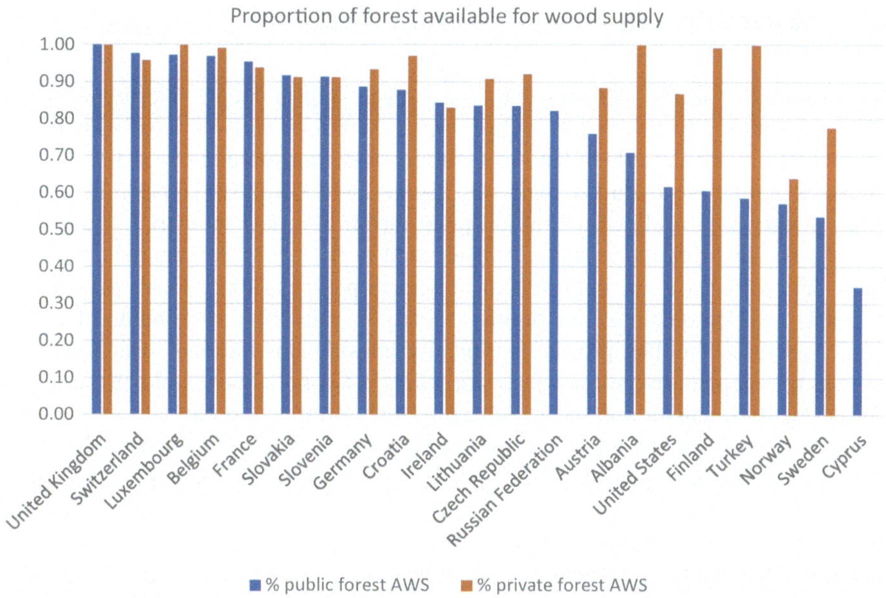

Fig. 6.7 Proportion of public and private forest available for wood supply. *Source* UNECE/FAO (2020)

More detailed analyses, for instance, concerning other ownership categories or for other FES, can hardly be conducted across countries due to very limited data sets (Aggestam, 2024). Many countries do not report on sub-categories of public or private forests because the data are not available, or ownership classifications differ. National forest inventories still focus mostly on wood production, which makes the analysis of productive functions of forest resources easier than biodiversity or other FES (Aggestam et al., 2020b).

From single-country data, indications suggest pronounced differences between different types of public forests and between large and small private forests (Aggestam, 2024). Based on scientific case studies, large private forests are usually more intensively managed than small private forests. Management practices such as clear-felling, conifer plantations, and shorter rotation cycles are more prevalent in large privately owned forests than in public forests (Mölder et al., 2021). Systematic, quantitative comparisons, however, have not yet been done. Better databases from forest inventories, including ownership information, would therefore be highly desirable (Aggestam et al., 2020a).

6.5 Ownership and Policy—Mutual Relations

The information so far indicates that forest ownership is deeply intertwined with policy in a mutually beneficial relationship. Few policy analysis approaches can ignore the fundamental role of land ownership. As demonstrated in Sect. 6.2, historical forest regulations included ownership aspects, with ownership types influencing management, e.g., traditional multipurpose farming versus industrial, "rational" forestry. The institutional analysis, as presented in Sect. 6.3, explains which forest goods and services align with specific ownership types (state, local, private, or common) and demonstrates that tenure security is crucial for sustainable management. Analyses of policy processes typically consider institutional structures and actors as key elements of policy systems or policy networks, for example, in the actor-centred institutionalism approach (Scharpf, 1997). In this approach, forest owners and their interest groups are key actors in both the policy formulation and implementation processes (Pülzl & Treib, 2017). Forest ownership structures are thus reflected in a country's forest policies and characterise their implementation (Weiss, 2004). Owner organisations are powerful actors in the political negotiations on the regulation, protection, management and use of forest resources (Krott, 2005).

Policies relate to ownership in two main ways: defining ownership structures and assigning property rights and responsibilities (Lawrence, 2020; see also Sect. 6.3). Ownership, in turn, shapes policymaking through the representation of owner interests in both formal and informal ways (Krott, 2005). In most countries with private forest owners, the policymaking system acknowledges their legitimate interests and includes forest owner organisations in policymaking processes (Sarvašová et al., 2015).

6.5.1 Policy Influence on Forest Ownership

There are several ways in which policy currently influences the structure of forest ownership, including land reform, land consolidation and afforestation (Lawrence, 2020; Nichiforel & Hujala, 2020). **Land reform** most prominently occurred in the post-socialist countries of the ESEE, with the restitution of forest properties to their former owners, and in some Baltic countries, the privatisation of state forests. There were also interesting steps in land reform in the United Kingdom, such as offering state forests to local communities in Scotland, following the devolution of UK forest laws to the country level. The details on these land reform processes and outcomes are reported in Sect. 4.2.

Various attempts have been made to **consolidate fragmented ownership structures** in Central European countries (Lidestav et al., 2020; Nichiforel & Hujala,

2020). They include changes to inheritance laws but mostly offer voluntary mechanisms through land exchange or loose collaboration of landowners in forest management and timber sales. In several countries, defragmentation policies for agricultural and forest land exist, often through preemptive rights granted to neighbouring landowners (e.g., Austria, France, Lithuania, Slovenia, and Romania) and/or to buyers who demonstrate knowledge and experience in forest management (e.g., Austria and Estonia). In Western European countries, those regulations have a longer tradition. In several countries, the splitting of forest land among heirs is prohibited (below 5 ha in Slovenia) or taxed (below 2 ha in Slovakia). In some countries, specific property regimes have been created that allow the pooling of forest parcels for joint management (Portugal, Belgium, Finland). In Hungary and Slovakia, shared ownership is possible since the forest restitution laws. Initiatives for voluntary land consolidation have been reported for Sweden, Finland and Germany. For Bavaria, it is reported that the land consolidation not only led to larger parcels but also had the desired side-effect that it initiated the selling of parcels of urban owners to "traditional" farm forest owners (Koch & Gaggermeier, 2011).

Afforestation programs on agricultural lands also led to changes in forest ownership structures, e.g., in Mediterranean countries, the UK, or Ireland. The EU Common Agricultural Policy has mostly fueled such afforestation measures through Rural Development Programmes in the Member States, implemented since 1990, and sometimes national subsidy schemes that support afforestation. In Eastern Europe, similar measures were included in their accession programmes to the EU. These were particularly utilised in the Czech Republic, Hungary and Poland (Nichiforel & Hujala, 2020).

With few exceptions reported here, progressive measures that would fundamentally alter ownership patterns or governance settings are rare; instead, most measures address short-term timber market issues arising from increasing fragmentation. Most reported measures are conservative: they do not alter ownership structures but aim to incentivise owners toward active management (see below).

6.5.2 Policy Influence on Forest Management

State influence on forest management primarily falls into two major domains: state-owned and non-state-owned forests. The basic legal provisions are outlined in the forest laws, specifically in the definitions of property rights (see Sect. 3.3). Although these forest law regulations are usually valid for both public and private properties, State forests have a special status, as state governments may directly prescribe their management goals.

Currently, two opposing trends are evident in forest-related policy (Nichiforel et al., 2020; Winkel & Sotirov, 2015): One, based on environmental policies, is oriented toward the **stronger conservation** of natural resources and biodiversity, resulting in management restrictions in managed forests and a call for more protected areas. The current EU Forest Strategy, together with the EU Nature Restoration Regulation, focus strongly on protecting and restoring the European forests. In practice, though, nature conservation in forests is predominantly achieved through voluntary financial instruments. Another trend, mostly found in national forest policies, aims to liberalise and incentivise forest management, e.g., with the aim of "wood mobilisation".

Forest policies of many European countries include both aims, a more environmentally oriented management and an increase in timber harvests. For both aims, the primary policy instrument is **financial incentives**. For the former aim, silvicultural methods of close to nature management are often subsidised, such as the conversion of coniferous monoculture forests to mixed and site-adapted species. Today, this is often connected with adapting forests to the changing climate and thus to the use of climate-adapted species. For the latter aim, a range of instruments is used, ranging from education and advisory services to support for forest owner cooperation and management activities, such as creating management plans. Those instruments are often specifically oriented towards small-scale owners. As part of the EU Rural Development Programmes, for instance, several countries utilised the defined measures for cooperation in the development of new products and techniques, or for forest management planning among small-scale forest holders (Austria, Germany, Spain, Italy, and the United Kingdom; Nichiforel & Hujala, 2020).

Many European countries have had policy programmes for **wood mobilisation**, focused on small-scale forest owners. These included awareness-raising campaigns and the support of cooperation. At the core, most campaigns aimed to motivate owners by highlighting the increased income and improved efficiency in forest management. The measures had limited effects. Traditional farm owners often respond to the support of cooperation, but their market participation ultimately depends primarily on market signals—something that the campaigns cannot change. Non-traditional owners rarely have economic goals for their forest properties and are therefore not effectively reached with conventional campaigns (EC, 2010; Wilkes-Allemann et al., 2021). Among other things, the limited impact of these policy programmes is caused by a disregard for the real situations of forest owners and the neglect of the goals and preferences of different owner types. Measures seem to be typically oriented toward an ideal type forest owner and constraints for rational forest management, rather than real-life situations (Lawrence, 2018). Concerning non-traditional owners, studies recommend developing new types of forest owner associations, particularly focusing on fragmented non-traditional ownership (EC, 2010), which would serve as intermediary actors, connecting and building trust between policymakers and forest owners (Wilkes-Allemann et al., 2021). Those approaches would help break the conventional "normal forest owner" ideal among policymakers and advisors by developing different incentives, using alternative channels, and employing less technical language. This would be particularly relevant for traditional owners and even more so for new, non-traditional owner types. Since profit rarely ranks first among the goals of forest owners, price-only incentives have limited effects. However, tailored advisory or cooperative models for different owner types could be more effective.

Such policy innovations, however, would face resistance by established structures (Buttoud et al., 2011; EC, 2010). Governments do not act in isolation to formulate their goals—policy goals are the result of policy processes that involve different steps and various public and private actors, including, for example, the landowners (Krott, 2005). This is briefly discussed in the following.

6.5.3 Forest Owners' Influence on Policies

Forest ownership patterns in a country not only matter, as they are the ones actually responsible for managing the forest resources. Also, in the policy process, they are not solely the "target group" of forest-related policies who would follow legal provisions or not. Through their interest organisations, they participate in the entire policymaking process, including formulation (Krott, 2005). Ownership patterns thus establish which interests are actually those of the owners, but also how easily their interests are organised and eventually fed into policymaking by their associations.

The effects of EU policies on forest management and conservation have been limited and fragmented. While the EU has set ambitious goals for forest protection and restoration, challenges remain in implementation and coordination across Member States (Aggestam & Pülzl, 2018; Sotirov, 2017). An inter- and transdisciplinary evaluation of the implementation of the **EU Natura 2000** policy in forests identified challenges related to tensions between different policy goals, between scientific knowledge and stakeholder demands, regarding new risks from climate change. It is connected to the effective allocation of funding (Winkel et al., 2015). Many of those challenges are directly linked to the fact that the landowners targeted with those policies are mostly private. Since the policy measures aim to direct their behaviours, that is, directly affect their property rights, their interest organisations try to influence the policy process for their interests, at local, national and EU levels. The EU Natura 2000 policy nicely illustrates the power dynamics of different interest groups. The establishment of the EU's Habitats Directive in 1992 was enabled by a supportive window of opportunity related to the growing environmentalism connected with the UNCED in Rio de Janeiro, and the involvement of Environmental NGOs and conservation biology scientists, in the formulation of the Habitats Directive and in the implementation process, through the European Habitat Forum. The often top-down and science-driven approach towards the designation of Natura 2000 sites caused resistance from affected landowners and authorities. The Member States' strategies to manage these conflicts led to a more flexible approach, taking into account local groups and land users (Weiss & Sotirov, 2017). More detailed studies at the national level reveal the various dynamics that led to significantly different policy outcomes and highlight the importance of ownership type in effective practice examples (Sotirov, 2017).

The impact of national ownership patterns on forest policies and how they may hinder or facilitate new solutions has been illustrated by the example of **mountain forest policies in Alpine countries** (Weiss, 2004). Depending on the distribution of forest ownership and administrative arrangements, different restoration approaches have been developed in Switzerland, Bavaria (Germany), Austria and Slovenia, for degraded mountain forests. According to their different institutional settings, various actors are involved in the policy networks, and with their distinct interests and value systems, they determine the goals, design, and outcomes of the restoration programmes. While, for instance, the prevailing collective ownership of forests in Switzerland allowed the development of non-intervention policies, the prevailing private ownership of forests in Austria led to either timber-oriented restoration concepts (by the forest authorities) or protection-oriented restoration concepts (by the services responsible for controlling torrents and avalanches).

The former socialist countries in East and Southeast Europe present an interesting case for studying forest owners' interest groups, as private ownership was nationalised in most of these countries during the socialist era. Subsequently, these interest groups were newly established. The formation of **forest owners' associations in Central and Eastern Europe** constituted one of the opportunities for interest representation in the land restitution processes (Sarvašová et al., 2015). Later, they gained a growing role in influencing forest policymaking. Although many owners, following the socialist experience, are rather restrained towards the idea of joining organisations, they seem to be able to consolidate themselves as interest groups and offer services to their members (Glück et al., 2010; Sarvašová et al., 2015). It also shows that policy makers seem to support the formation of forest owners' associations because they provide them with a clear addressee to communicate with (Glück et al., 2010; Weiss et al., 2012). Although they have thus become a significant factor in political power, it does not mean they receive great support from the political system (Šálka et al., 2016).

6.6 The Implications of Forest Ownership for Transformative Change

What implications do these findings about the diversity and trends of forest ownership have for forest governance, particularly with regard to the transition to a sustainable economy and society?

6.6.1 The (Growing) Diversity of Forest Owners Needs to Be Recognised by Policymakers

The main finding is that land ownership structures represent a fundamental institutional framework for forest governance that must be considered in policy-making. There should be greater awareness of the **wider variety of ownership types beyond just public and private**. Given the ongoing changes in forest ownership patterns, this diversity is expected to continue growing. The COST Action FACESMAP concluded from these observations that policy instruments and forest management approaches need to be revised, such as forest owners' associations, advisory and innovation systems, and the types of support and management services offered to owners (Weiss et al., 2017). This is particularly true in view of how forests could effectively contribute to societal and economic transformation.

The growing diversity among forest owners is largely viewed as a threat to ensuring a sufficient timber supply for forest-based industries. However, they could also be seen as an asset in terms of achieving resilience in a changing climate and meeting the evolving societal expectations of forests. Furthermore, effective policy change may require not only different policy instruments, but also new organisational and institutional solutions, such as specific cooperations or associations for different owner types, new ownership forms, or new service organisations (Weiss et al., 2017).

6.6.2 The Misconception of an Ideal Type "Normal Forest Owner"

A central insight from this chapter is that forest sciences and forest policy generally and unconsciously apply the ideal figure of the "normal forest owner". The basic scientific model of forest management assumes a standardised forest and policy measures assume landowners who are rational and economically oriented. These landowners are expected to be interested in placing a steady supply of valuable timber on the market, which is welcomed by industry actors and forest policy but viewed critically by environmentalists and conservation policy. Small-scale, urban and absentee owners are also assumed to have the same attitude, but with limited skills, knowledge and capacities. This demonstrates how difficult it is for traditional forestry stakeholders to accept that the growing number of non-traditional owners often have completely different objectives and preferences for their forests.

While the "normal forest" may be a useful planning concept, it rarely reflects reality. A large cross-country study of actual forest management and harvesting patterns across Europe revealed that "that there is no such thing as yield table harvesting in Europe" (Schelhaas et al., 2018). Likewise, **there is no such thing as a typical or standard forest owner** (cp. Section 4.3).

Large forest holdings (whether public or private) may approach the ideal type of ownership, as they are organised professionally for the specific purpose of managing their land, setting aside any secondary goals of their owners. However, most forest holdings are less profit-oriented, focusing instead on maintaining the resource (Rametsteiner et al., 2005). Non-industrial private forest owners, in particular, tend to be less profitable, less innovative and less responsive to market signals (Beach et al., 2005). Since money is not the main motivator for the behaviour of forest owners, forest management and harvesting should be viewed as more of a social issue than an economic one (Lawrence, 2018). For example, these findings suggest that personal, social and cultural factors, such as traditions, local networks and trust, are often more important than economic incentives or neutral "information".

6.6.3 Is a Different Ownership Structure Needed?

In light of the multiple global crises facing the world, it is worth considering bold solutions, such as rethinking forest ownership structures. Is there an ideal ownership model that could enable transformative societal change?

Research does not clearly indicate which ownership type best supports sustainability. Although evidence often associates larger (public or private) forest holdings with higher efficiency, this relationship is highly context-dependent and contingent on institutional support. It is also sometimes constrained by systemic barriers. Relevant examples include a variety of new forest products and services, such as bioenergy, district heating, non-wood products, recreational and cultural services, and social innovations.

Policy and innovation research suggests that, from an institutional perspective, the ideal ownership framework is one in which beneficiaries of goods and services play a direct role in management. Since many forest ecosystem services benefit local populations, **locally based common ownership** appears to be the most promising option (Glück, 2000). Community woodlands in Wales and Scotland exemplify this approach, integrating the multiple benefits of forests, including non-market benefits, with local, integrated economic development (Lawrence & Ambrose-Oji, 2015; Ludvig et al., 2019). Local government ownership with participatory governance is a close alternative. Various examples of common, charitable or alternative forms of forest ownership or management oriented towards multiple societal benefits have been described as social innovations (Weiss et al., forthcoming).

While these examples focus on services that primarily benefit local communities, many forest ecosystem services, such as biodiversity and carbon sequestration, are global public goods. As global forest ownership is impractical, securing these services requires alternative tools, such as regulation or market mechanisms. To be effective, these tools must be tailored to landowners and supported by suitable implementation systems.

6.6.4 Fighting Global Crises

When examining global changes, such as the climate and biodiversity crises, large-scale solutions are required, ideally on a global level. In theory, this would require large-scale public ownership of forests. **However, it is doubtful that state forests would be primarily managed for biodiversity or climate goals**. Specific examples and a certain trend indicate that state forests tend to be managed for multiple purposes or to create protected areas. However, most remain commercially focused on timber. As governments are mostly inactive or unable to implement effective measures to reduce biodiversity loss or slow down climate change, major shifts in forest management, even on public land, seem unlikely. Furthermore, private forest owners are not the only group with a primary interest in timber production. Industry representatives certainly have a stronger interest in securing the wood supply and are much more powerful.

If governments aimed to change forest management to restore ecosystems and mitigate climate change, they would have the policy instruments to direct both public and private owners towards these goals. However, they have been shown to act late and maintain timber production goals. Forest sciences have highlighted the need for forest restoration—transitioning from risky monoculture stands to more diverse, site-adapted ones—for decades, but this has only gained traction following large-scale forest damage from climate impacts in Central Europe.

The European Green Deal called for more progressive measures to combat the biodiversity and climate crises. However, the related Forest Strategy was met with controversy from forestry stakeholders and civil society. Environmental NGOs called for a stronger focus on biodiversity and climate change. Conversely, forest owners feared that reduced wood harvesting would negatively impact forest-based value chains and employment opportunities. Scientific studies often lack balance. In order to fully understand the mitigation potential of forests and forestry, a holistic approach is necessary that considers all forest and wood use options (Verkerk et al., 2022). Carbon can be sequestered in forests either by increasing the standing volume or by harvesting it and storing it in long-lived wood products. Since these strategies have trade-offs and are interlinked, they must be assessed together in both scientific research and policy. Neither strategy is a one-size-fits-all solution. Nevertheless, such integrated perspectives remain rare. Integrated management options need to be developed that combine the benefits of forest protection, restoration and utilisation. Climate-Smart Forestry is an emerging scientific approach that incorporates adaptation, mitigation, and social aspects into forest management (Bowditch et al., 2020). It considers wood production, biodiversity protection, and ecosystem services, with the aim of achieving effective carbon removal and enhancing forest resilience (Verkerk et al., 2020). However, such integrative thinking is largely absent from global decarbonisation strategies and national forest policies. At the policy level, the long-standing conflict between environmental and forestry perspectives remains challenging to resolve.

References

Aggarwal, S., Reeb. D., & Walji, K. (2020). Forest ownership and tenure in the world and the ECE region. In *UNECE/FAO*. (pp. 31–42)

Aggestam., F. (2024). *European forest policy and governance: An integrative analytical framework*. Routledge.

Aggestam, F., & Pülzl, H. (2018). Coordinating the uncoordinated: The EU forest strategy. *Forests, 9*(3), 125.

Aggestam, F., Krott, M., & Hauck, M. (2020a). *Municipal forests* (pp. 155–159). UNECE/FAO.

Aggestam, F., Quiroga, S., Suárez, C., Martínez, P., & Sarvašová, Z. (2020b). *Forest ownership and the provision of wood and other forest ecosystem services* (pp. 73–85). UNECE/FAO.

Beach, R. H., Pattanayak, S. K., Yang, J. C., Murray, B. C., & Abt, R. C. (2005). Econometric studies of non-industrial private forest management: a review and synthesis. *Forest Policy and Economics, 7*(3), 261–281.

Bouriaud, L. (2005). Causes of illegal logging in Central and Eastern Europe. *Small-Scale Forest Economics, Management and Policy, 4*, 269–291.

Bouriaud, L., & Schmithüsen, F. (2005). Allocation of property rights on forests through ownership reform and forest policies in Central and Eastern European countries. *Schweizerische Zeitschrift fur Forstwesen, 156*(8), 297–305.

Bowditch, E., Santopuoli, G., Binder, F., Del Rio, M., La Porta, N., Kluvankova, T., & Tognetti, R. (2020). What is climate-smart forestry? A definition from a multinational collaborative process focused on mountain regions of Europe. *Ecosystem Services, 43*, 101113.

Buttoud, G., Kouplevatskaya-Buttoud, I., Slee, B., & Weiss, G. (2011). Barriers to institutional learning and innovations in the forest sector in Europe: Markets, policies and stakeholders. *Forest Policy and Economics, 13*(2), 124–131.

Carlowitz, H. C. V. (1713). Sylvicultura oeconomica, oder, Hausswirthliche Nachricht und naturmäßige Anweisung zur wilden Baum-Zucht (nebst gründlicher Darstellung, wie zuförderst durch Göttliches benedeyen dem allenthalben und insgemein einreissenden grossen Holtz-Mangel, vermittelst Sae-Pflantz-und Versetzung vielerhand Bäume zu rathen, auch also durch Anflug und Wiederwachs des so wohl guten und schleunig anwachsend, als andern gewüchsig-und nützlichen Holtzes, ganz öde und abgetriebene Holtz-Ländereyen, Plätze und wiederum …). Leipzig.

Deuffic, P., Sotirov, M., & Arts, B. (2018). "Your policy, my rationale", How individual and structural drivers influence European forest owners' decisions. *Land Use Policy, 79*, 1024–1038.

Dobšinska, Z., Živojinović, I., Nedeljkovic, J., Petrovic, N., Jarsky, V., Oliva, J., Salka, J., Sarvasova, Z., & Weiss, G. (2020). Actor power in the restitution processes of forests in three European countries in transition. *Forest Policy and Economics, 113*, 102090.

EC. (2010). *Prospects for the market supply of wood and other forest products from areas with fragmented forest-ownership structures*. Final study report to the European Commission.

Elomina, J., Živojinović, I., Lidestav, G., Sandström, P., Sandström, S., & Pülzl, H. (2025). Local stakeholder's perspectives on development of economic activities: The Gällivare case. *The Extractive Industries and Society, 23*, 101664.

FAO. (2006). Time for action—Changing the gender situation in forestry. *Report of the UNECE/FAOI team of specialists on gender and forestry*. FAO.

FAO. (2018). *FRA 2020 terms and definitions* (Forest resources assessment working paper 188). Food and Agriculture Organization of the United Nations.

Feeny, D., Berkes, F., McCay, B. J., & Acheson, J. M. (1990). The tragedy of the commons: Twenty-two years later. *Human Ecology, 18*(1), 1–19.

Feliciano, D., Bouriaud, L., Brahic, E., Deuffic, P., Dobsinska, Z., Jarsky, V.,... & Ficko, A. (2017). Understanding private forest owners' conceptualisation of forest management: Evidence from a survey in seven European countries. *Journal of Rural Studies, 54*, 162–176.

Ficko, A., Lidestav, G., Dhubhain, A. N., Karppinen, H., Živojinović, I., & Westin, K. (2019). European private forest owner typologies: A review of methods and use. *Forest Policy and Economics, 99*, 21–31.

Follo, G., Lidestav, G., Ludvig, A., Vilkriste, L., Hujala, T., Karppinen, H., & Mizaraite, D. (2017). Gender in European forest ownership and management: Reflections on women as "New forest owners." *Scandinavian Journal of Forest Research, 32*(2), 174–184.

Forest Europe. (1993, June). Resolution H1: general guidelines for the sustainable management of forests in Europe. In *Second ministerial conference on the protection of forests in Europe* (pp. 16–17).

Forest Europe (2003). State of Europe's Forests 2003. The MCPFE Report on Sustainable Forest Management in Europe. Ministerial Conference on the Protection of Forests in Europe, Liaison Unit Vienna.

Forest Europe. (2020). *State of Europe's forests report 2020.* Liaison Unit Bratislava.

Glück, P. (1987). Social values in forestry. *Ambio, 16*(2/3), 158–162.

Glück, P. (2000). Policy means for ensuring the full value of forests to society. *Land Use Policy, 17*(3), 177–185.

Glück, P. (2002). Property rights and multipurpose mountain forest management. *Forest Policy and Economics, 4*(2), 125–134.

Glück, P., Avdibegović, M., Čabaravdić, A., Nonić, D., Petrović, N., Posavec, S., & Stojanovska, M. (2010). The preconditions for the formation of private forest owners' interest associations in the Western Balkan Region. *Forest Policy and Economics, 12*(4), 250–263.

Górriz-Mifsud, E., Govigli, V. M., & Bonet, J. A. (2017). What to do with mushroom pickers in my forest? Policy tools from the landowners' perspective. *Land Use Policy, 63*, 450–460.

Hardin, G. (1968). The tragedy of the commons. Science 162: 1243–1248.

Hehn, M., Westermayer, T., & Lewark, S. (2006) FAO (pp. 144–146).

Huss, J., & von Gadow, F. (2012). Einführung in das Faksimile der Erstausgabe der Sylvicultura oeconomica von H. C. von Carlowitz, 1713. Kessel.

IFSA. (2022). *Gender and forest education: An open letter by IFSA.* Retrieved July 25, 2025, from https://ifsa.net/gender-open-letter/

Jeanrenaud, S. (2001). *Communities and forest management in Western Europe.* WG-CIFM.

Katila, P., McDermott, C., Larson, A., Aggarwal, S., & Giessen, L. (2020). Forest tenure and the sustainable development goals—A critical view. *Forest Policy and Economics, 120*, 102294.

Kissling-Näf, I., Volken, T., & Bisang, K. (2002). Common property and natural resources in the Alps: The decay of management structures? *Forest Policy and Economics, 4*(2), 135–147.

Koch, M., & Gaggermeier, A. (2011). Forest land consolidation—Who benefits? In C. Hartebrodt, & K. Howard (Eds.), *IUFRO small scale forestry conference: Synergies and conflicts in social, ecological and economic interactions.* Universität Freiburg and FVA Baden-Württemberg.

Koch, N. E., & Kennedy, J. J. (1991). Multiple-use forestry for social values. *Ambio, 20*(7), 330–333.

Krott, M. (2005). *Forest policy analysis.* Springer.

Kuemmerle, T., Kozak, J., Radeloff, V. C., & Hostert, P. (2009). Differences in forest disturbance among land ownership types in Poland during and after socialism. *Journal of Land Use Science, 4*(1–2), 73–83. https://doi.org/10.1080/17474230802645857

Lähdesmäki, M., Matilainen, A., & Lehto, P. (2023). Forest owner or shareholder? Ownership feelings in a jointly-owned forest. *Scandinavian Journal of Forest Research, 38*(4), 254–264. https://doi.org/10.1080/02827581.2023.2216945

Lawrence A. (2020). *Overview* (pp. 1–17). UNECE/FAO.

Lawrence A., Hujala, T., Deuffic, P., & Nichiforel, L. (2020). *Forestry extension and advice: diversity and change across Europe* (pp. 148–150). UNECE/FAO.

Lawrence, A. (2018). Do interventions to mobilise wood lead to wood mobilisation? A critical review of the links between policy aims and private forest owners' behaviour. *Forestry: An International Journal of Forest Research, 91*(4), 401–418.

Lawrence, A., & Ambrose-Oji, B. (2015). Beauty, friends, power, money: Navigating the impacts of community woodlands. *The Geographical Journal, 181*(3), 268–279.

Lawrence, A., Gatto, P., Bogataj, N., & Lidestav, G. (2021). Forests in common: Learning from diversity of community forest arrangements in Europe. *Ambio, 50*, 448–464.
Leipold, S. (2014). Creating forests with words—A review of forest-related discourse studies. *Forest Policy and Economics, 40*, 12–20.
Leslie, A. J. (1966). A review of the concept of the normal forest. *Australian Forestry, 30*(2), 139–147.
Lidestav, G. (2020). *Gender by numbers—And beyond* (pp. 151–154). UNECE/FAO.
Lidestav, G., & Egan Sjölander, A. (2007). Gender and forestry: A critical discourse analysis of forestry professions in Sweden. *Scandinavian Journal of Forest Research, 22*(4), 351–362.
Lidestav, G., Weiss, G., & Zivojinovic, I. (2020). *Changes in forest ownership* (pp. 43–58). UNECE/FAO.
Louda, J., Dubová, L., Špaček, M., Brnkaľáková, S., & Kluvánková, T. (2023). Factors affecting governance innovations for ecosystem services provision: Insights from two self-organised forest communities in Czechia and Slovakia. *Ecosystem Services, 59*, 101497. https://doi.org/10.1016/j.ecoser.2022.101497
Ludvig, A., Öllerer, B., & Aubram, T. (2024). Connecting gender balance, crisis resistance and innovativeness in the forestry sector: Women in leadership and management. *Environmental Science & Policy, 161*, 103890.
Ludvig, A., Rogelja, T., Asamer-Handler, M., Weiss, G., Wilding, M., & Zivojinovic, I. (2020). Governance of social innovation in forestry. *Sustainability, 12*(3), 1065.
Ludvig, A., Zivojinovic, I., & Hujala, T. (2019). Social innovation as a prospect for the forest bioeconomy: Selected examples from Europe. *Forests, 10*(10), 878.
Mai, Y. H., Mwangi, E., & Wan, M. (2011). Gender analysis in forestry research: Looking back and thinking ahead. *International Forestry Review, 13*(2), 245–258.
Mantau, U. (2001). *Recreational and environmental markets for forest enterprises: A new approach towards marketability of public goods*. CABI.
McKean, M. A. (1998). Common property: What is it, what is it good for, and what makes it work? In C. Gibson, M. A. McKean, & E. Ostrom (Eds.), *Forest resources and institutions* (pp. 23–47). Center of the Study of Institutions, Population, and Environmental Change (CIPEC), Workshop in Political Theory and Policy Analysis. Forests, Trees and People Programme. Working Paper No. 3, Bloomington, Indiana.
Miller, D., & Le Breton-Miller, I. (2006). Family governance and firm performance: Agency, stewardship, and capabilities. *Family Business Review, 19*(1), 73–87.
Mitscherlich, G. (1963). Zustand, Wachstum und Nutzung des Waldes im Wandel der Zeit. Schulz.
Mölder, A., Tiebel, M., & Plieninger, T. (2021). On the interplay of ownership patterns, biodiversity, and conservation in past and present temperate forest landscapes of Europe and North America. *Current Forestry Reports, 7*, 195–213. https://doi.org/10.1007/s40725-021-00143-w
Nichiforel, & Hujala. (2020). *Policy instruments and legislation to govern forest ownership* (pp. 87–100). UNECE/FAO.
Nichiforel, L., Deuffic, P., Thorsen, B. J., Weiss, G., Hujala, T., Keary, K., & Bouriaud, L. (2020). Two decades of forest-related legislation changes in European countries analysed from a property rights perspective. *Forest Policy and Economics, 115*, 102146.
Nichiforel, L., Keary, K., Deuffic, P., Weiss, G., Thorsen, B. J., Winkel, G., & Bouriaud, L. (2018). How private are Europe's private forests? A comparative property rights analysis. *Land Use Policy, 76*, 535–552.
Ostrom, E., Gardner, R., & Walker, J. (1994). *Rules, games, and common-pool resources*. Michigan University Press.
Plieninger, T., & Schaich, H. (2014). Socialist and post-socialist land-use legacies determine farm woodland composition and structure: Lessons from Eastern Germany. *European Journal of Forest Research, 133*, 597–610.
Plochmann, R. (1985). Bemerkungen zur Waldkultur Mitteleuropas. *Zeitschrift für Politik*, 195–207.
Pulla, P., Schuck, A., Verkerk, P. J., Lasserre, B., Marchetti, M., & Green, T. (2013). *Mapping the distribution of forest ownership in Europe*. EFI technical report 88 (p. 92).

Pülzl, H., & Treib, O. (2017). Implementing public policy. In *Handbook of public policy analysis* (pp. 115–134). Routledge.

Radkau, J. (1986). Zur angeblichen Energiekrise des 18. Jahrhunderts: Revisionistische Betrachtungen über die “Holznot”. *VSWG: Vierteljahrschrift für Sozial-und Wirtschaftsgeschichte, 73*(H. 1), 1–37.

Rametsteiner, E., & Weiss, G. (2006). Assessing policies from a systems perspective—Experiences with applied innovation systems analysis and implications for policy evaluation. *Forest Policy and Economics, 8*(5), 564–576.

Rametsteiner, E., Weiss, G., & Kubeczko, K. (2005). *Innovation and entrepreneurship in forestry in central Europe* (Vol. 19). Brill.

Robinson, B. E., Holland, M. B., & Naughton-Treves, L. (2014). Does secure land tenure save forests? A meta-analysis of the relationship between land tenure and tropical deforestation. *Global Environmental Change, 29*, 281–293.

Šálka, J., Dobšinská, Z., & Hricová, Z. (2016). Factors of political power—The example of forest owners associations in Slovakia. *Forest Policy and Economics, 68*, 88–98.

Sarvašová, Z., Zivojinovic, I., Weiss, G., Dobšinská, Z., Drăgoi, M., Gál, J., Jarsky, V., Mizaraite, D., Pollumae, P., Salka, J., Schiberna, E., Sisak, L., Wolfslehner, B., Zalite, Z., & Zalitis, T. (2015). Forest owners associations in the Central and Eastern European region. *Small-Scale Forestry, 14*(2), 217–232.

Scharpf, F. W. (1997). *Games real actors play actor-centered institutionalism in policy research.* Routledge.

Schelhaas, M.-J., Fridman, J., Hengeveld, G. M., Henttonen, H. M., Lehtonen, A., Kies, U., et al. (2018). Actual European forest management by region, tree species and owner based on 714,000 re-measured trees in national forest inventories. *PLoS ONE, 13*(11), e0207151. https://doi.org/10.1371/journal.pone.0207151

Schlager, E., & Ostrom, E. (1992). Property-rights regimes and natural resources: a conceptual analysis. *Land Economics*, 249–262.

Scott, J. C. (1998). *Seeing like a state: How certain schemes to improve the human condition have failed*. Yale University Press.

Sieferle, R. P. (1982). Der unterirdische Wald. Energiekrise und industrielle Revolution. Beck. English edition: Sieferle, R. P. (2001). *The subterranean forest: Energy systems and the industrial revolution.* Translated from the German by Michael P. Osman. The White Horse Press.

Sotirov, M. (Ed.). (2017). *Natura 2000 and forests: Assessing the state of implementation and effectiveness. What science can tell us 7*. European Forest Institute.

Stojanovski, V. (2022). Policy processes in the institutionalisation of private forestry in the Republic of North Macedonia. *Sustainability, 14*(7), 4018. https://doi.org/10.3390/su14074018

Tiebel, M., Mölder, A., & Plieninger, T. (2022). Conservation perspectives of small-scale private forest owners in Europe: A systematic review. *Ambio, 51*, 836–848. https://doi.org/10.1007/s13280-021-01615-w

UNECE/FAO. (2020). *Who owns our forests? Forest ownership in the ECE region.* Report published by the United Nations Economic Commission for Europe (UNECE) and the Food and Agriculture Organization of the United Nations.

United Nations. (1992) Non-Legally Binding Authoritative Statement of Principles for a Global Consensus on the Management, Conservation and Sustainable Development of all Types of Forests. A/CONF.151/6/Rev1. United Nations, Rio de Janeiro.

Verkerk, P. J., Costanza, R., Hetemäki, L., Kubiszewski, I., Leskinen, P., Nabuurs, G. J., & Palahí, M. (2020). Climate-smart forestry: The missing link. *Forest Policy and Economics, 115*, 102164.

Verkerk, P. J., Delacote, P., Hurmekoski, E., Kunttu, J., Matthews, R., Mäkipää, R., Mosley, F., Perugini, L., Reyer, C. P. O., Roe, S., & Trømborg, E. (2022). *Forest-based climate change mitigation and adaptation in Europe. From Science to Policy 14*. European Forest Institute. https://doi.org/10.36333/fs14

Weiss et al. (forthcoming). The Role of Social Innovation for Transformative Change in Forestry.

Weiss, & Nichiforel. (2020). *Concepts and definitions of forest ownership* (pp. 19–28). UNECE/FAO.

Weiss, G. (2001). Mountain forest policy in Austria: A historical policy analysis on regulating a natural resource. *Environment and History, 7*(3), 335–355.

Weiss, G. (2004). The political practice of mountain forest restoration—Comparing restoration concepts in four European countries. *Forest Ecology and Management, 195*(1–2), 1–13.

Weiss, G., & Sotirov, M. (2017). Implementation of Natura 2000 in forests. In M. Sotirov (Ed.), *Natura 2000 and forests: Assessing the state of implementation and effectiveness. What science can tell us 7* (pp. 39–64). European Forest Institute.

Weiss, G., Gudurić, I., & Wolfslehner, B. (2012). *Review of forest owners' organisations in selected Eastern European countries* (Working Paper No. 30, p. 46). FAO Forestry Policy and Institutions. http://www.fao.org/docrep/016/me171e/me171e00.pdf

Weiss, G., Lawrence, A., Hujala, T., Lidestav, G., Nichiforel, L., Nybakk, E., Quiroga, S., Sarvašová, Z., Suarez, C., & Živojinović, I. (2019). Forest ownership changes in Europe: State of knowledge and conceptual foundations. *Forest Policy and Economics, 99*, 9–20.

Weiss, G., Lawrence, A., Lidestav, G., Feliciano, D., & Hujala, T. (2017). *Changing forest ownership in Europe—Main results and policy implications, COST Action FP1201 FACESMAP Policy Paper. EFICEEC-EFISEE research report* (25 pages). University of Natural Resources and Life Sciences, Vienna (BOKU).

Weiss, G., Ludvig, A., & Živojinović, I. (2023). Embracing the non-wood forest products potential for bioeconomy—Analysis of innovation cases across Europe. *Land, 12*(2), 305.

Wilkes-Allemann, J., Deuffic, P., Jandl, R., Westin, K., Lieberherr, E., Foldal, C., & Jarský, V. (2021). Communication campaigns to engage (non-traditional) forest owners: A European perspective. *Forest Policy and Economics, 133*, 102621.

Winkel, G. (2012). Foucault in the forests—A review of the use of 'Foucauldian' concepts in forest policy analysis. *Forest Policy and Economics, 16*, 81–92.

Winkel, G., & Sotirov, M. (2015). Whose integration is this? European forest policy between the gospel of coordination, institutional competition, and a new spirit of integration. *Environment and Planning C: Government and Policy, 34*(3), 496–514. https://doi.org/10.1068/c1356j (Original work published 2016).

Winkel, G., Blondet, M., Borrass, L., Frei, T., Geitzenauer, M., Gruppe, A., & Turnhout, E. (2015). The implementation of Natura 2000 in forests: A trans-and interdisciplinary assessment of challenges and choices. *Environmental Science & Policy, 52*, 23–32.

Wong, J., Lawrence, A., & Feliciano, D. (2020). The impact of ownership type on the implementation of forest management. In *Who owns our forests? Forest ownership in the ECE region* (pp. 60–72). United Nations Publications.

Živojinović, I., Elomina, J., Pülzl, H., Calanasan, K., Dabić, I., Ólafsdóttir, R., Siikavuopio, S., Iversen, A., Robertsen, R., Bjerke, J., Engen, S., Tommervik, H., Bogadóttir, R., Moioli, S., Tuulentie, S., Rautio, P., Lynge-Pedersen, K., Lidestav, G., Edvardsdòttir, A. G., & Nygaard, V. (2024). Exploring land use conflicts arising from economic activities and their impacts on local communities in the European Arctic. *Journal of Land Use Science, 19*(1), 186–210. https://doi.org/10.1080/1747423X.2024.2382676

Živojinović, I., Nedeljković, J., Petrović, N., Nonić, D., Weiss, G., & Nichiforel, L. (2025). The forest restitution process in Serbia: The role of the Serbian Orthodox Church in shaping an adapted forest governance framework. *Forest Policy and Economics, 177*, 103538.

Živojinović, I., Weiss, G., Lidestav, G., Feliciano, D., Hujala, T., Dobšinská, Z., & Lawrence, A. (2015). Forest land ownership change in Europe. In *COST action FP1201 FACESMAP country reports, Joint volume, EFICEEC-EFISEE research report*. University of Natural Resources and Life Sciences, Vienna (BOKU).

Chapter 7
The Missing Link: Leveraging Communication Science Insights for Forest Policy and Governance Research

Lena Riemann, Alexandru Giurca, Sandra Liebal, Philipp Mack, and Rina Tsubaki

Abstract This chapter makes the case for closer integration of communication science in forest policy and governance research, emphasising the need for theoretically robust and empirically sound approaches. The existing literature emphasises the pivotal role of communication in forest governance, particularly with regard to stakeholder engagement, conflict resolution, the interaction between science andpolicy, public participation, and the implementation of policies. However, despite acknowledging its importance, much of this research lacks theoretical clarity. While many studies advocate for improved communication, they often fail to substantiate claims of effectiveness, particularly in navigating divergent worldviews, political beliefs, and competing stakeholder interests. At the same time, the use of digital technology to improve communication is being addressed more and more, but without detailing not just the practical implications and challenges, or the overall impact on the democratic process. This chapter bridges this gap by shedding light on the key theoretical insights and methodological approaches from communication science that inform forest policy research. It critically analyses the intersection between communication science and forest policy, highlighting the untapped potential of communication theories and approaches in addressing the complex challenges of forest governance. Through a review of current research and the provision of additional conceptual insights, the discussion in this chapter will guide future research and

L. Riemann (✉) · P. Mack
Chair of Forest and Environmental Policy, University of Freiburg, Freiburg, Germany
e-mail: lena.riemann@fp.uni-freiburg.de

A. Giurca · R. Tsubaki
European Forest Institute, Governance Programme, Bonn, Germany

S. Liebal
Chair of Forest Policy and Forest Resource Economics, TU Dresden, Tharandt, Germany

R. Tsubaki
Forest and Nature Conservation Policy Group, Wageningen University and Research, Wageningen, The Netherlands

F. Aggestam (ed.), *Governing Forests Through Transformative Change*, Advances in Global Change Research 70, https://doi.org/10.1007/978-3-032-19072-7_7

practice towards the better integration of communication science principles in forest and policy governance research.

7.1 Introduction

Communication is central to human interactions and shapes every part of society and daily life—from interpersonal interactions to political negotiation processes. Communication science is a multidisciplinary field that examines how people seek, utilise, and process information, and ultimately how these processes impact decision-making and behaviour. It is a cross-cutting topic in scientific studies, covering both applied and basic research, and integrating insights from social, behavioural, and decision sciences to develop effective communication strategies (Sherry, 2010). This rich field incorporates various theoretical frameworks ranging from rhetorical and semiotic to phenomenological, cybernetic, socio-psychological, socio-cultural, and critical traditions. Each of these traditions is characterised by its own approach to defining communication and communication problems (Craig, 1999).

Communication is a vital aspect of human behaviour, encompassing more than just speech. In fact, models from communication psychology models argue that the verbal content accounts for less than 10% of a statements impact; the rest depends on the speaker's voice, body language, and other non-verbal attributes (Mehrabian & Wiener, 1967). An important yet challenging question is where communication begins and where it ends. There is no universally accepted definition of communication (Littlejohn & Foss, 2011). Therefore, the understanding of communication can vary significantly based on the context, goals, and disciplinary approaches to the study, as well as the research methods employed. Given the scope of this volume, which aims to discuss various perspectives on transformational forest governance, this chapter will explore different forms of communication that could serve as transformative levers for forest policy and governance. To this end, the focus will be on two branches of communication that are particularly relevant to forest policy and governance: public and political communication.

Political communication can be understood as the "purposeful communication about politics" (McNair, 2011: 4). According to McNair's definition (ibid.), political communication encompasses: (a) all communication by politicians or political actors that is intended to achieve specific goals, (b) communication directed at political actors by non-politicians, and (c) discussions about political actors and their actions, as found in news articles, editorials, and other media commentary on politics. The objectives of analysing political communication are to understand how to communicate effectively with decision-makers to influence policy outcomes (a), to support policy implementation (b), and to gain insight into how public opinion on political issues is shaped by the media (c).

Public communication includes all communication processes and structures that take occur in public, which are frequently—though not invariably—mediated by mass media. This includes both "traditional" media, such as radio or newspaper, and "new" online media (Bentele et al., 2003: 7). Public communication involves sharing information or messages with the public or its specific audiences and can cover a variety of topics and goals. One subcategory of public communication is the broad field of "public relations" (PR), which Grunig and Hunt (1984: 6) define as "the management of communication between an organisation and its publics". Another definition, by van Woerkum (1982: 39), describes PR as a means of "influencing the knowledge, attitude, and behaviors of target groups in a direction that the organisation predefines." Trust and credibility typically play a crucial role in successful public communication (Bentele et al., 2003).

However, these two branches of communication have been significantly influenced by the introduction of new technologies, such as the Internet and social media. Nowadays, political and public communication both take place publicly online through interactive and participatory media. Social media platforms such as YouTube, Twitter, Instagram, Facebook, and LinkedIn are widely used for both political and public communication by institutions, civil society, and traditional (or legacy) media outlets, enabling direct interaction between a diverse range of participants. These commercially owned digital platforms play a crucial role, setting the rules of the game through their own policies, algorithms and interface design (Nieborg & Poell, 2018). Scholars have therefore emphasised the indispensable role of these platforms in transforming socio-cultural practices and "affecting the very core of democratic processes and political communication" (van Dijck, 2019: 1). Some of the opportunities and challenges of social media communication will be explored in Chap. 8.

Broadly speaking, communication channels can be divided into "old" and "new" media, each of which shapes political and policy communication in a different way. Traditional or "old" media, such as newspapers, radio, and television, have historically served as gatekeepers of information, facilitating a top-down, one-to-many model of communication (McQuail, 2010). These legacy media outlets continue to play a central role in shaping public opinion and setting the agenda, largely due to their institutional authority and journalistic norms (Strömbäck, 2008). In contrast, "new" media, including digital platforms, blogs, and social networking sites, enable decentralised, many-to-many communication, blurring the boundaries between producers and consumers of information (Castells, 2009). This shift towards participatory media has transformed the way political actors disseminate messages, engage with audiences, and construct narratives in real-time. While old and new media often coexist and interact in hybrid communication systems, their affordances, logics, and impacts on democratic discourse differ significantly (Chadwick, 2013). Understanding these differences is essential for analysing how communication flows within the contemporary policy and political landscape.

As more and more societal interactions occur online among various groups of actors, communication is becoming an increasingly important aspect of forest policy and governance. Effective forest policy and governance require more than just sound scientific knowledge and well-designed policies. They also require strategic communication to foster public participation, stakeholder engagement, collaboration, co-creation, and broader societal support. Incorporating insights from communication science into forest policy and governance research can enhance these efforts by revealing how information is framed, transmitted, received, and acted upon within different social and political contexts. A deeper understanding of communication processes can shed light on how forest policies influence decision-making, shape public perceptions, and affect forest governance. However, the current disconnect between communication science and forest policy and governance research prevents researchers and decision-makers from fully grasping and addressing the complexities of contemporary forest governance.

7.2 Why Does Communication Matter for Forest Policy and Governance?

It is beyond the scope of this chapter to deliver a comprehensive overview of forest policy and governance research, especially since other contributions, such as the initial chapters in this book (see Chap. 3–5), will do justice to this rich field. It is, however, useful to revisit some key conceptual clarifications relevant to the discussion about potential complementarities between the two fields of communication science, forest policy and governance.

According to Arts (2021), the terms "policy" and "governance" are closely interlinked. The FAO (2010) defines a forest policy as: "a negotiated agreement between government and stakeholders (e.g., all those who depend on or benefit from forests or who decide on, control or regulate access to these resources) on the orientations and principles of actions they adopt, in harmony with national socioeconomic and environmental policies, to guide and determine decisions on the sustainable use and conservation of forest and tree resources for the benefit of society." If policy defines what is aimed for on paper, governance refers to how those aims might be achieved in practice (Arts, 2021). As this volume also illustrates, governance is a contested term, with multiple existing definitions (Ibid.).

Forest governance scholars prefer a broad and holistic analysis of the different modes of forest governance, extending beyond governments' policy instruments to include rules of the game (institutions), actors and coalitions (social interactions), and impacts (effects on forests) (Arts, 2021). Such definitions acknowledge the existence of classical approaches, such as public or formal governance, as well as newer ones, like private or informal governance (Ibid.). One such definition envisions forest governance as comprising "(a) all formal and informal, public and private regulatory structures, i.e. institutions consisting of rules, norms, principles, decision procedures, concerning forests, their utilisation and their conservation, (b) the interactions between public and private actors therein and (c) the effects of either on forests" (Giessen & Buttoud, 2014: 1).

The so-called governance triangle, namely, the state (i.e., governments), market (i.e., businesses), and civil society (i.e., communities), emphasises the mutual relationships between these three domains (Arts, 2021). The domains are linked through public–private partnerships, co-management, or voluntary standards governance

(Arts and Buizer, 2009; Lemos & Agrawal, 2006; Sergent et al., 2018). What this triangle overlooks is the crucial role that communication plays in connecting these domains. More specifically, the relationships shaped by communication between stakeholders are relevant to the discussion here. Because communication is about relations and interactions, understanding who the main actors involved in or influencing forest policy and governance is crucial.

In forest and environmental policy research, analysing power dynamics and structural imbalances among actors is of central importance. While power asymmetries are often rooted in historically embedded hierarchies and institutional structures, another—so far underexplored—dimension involves disparities in communication capacities. Even within hypothetical platforms designed to provide equal voice to all participants, those actors who possess superior communication skills tend to dominate. In this sense, communication skills constitute a form of power in its own right. This raises a new and compelling question for forest and environmental policy research: What role does communication play in shaping power relations between different actors and political contexts? While this chapter does not aim to answer this question in depth, it highlights an important direction for further inquiry—one that warrants more systematic exploration in future research.

Forests encompass a diverse range of stakeholders, each with distinct power positions and varying degrees of influence. Those actors often have different ideas (e.g., beliefs, values, perceptions, and mentalities), interests (e.g., the main drivers for (political) action), and act within different institutions (e.g., the rules of the game in a society) (Palier & Surel, 2005). Janse (2005: 36) identifies the following actors who make up the forest policy arena: forest policymakers: (a) in government and parliament, (b) in administration, forest industry; Forest owners, forest managers, and contractors; Forest scientists; Forest user/interest groups (e.g., NGOs); Society at large; The media. Janse (ibid.) further categorises those actors in the *forest sector core*, which comprises the forest industry, forest owners, national and international governmental bodies involved in forest policy, state forestry administration, forest science, and education. Those actors are the ones more directly involved in the policymaking process. The *forest cluster* comprises interest groups and international organisations focused on forests, including environmental protection agencies, as well as trade and social welfare organisations. The *media* includes specialised media as well as general and more popular mass media. *Society at large* refers to the public that holds opinions and interests on the protection, use, and function of forests.

Engaging public and private actors in policy design and governance-related issues is a central focus of a growing body of literature in both fields. Forest governance frameworks increasingly recognise that effective communication is essential for transparency and legitimacy, especially when policies directly impact different stakeholders. For example, "new" governance approaches encompass the coordination of collective action by public and private actors to address societal problems and opportunities (Arts, 2021; Hogl et al., 2012). New governance arrangements are characterised by the broad participation of non-state actors, increasing attempts to improve vertical and horizontal coordination, as well as efforts to integrate different types of expertise in a democratic fashion (Hogl et al., 2012).

In this context, Habermas' idea of a deliberative discourse has been held up as a democratic ideal in environmental governance studies (e.g., Cashore, 2002; Kleinschmit, 2012; Kleinschmit et al., 2018; Steffek, 2009). Habermas' theory of communicative action emphasises the importance of rational, inclusive discourse in democratic decision-making (Habermas, 2006). He emphasises the importance of including civil society actors in deliberative policy discourse, acknowledging them as valuable sources of expertise (ibid.). A core concept of this chapter relates to Habermas' notion of a deliberative discourse in the public sphere, where the essential elements for a democratic ideal are openness to speakers from the periphery of the political system, a discursive communication flow, and a public consensus or a result supported by the majority. Since today's political public sphere primarily comprises the media, these elements are constrained by the media's limitations.

Principles of engagement, transparency, and legitimacy are central to debates around "good governance" as well. "Good governance" is generally interpreted as being compliant with the rule of law, involving stakeholder participation, promoting democratic and transparent decision-making, and ensuring accountability (Arts, 2021; Gisselquist, 2012). Principles of transparency and accountability, or stakeholder participation, play an important role in planning and decision-making processes of good forest governance frameworks such as the Food and Agriculture Organization of the United Nations Program on Forests (FAO-PROFOR) framework, the World Bank Institute's Worldwide Governance Indicators (WGI), the UN University's World Governance Assessment (WGA) Index, the Governance of Forests Initiative (GFI) or the Forest Governance Diagnostics Tool by the Agriculture and Rural Development Department (ARD) of the World Bank (Secco et al., 2011).

For example, one of the FAO-PROFOR's framework objectives is to "enhance common understanding of and communication about forest governance and lead to improvements that strengthen the practice of governance" (FAO-PROFOR, 2011: 8). In other words, the framework aims to encourage dialogue between stakeholders in the forest sector. It is based on recognised principles of good (forest) governance such as accountability, effectiveness, efficiency, fairness, equity, participation, and transparency. These principles uphold three distinct pillars: (1) policy, legal, institutional, and regulatory frameworks; (2) planning and decision-making processes; and (3) implementation, enforcement, and compliance (ibid.). Similarly, the GFI framework refers to five "principles" of good governance: transparency, participation, accountability, coordination, and capacity (Brito et al., 2009). While it is beyond the scope of this discussion to detail all aspects of these good governance frameworks, it is clear that, communication plays a central role in fostering the different pillars and principles advocated by these frameworks.

7.3 What Has Been Done So Far: Mapping Communication Approaches in Forest Policy Research

7.3.1 Policy Analysis: Which Discourses Are Represented in Policy Processes?

Over the last two decades, the positivist paradigm in policy analysis has been challenged by a diverse range of approaches unified by a shared focus on language, meaning, rhetoric, and values as central elements in the examination of policymaking and planning. Considering this "argumentative turn" (Fischer & Gottweis, 2013), interpretative methods from social and political sciences have been applied in environmental policy studies (e.g., Behagel & Turnhout, 2011; Feindt & Oels, 2005; Hajer & Versteeg, 2005) and the field of forest policy, in particular (e.g., Arts et al., 2010). For instance, a recent IUFRO report (Kleinschmit et al., 2024) highlights the central role of discourses in forest policy analysis, emphasising their significance as a highly researched topic within the field.

Based on social constructivist approaches, it is assumed that perceptions of reality are subjective and shaped by societal practices as well as power structures and cultural contexts—all of which are constructed through linguistic and visual signs and sign systems (Berger & Luckmann, 1967). In this sense, language is a significant device for the construction and subjective (re-)production of reality. Policy researchers started to ask the question: "What if our language does not simply mirror […] the world but instead profoundly shapes our view of it in the first place?" (Fischer & Forester, 1993: 1). Inspired by discourse-based approaches in policy analysis (e.g., Steffek, 2009), social constructivist approaches in political ecology (e.g., Winkel, 2012) and new governance models (e.g., Kleinschmit et al., 2009), researchers now started to examine how narratives and discourses shape policymaking at all levels.

Several studies in the forest policy field have applied interpretive, argumentative, and critical discourse policy analysis. Though methods and foci differ for each approach, they are summarised here under the term "discursive approaches". The following sections outline the most prominent research methods employed under this umbrella term and provide examples of case studies from forest policy research.

7.3.1.1 Forest Discourses

Discourses have been extensively studied in the forestry field, and the vast body of discourse literature has been reviewed by Arts and Buizer (2009), Arts et al. (2010), Winkel (2012), and Leipold (2014). It can be distinguished between studies that analyse discourses in national policies and other state-produced documents (Asselin et al., 2022; Bidone, 2022; Fischer et al., 2020; Kelly & Crandall, 2022; Ofoegbu & Speranza, 2021; Papaspyropoulos et al., 2023; Thi et al., 2012) and those that analyse discourses found in interviews with policy actors (Aicher, 2003; de Koning et al.,

2014; Frei et al., 2022; Halonen et al., 2022; Kalaba, 2016). Many of those studies refer to Hajer and Versteeg (2005) for the definition of discourses as "ensembles of ideas, concepts and categories through which meaning is given to social and physical phenomena, and which are produced and reproduced through an identifiable set of practices" (ibid.: 175). Discourse analysis, in this sense, is about identifying patterns of thought and argumentation, concepts, and overarching rationalities that make discussions possible (ibid.: 301). Some studies more explicitly refer to Foucauldian concepts (e.g., Vainio & Paloniemi, 2012; Smallwood et al., 2023, see Winkel, 2012, for a review). The discourse itself is perceived as a powerful unit of stability and change (Winkel, 2012), and these analyses can reveal power dynamics and structural hierarchies in forest policy (e.g., Porter, 2007).

7.3.1.2 Forest Framings

Another prominent approach to analysing policy documents and stakeholder perceptions is through frame analysis. As Edelman (1993: 232) describes, the world is like a "kaleidoscope of potential realities," where shifting the framing alters how observations are understood. How a topic is framed influences how it is perceived and interpreted by the audience (Entman, 1993). Frames, as defined by Rein and Schön (1993: 146), provide a way to organise, interpret, and make sense of complex realities, serving as guideposts for analysis, persuasion, and action. Frames construct meaning and can mobilise collective action (Benford & Snow, 2000).

Several studies have applied frame analysis to identify how forest issues are presented in policy documents at the international level (Kleinschmit et al., 2017), the EU level (Elomina & Pülzl, 2021), and the national level (Ambroise et al., 2023; Andersson et al., 2022). Others have identified common frames of interest within groups and associations (Müller & Kruse, 2021; Riemann et al., 2022) and stakeholder perceptions (Lindahl & Westholm, 2012; Lindahl et al., 2013).

7.3.1.3 Forest Narratives

Narrative and discourse analysis have been used interchangeably in some cases, however, here narratives are understood in a more narrowly defined sense as causal stories with characteristic elements, such as setting, characters, plot, and solution or moral (Jones et al., 2014; Stone, 1988). They assign roles to different actors in complex forest-policy-making processes (Beymer-Farris & Bassett, 2012). The influence of narratives on the policy process has been studied by a few scholars in the forestry context (Häublein & Granzow, 2024; Kansanga et al., 2019; Tietze et al., 2023).

7.3.2 Media Analysis: Which Actors Communicate What in the Public Sphere?

Political and public communication can occur through mass media, and as such, it relates closely to mass communication research. The term mass refers not only to the large size of the audience reached by such media but also to ideas of uniform consumption, uniform effects, and audience anonymity—concepts increasingly at odds with contemporary media usage (McQuail, 2010). Mass media serve not only as channels of communication but also represent an important field of academic inquiry in its own right.

Mass media shape the public sphere (Kepplinger, 2002; Mazzoleni & Schulz, 1999; Schulz, 2004). For most individuals in modern and complex societies, media content serves as the primary source of information. Notably, the less an individual knows about a particular issue, the more likely they are to adopt the opinions presented in the media (Neuman et al., 1992; Krott, 2005; Kleinschmit & Sjöstedt, 2014). Various scholars argue that the media influence opinion formation (Curran, 2002; Kingdon, 2003). Alongside opinion polls, media coverage also serves as a key indicator of public opinion (Kleinschmit & Krott, 2008), which is of interest for political actors. In this way, mass media shape the public debate, serve as a platform for political discussion, and influence the political agenda (Kleinschmit, 2012; Reese, 1991).

In forest policy, both public and private stakeholders utilise the media as a means of communication to advance their interests (Krott, 2005). Media attention is highly sought after—but only a limited number of actors gain media prominence and have their messages reproduced. The media industry has established routines and criteria to filter significant events and present them in media-tailored formats. In the news selection process, journalists act as both gatekeepers and storytellers (Gerhards, 1994). Those seeking access to the public sphere understand the need to conform to the media's selective criteria and adjust their messaging accordingly (Schulz, 2004). "Media standing" refers to the position in which actors can actively place their perspectives in the public discourse via mass media (Gerhard & Schäfer, 2007).

Besides questions about which actors and topics enter the media discourse, it is also of interest to ask how these topics are presented. Mass media can assign specific meanings to issues, which audiences may later adopt for discussing and interpreting these topics. This is especially relevant for issues that cannot be directly experienced, such as climate change (Kleinschmit & Sjöstedt, 2014; Rollhansen, 1994).

Forest policy researchers have used discourse and frame analysis of media coverage to identify influential actors and the messages they communicate (Feindt & Kleinschmit, 2011; Hallberg-Sramek et al., 2020; Kleinschmit & Sjöstedt, 2014; Mack et al., 2023; Park & Kleinschmit, 2016). Some studies have focused specifically on the portrayal of the forest industry in the media (Ekayani et al., 2016; Fabra-Crespo & Rojas-Briales, 2015; Krumland & Krott, 2004; Nhem et al., 2017; Takala et al., 2019).

The research focus in forest policy research has been on traditional media types, such as newspapers, television, and radio, as these have been perceived to be the preferred channels for opinion leaders and political decision-makers seeking indicators of public opinion on specific issues (Kleinschmit & Krott, 2008). "New media," including blogs and social media, create open, interactive, and discursive spaces that accommodate multiple stakeholders. However, they lack the gatekeeping function of journalistic media (Singer, 2005). Instead, interactions occur within the infrastructural frameworks and rules established by the platforms. Recently, the potential and challenges of new media for deliberate communication between forest interest groups have been researched (Calibeo & Hindmarsh, 2022; Mäder et al., 2025). Nevertheless, based on the social media data, research has been conducted to examine discourses and framings (Gerosideris & Ferra, 2020; Mirza & Kusumasari, 2022; Skill et al., 2021) as well as emerging topics and sentiments (Pellicer-Bhatt & Pickering, 2021; Chenoll et al., 2023).

7.3.3 Persuasion Research: How Can Communication Lead to Certain Effects?

Throughout history, various methods of influence through communication and mass communication (e.g., rhetoric and propaganda) have attracted the interest of different societal actors (Bonfadelli & Friemel, 2015). Studies that explore the conditions under which communication can have a specific effect on audiences are referred to as "persuasive communication." It is not a unified research field, but rather a combination of approaches from psychology, media effectiveness studies, philosophy, and social science (Druckmann, 2021). Persuasion is "a successful intentional effort at influencing another's mental state through communication in a circumstance in which the persuadee has some measure of freedom" (O'Keefe 2016: 4). Mutz et al. (1996: 146) specify the type of intended influence as a change in the audience's behaviour, beliefs, or attitude. Those effects usually overlap and often, one effect leads to another (Ajzen & Fishbein, 2011). Recently, emotional change has been included as an outcome of communication (Jäckel, 2019). The term "persuasion" is rarely mentioned in forest-related studies (exceptions include Chung et al., 2020; Tarrant et al., 1997; Wilkes-Allemann et al., 2021), despite an increasing number of studies raising questions about communication-induced behaviour and attitude change in the context of targeted campaigns.

Relevant research questions are: How to communicate in order to raise public support and social acceptance for policies (change in beliefs and attitudes; Mylek & Schirmer, 2020; Novak et al., 2023; Setiono, 2024, in all those cases: forest fire prevention measurements) or how to communicate to get forest owners to implement certain management measures or enrol in certification schemes (implying a behaviour change; Graves et al., 2022; Rubino et al., 2022; Vulturius, 2019; Wilkes-Allemann et al., 2021).

A common assumption of communication efforts is that individuals will change their attitudes and behaviours in response to (scientific) information. This "information deficit" approach assumes that people need to be better educated to take "better" actions (Ma et al., 2012; Nisbet & Scheufele, 2009). Wilkes-Allemann et al. (2021), based on Atkin and Rice (2013) differentiated communication campaigns directed at forest owners according to their messaging style: either informational ("seeking to create awareness or provide instruction", Wilkes-Allemann et al., 2021: 5) or persuasive ("seeking to emphasise reasons why the audience should adopt the advocated action or avoid the proscribed behavior", ibid.). The study found that messages intended to serve multiple purposes (e.g., inform, engage, and educate simultaneously) are common but inefficient, and that for efficiency, communication styles should ideally match one clearly defined purpose (Wilkes-Allemann et al., 2021).

However, research indicates that the progression from learning about a topic to altering attitudes or behaviour is not linear but rather influenced by a complex interplay of factors, such as values, politics, and expert knowledge (Ardoin et al., 2013; Kollmuss & Agyeman, 2002; Nisbet & Scheufele, 2009).

More in-depth insights are provided by a study conducted by Novak et al. (2023), which bases its findings on raising social acceptance of fire prevention measures, such as prescribed burning, on the stages of change model (Prochaska et al., 2015). The model suggests that before a behaviour change occurs, various steps or changes lead towards behavioural adaptation. In Novak et al. (2023), the audience is segmented according to their respective stage in the change process, and findings show that information preferences differ depending on the stage. Chung et al. (2020) applied the stages of change model to persuade forest owners to adapt a tree planting behaviour. They came to very similar conclusions: informational advice messages are more likely to persuade landowners at a later stage of behaviour change, while those in early stages of behaviour change are better persuaded by narrative advice messages (ibid.). Findings from Novak et al. (2023) and Mylek and Schirmer (2020) suggest that information does affect acceptance and behaviour change. However, the change is rather progressive than linear, and adaptive messaging can support this progression.

Studies that research the effectiveness of persuasive communication in experimental settings (Chung et al., 2020; Mylek & Schirmer, 2020; Novak et al., 2023) are highly significant, as they not only provide interesting findings but also introduce new theoretical approaches to forest policy science. Other studies on attitude and behaviour change are usually ex-ante or ex-post preference surveys (Graves et al., 2022; Rubino et al., 2022; Setiono, 2024) or segmentation studies (Butler et al., 2007; Kuipers et al., 2013).

7.3.4 Interpersonal Communication: How Do Stakeholders Communicate with Each Other?

Another aspect of communication is in-person interaction. Interpersonal communication is essential for improving forest resource governance (Kubo & Supriyanto, 2010), managing and mediating conflicts (Nousiainen & Mola-Yudego, 2022; Yasmi et al., 2007), fostering participatory processes (Dragoi et al., 2011), and meeting public expectations (Bethmann et al., 2018). However, participatory and co-management approaches often rely on ineffective stakeholder communication and weak communication networks (Baten, 2024; Ullah et al., 2023). Infrequent communication can exacerbate conflicts between stakeholders (Fabricius et al., 2001). Furthermore, Blicharska et al. (2020) found that evidence-based knowledge alone is insufficient to address complex conservation conflicts—values, trust, and power relations significantly influence stakeholder communication and interactions, along with careful consideration of the local and historical context (Blondet et al., 2017).

For participatory processes to be successful, besides the frequency of communication, process design also matters. Participation spans a spectrum from one-way dissemination of information to collaborative, co-design, processes that actively engage stakeholders, including through citizen science approaches (Leskinen, 2004). Maier et al. (2014) illustrate how forest policy processes are shifting along this participatory spectrum—from traditional, expert-driven models toward more inclusive formats that seek to involve a broader range of civil society actors in the decision-making process. However, best practice examples for collaborative modes, such as co-designed processes or citizen science, are still rare, indicating that participation remains limited in its transformative potential (ibid.).

As Aasetre (2006) notes, formal participation is no guarantee of real influence and communication patterns between stakeholders in forestry conflicts are usually greatly influenced by power distributions. Parkins and Sinclair (2023) further criticise that public participation in certain parts of the Western world is often treated as a tool to legitimise already-made policy decisions. In cases of genuine inclusion and consultation with local communities, contested management plans are more likely to be adopted and socially accepted (Fienitz et al., 2022; Grundel et al., 2022). Collaborative learning (e.g., Daniels & Walker, 2001) and social learning (e.g., Romina, 2014) are two approaches proposed to foster a deeper understanding among policymakers, practitioners, and the broader public. Nevertheless, even during these processes, Esselin and Ljung (2010) emphasise the importance of communication and outreach efforts in supporting various steps to make it work.

Trust is essential for successful stakeholder interactions (Kubo & Supriyanto, 2010) and also influences whether forest owners seek out professional management advice or not. In this context, in-person communication is important, as personal relationships are best suited to foster trust and mutual understanding (Hujala & Tikkanen, 2008; John et al., 2024). Pregernig (2002) finds that meanings arise from the interaction of the individual with others, therefore, the relevance and meaning of a message often need to be discussed between peers before being adopted. For forest advisory practice, internet-based services can serve a supportive role, however, the interactive elements provided in face-to-face advisory encounters cannot be fully replaced by digital services (Virkkula & Hujala, 2014).

Social network studies find that social networks can operate as creative learning environments in local contexts (Hamunen et al., 2015)—yet some scholars argue they are meaningless, if the members of the social network interact infrequently (André et al., 2017) or communicate opposing interests, thereby reproducing competing narratives that fuel a "us versus them" mentality (Koch, 2023). Recently, marteloscopes have been studied as sites for deliberative communication and cooperation between stakeholders (Joa et al., 2020; John et al., 2024). Citizen science approaches are being tested, especially in the context of forest monitoring (De Groot et al., 2022; Hulbert et al., 2023).

7.3.5 Pedagogic Approaches: How to Educate About Forests?

Since several studies have revealed that children and young people in developed countries are increasingly losing their awareness of and connection to nature (Brämer, 2006; Louv, 2011), pedagogical approaches such as nature, wilderness, and forest pedagogics, as well as environmental education, have been put back into focus over the last few decades. There are dozens of communication methods applied in this field, ranging from one-way communication approaches, which are mainly aimed at imparting knowledge (e.g., articles on websites and in journals, training courses e.g., for the work with chainsaws, and guideline brochures), two-way communication approaches (e.g., experiments with nature materials, guided walks through the forest, discussion forums) to participatory methods (e.g., group and collaborative games, joint timber constructions or tree planting events). Even "silent" communication methods such as land art, Shinrin Yoku (forest bathing), or hugging a tree are commonly used. As broad as the methods are, the specific objectives and addressed learning outcomes of the above-mentioned educational approaches (Bolay & Reichle, 2015; Kohler, 2000; Liebal, 2011) are also noteworthy. However, there is a unifying main goal of supporting the transition of society towards a higher ecological awareness, a closer human-nature relationship, and—along with this—more environmentally friendly behaviour (Liebal, 2013). Environmental awareness and attitudes can in turn influence political decisions such as the legislation on forest and nature conservation laws and regulations.

Forest pedagogics can be considered as forest-related environmental education (Deutsche Forstchefkonferenz, 2007), but differ in the values and beliefs associated with pro-environmental behaviour. While environmental education focuses on nature protection and often follows a conservationist perspective on nature, forest pedagogics puts more emphasis on the sustainable use of nature and forests, aims to raise acceptance of sustainable forest management, and—at least sometimes—strives to improve the image of forestry. In this regard, forest pedagogics overlap with public relations (Kohler, 2000), while this interpretation is partly seen as critical (Dobler, 2013: 88). Indisputably, however, an increasing influence of sustainability aspects catalysed by the Rio Declaration 1992 and the following pedagogical approaches, such as the 'education for sustainable development' can be recognised both in environmental education and forest pedagogics. Therefore, "for almost 20 years, the goal of education has changed from the mere acquisition of factual knowledge in individual subjects to mastering the world around us and the ability to solve problems in everyday situations" (Moormann et al., 2021: 59). This results in a transition from observational learning to competence-oriented education, which, in turn, highlights the need for alternative communication and education methods that teachers and instructors can apply.

7.3.6 Science Communication

Science communication covers a "spectrum of issues from the factual dissemination of scientific research to new models of public engagement whereby lay persons are encouraged to participate in science debates and policy" (Bubela et al., 2009: 514). It is widely acknowledged that trust in scientific methods and results plays a crucial role in science communication with political actors and the general public (Hajdu & Simoneau, 2020; Weingart & Guenther, 2016). In forestry contexts, especially, trust may be challenged by the high degree of uncertainty surrounding future forest development and management, caused by climate change as well as social and economic transformations. Science communication thus strives to promote understanding of uncertainties on the one hand and "actively cultivates its image as society's major tool for reducing uncertainty" on the other hand (Dunwoody, 2014: 33).

Science communication, whether explicitly or implicitly, often assumes that a specific target group lacks knowledge, which can lead to undesirable or problematic situations. There is an assumption that "if members of the public [or: politics] only understood the scientific facts, they would be more likely to see the issues as experts do" (Bubela et al., 2009: 515). This perception raises questions about the legitimacy and hierarchy of expert knowledge, a topic that has been explored by the forestry policy research community (Giessen et al., 2009; Kleinschmit et al., 2018). Further, science communication has been analysed in the context of improving the science-policy knowledge transfer (Pregernig, 2000), evaluating and assessing changing approaches, methods, and quality criteria of science communication concerning communication success (Fähnrich et al., 2023; Metcalfe, 2019) and developing practical guidelines to improve the communication of scientific messages to the public or policymakers (Janse & Schifferdecker, 2024).

7.4 Research Gaps: What Is Missing?

Despite the vast amount of research on forest policy and governance relating to communication, most studies conduct policy or media analysis using discursive approaches. The importance of discursive approaches has been highlighted in a recent IUFRO report (Kleinschmit et al., 2024), reflecting an interest in understanding the distribution, dynamics, and manifestation of power in decision-making, policy outcomes, and the relationships between different stakeholders. In this context, communication is analysed as a concept that shapes and is shaped by the broader social context. What is less studied is the communicative process itself—and related questions, such as: How does communication influence people's attitudes and behaviours? How does interpersonal communication "work" and what factors need to be considered for it to be effective? How do people learn about forests in an engaging way? How can scientists communicate to policymakers and the general public without reproducing hierarchies of expert knowledge but instead fostering a deliberative discussion that includes other forms of knowledge (e.g., local or indigenous knowledge)? Moreover, more broadly, how do new technologies and digital platforms influence communication and information-gathering processes?

The following sections introduce key concepts from communication studies alongside empirical case studies that link communication to forest and environmental policy research. By drawing on insights from communication science and related fields, the aim is to encourage more diverse methodological and theoretical approaches in forest policy and governance research—approaches that not only analyse transformation processes but also actively promote them.

7.4.1 The Communicative Process

In forest policy science, few studies consider the fundamentals of the communicative process (exceptions include, for example, Janse, 2005; Rubino et al., 2022; Wilkes-Allemann et al., 2021). However, a closer examination of the components of the communication process enables a better understanding and conceptualisation of communication and its effectiveness. The communication process is often broken down into five components: the sender (communicator), the content of the message, the channel of communication (medium), the receiver of the message (audience), and the impact of the communication on the audience (effect) (Wilkes-Allemann et al., 2021). One-way communication models conceptualise messages as flowing from senders to receivers in a single direction (Lähtinen et al., 2017). Often, one-way communication models correlate with media effectiveness theories that assume high media effects, as the messages are supposed to hit the audience like a "magic bullet" and produce an effect (Berger, 1995). For example, modern media effectiveness theories assume moderate media effects and conceptualise a heterogeneous audience that actively participates in the communication process by selecting, interpreting, and possibly rejecting senders, mediums, and messages. This perception aligns with two-way communication models that consider specific target audiences and socio-psychological characteristics of the individuals belonging to those target groups. Specifically, trust is a key element for two-way communication (Wilkes-Allemann et al., 2021).

There is untapped potential for more efficient public communication in the forest-based sector. In an analysis of communication campaigns targeting non-industrial forest owners, Wilkes-Allemann et al. (2021) found that one-way communication models are predominantly employed. While one-way communication is often aimed at informing and can be effective for short-term awareness raising, the potential of two-way communication, which aims to persuade and is essential for triggering forest management, has not yet been exhausted. More engaging two-way communication concepts can be applied not only to forest owner communication but also in other contexts.

While one-way communication models might define communication as the "transmission of information, ideas, attitudes, or emotion from one person or group to another (or others)" (Van Ruler, 2004: 128), modern, interactional communication approaches with active recipients (Bucher & Schumacher, 2012) that take into consideration specific target audiences and socio-psychological characteristics of the individuals belonging to those target groups understand communication as a social process of meaning-making, in which both sender(s) and receiver(s) of information are involved (Van Ruler, 2018). The latter links to Pearson et al. (2011)'s understanding of communication as a social and dynamic process that evolves, and the actors involved may also change accordingly.

Various communication scholars view communication as a process of meaning exchange (Burkart, 1995; Littlejohn & Foss, 2011; Pearson et al., 2011; Rosengren, 2000). Only when meanings overlap do individuals understand each other, which makes communication successful (Burkart, 1995). In this case, the sender and receiver exchange meaning and stimulate each other to construct new meanings (Rosengren, 2000; Schoeneborn & Tittin, 2013; van Ruler, 2018). Otherwise, misunderstandings occur, hindering effective communication and policy formulation processes.

7.4.2 Socio-psychological Factors Affecting Communication

The understanding of communication as a social interaction implies that both sender and receiver play an active part in the communicative process. The audience does not simply partake in everything that is offered; instead, they actively select, interpret, and either accept or reject information (Bonfadelli and Friemel, 2015; Stoeber, 2008). So-called audience-oriented approaches to communication investigate individuals' inner and outer worlds and how those influence the understanding, interpretation, and internalisation of communicative offers (McQuail, 2010). For this, values, trust and emotions must be taken into consideration.

- **Values**: Audience characteristics can explain why (or whether) communication is effective. However, socio-psychological barriers to communication have been largely overlooked in the field of forest policy and governance research. Selective Exposure Theory (Klapper, 1960) assumes that individuals tend to select information that aligns with their existing beliefs, attitudes, and values while avoiding or ignoring information that contradicts or challenges them. Selective Exposure Theory draws upon Cognitive Dissonance Theory, coming from psychology (Festinger, 1957; Stroud, 2008), and applies it to explain media consumption behaviour and how people choose media content that reinforces their pre-existing beliefs. It is, moreover, assumed that individuals are more likely to trust senders with whom they share values and other similarities (Burke, 1969). Takala et al. (2022) found in a study with Finnish forest owners that selective hearing, cognitive dissonance, and reinterpretation of information were discursive barriers that made the ongoing communication about forest management measures for biodiversity conservation ineffective for the majority of those forest owners. This study demonstrates that mere information (one-way communication) can be beneficial in certain cases, however, to effectively overcome such discursive barriers, two-way communication models and methods must be used. Furthermore, the findings largely challenge the information deficit model, which assumes that more information leads to better-informed actions.
- **Trust**: A complex socio-psychological variable that has challenged forest-related research. A study in the field of environmental conservation reveals that trust is crucial in establishing social relationships and often serves as the decisive factor in determining whether people take action or not (e.g., seeking management advice; Wilson, 2008). Trust is relevant for in-person interactions (Beck, 2012) as well as for the intact functioning of environmental governance systems (e.g., Ruppert-Winkel & Winkel, 2011; Saarikoski et al., 2013). Bentele and Seidenglanz (2008) argue that trust in public actors and institutions is carefully constructed and gains increasing relevance concerning the credibility of communication—thereby challenging academia to investigate the factors affecting trust further. Ford et al. (2020) provide a review of the factors that influence the development and maintenance of trust in natural resource governance.
- **Emotions**: Forestry has long been perceived as a rational natural science, reflecting the rationalisation and objectification of nature since the eighteenth century (Jardine et al., 1996; Rientjes, 2002; Wilber, 1998). However, more recent studies have found that emotions play a crucial role in shaping stakeholder behaviour and are key drivers of forestry-related conflicts (Bujis & Lawrence, 2013; Halla & Laine, 2022; Jakobsson et al., 2021). Related disciplines, particularly biodiversity conservation research, have revealed that natural resource-related decision-making is highly emotional and value-driven (Mayer, 2006; Takacs, 1996; Trudgill, 2001). As such, the often-perceived dichotomy between rationality and emotionality in natural resource conflict, management, and policymaking should be critically examined. Wahl-Jorgensen (2019: 28) argues that

this dichotomy does not exist at all: "rational decision-making is inseparable from and inextricably linked to emotion". These insights are also of great significance for media analysis, as media are considered the means by which shared emotions are generated among the masses. Media stories that evoke emotions are generally more effective in drawing the audience's attention towards complex topics. As such, emotions are a powerful means that can be performatively constructed. This practice is widely used by activists and non-governmental organisations (Wahl-Jorgensen, 2019). Especially in today's networked society, emotions can mobilise a collective action among different actors who may not necessarily share the same ideologies, or what Papacharissi (2015) refers to as "affective publics". Social media provides platforms and channels for expressing and sharing emotions, and therefore could contribute to forming collective sentiments. This is one of the rationales for researching what happens on participatory media platforms like YouTube, Instagram, Twitter, Facebook, and LinkedIn to explore how multi-actor networks may assemble themselves to initiate a movement using hashtags and even profile photos (e.g., a blacked-out profile picture to show solidarity for a specific global event).

7.4.3 Audience Segmentation

The basic idea of segmentation is to divide a population or audience into groups whose members are more similar to each other than to members of other segments (Grunig & Grunig, 1989). Moscardo et al. (2001) categorise segmentation approaches into demographic and geographic segmentation based on registry data, psychometric segmentation based on an individual's attributes, such as values, attitudes, or personality traits, and behavioural segmentation. Audience segmentation is typically the first step in a communication strategy, and having a clear target group in mind can help facilitate trust, communicate shared values, and evoke emotions. Additionally, knowledge about the target group will also influence the channel through which the information is delivered, as well as the style, format, and topic (e.g., Cutlip et al., 2000; Wells et al., 1989).

Forest owners have been grouped according to their values, attitudes, and motivations for owning land (Arnould et al., 2021; Butler et al., 2007; Ficko et al., 2019; Kuipers et al., 2013; Stockmann et al., 2024; Tiebel et al., 2024). To make these target groups more tangible and relatable, the development of *personas*—a method originally rooted in marketing to represent user segments through fictional yet evidence-based characters—can be promising, but their adoption in the context of forest and policy science has so far been limited. One example comes from Finley and Kittredge (2006), who applied the persona concept to forest ownership, identifying distinct forest owner personas to understand better and address their needs and behaviours.

Another innovative approach is to segment audiences based on their level of complex thinking (Mylek & Schirmer, 2020) and their stage of behavioural adaptation

(Chung et al., 2020). In both cases, the style of audience segmentation is directly linked to and serves a purpose for the desired communication goal (Chung et al., 2020; Mylek & Schirmer, 2020). There is much potential for further developing those approaches in forest policy and governance research.

7.4.4 Social Networks Studies

Social network analysis is closely linked to communication, as it examines the structure of relationships and interactions among individuals or groups within a social network. Communication is both a condition and a consequence of these social ties—it establishes, maintains, or disrupts relations, and shapes how meaning and trust are negotiated and established. In this sense, communication forms the connective tissue of networks, making its analysis indispensable for understanding the dynamics within social and political systems related to environmental governance.

Pregernig (2002) finds that the relevance and meaning of a message often need to be discussed among peers before being adopted. In social interactions, meanings are shaped and negotiated through mutual understanding and comprehension. Moreover, trust is more easily developed via face-to-face interactions. Social network analysis has been employed in environmental science to examine the significance of social connections in the management of natural resources. Social network analyses indicate that environmental management and collaborative decision-making are greatly facilitated when actors trust each other (Ardoin et al., 2017; Ghorbani & Azadi, 2021; Gorris & Koch, 2024). Koch et al. (2023) additionally find a link between communication and trust: A common narrative tends to emerge in sub-groups that is characterised by frequent member interaction and a highly trusted leader.

7.4.5 Social Media Interactions

In modern society, many social interactions occur through electronic media. While several existing forest policy research studies have referred to social media as a potential avenue to bridge the gap between forest policymaking and society (e.g., Parkins & Sinclair, 2023), this area is relatively understudied compared to other areas highlighted in this section. However, several studies have been conducted using social media data to explore cultural ecosystem services, e.g., via geotagged Flickr image data (Fox et al., 2022; Lingua et al., 2022) and public perceptions and semantic networks using data from X (previously Twitter) (e.g., Bruzzese et al., 2022; Mack et al., 2025).

However, some research applies conventional methods and theoretical frameworks that were developed, for example, in mainstream journalistic media, which significantly underestimates the networked nature of social media. At the same time, several existing studies overlook the "research affordance" of social media platforms (Weltevrede & Borra, 2016). In particular, they have overlooked the issue of "technicity", which refers to the inextricable link between content and the platforms, as well as the social and cultural practices that evolve within each platform (Niederer, 2019). Another issue that Weltevrede and Borra (2016) points out is the medium dependency, e.g., the decline of API data access significantly influences how the research can be conducted. Nevertheless, online engagement around forest policy-relevant issues could provide additional insights into which actors tend to echo similar voices and thereby support the development of effective communication strategies. Which new theories and methods (e.g., from new media studies) can be used to understand the various ways in which social media shapes our behaviours, ideas, emotions, interactions, and understanding of the world? What are the risks, challenges, and opportunities? These research questions are still to be explored.

7.4.6 *Methodological Diversity*

Communication science is a broad field covering a range of methods and approaches from different scientific disciplines. At the level of syntax and lexis, linguistic methods such as language intensity analysis may offer explanations for why certain messages attract more attention than others or whether conflicts are more likely to be successfully solved (Köhler et al., 2024). However, linguistic methods have rarely been applied to research on forest-related communication. This points out a clear research gap. Another approach that has so far received little attention comes from the direction of design and user experience—it is about finding out what end users of a system or product need and want. These findings are then used to improve the design process for products, services or software. A research example, including user design, is presented in this chapter.

7.5 Future Perspectives: Transformative Research Approaches

Communication can be a fundamental driver of social and political change—or a means to reproduce and stabilise the status quo. Emerging modes of communication, driven by technological advancements and shifting societal expectations, are transforming the way stakeholders interact, share knowledge, and influence decision-making. However, recent work in communication science shows promising approaches to tackle questions also relevant to research on forest communication. This is the case not only because communication systems themselves are undergoing deep transformation. Even more, new digital technologies also underscore the transformative potential for participation. The ultimate aim is to shed light on the evolution of different perspectives in media research. In contrast to deliberative ideals of communication, scholars criticise these approaches for being oblivious to power relations. Instead, informed by feminist scholars, considering the role of emotions in heated debates, as well as the materiality of communication devices, is presented as crucial for a better understanding of agency in communication processes. In this context, the role of how content is presented should also be considered, taking into account the visual characteristics of websites and other forms and objects that could intervene in or facilitate communication. The following sections will elaborate on these different perspectives, drawing on the broader context of the book.

7.5.1 Transformation of Communication

Just like past media systems, the current political and public communication is under constant transformation. Yet, the current transformation exhibits particular dynamics, mainly driven by digital media. These dynamics include an increased range and speed of distribution facilitated by mobile devices as well as new forms of commercialisation. Further, (news) communication occurs in real-time and is almost accessible everywhere. However, instead of conceiving these new forms of media as complementary or concurrent with legacy media, Chadwick (2013, 2017) and Bruns (2023) emphasise the need to consider the hybrid nature of media systems. In this understanding of Hybrid Media Systems, political communication is characterised by the blending of traditional and social media logics. Both communication spaces mutually adapt and accommodate formats, genres and norms to maximise their attention. Consequently, the discursive power of actors is understood as their "ability to introduce, amplify, and maintain topics, frames, and speakers" (Jungherr et al., 2019, p. 412), not only in a specific communication space but also across the blending of traditional and new media outlets. Further, a differentiated perspective on who introduces topics and who is able or required to amplify and maintain those in the public sphere "reduce[s] the risk of misattributing power" (Jungherr et al., 2019: 410). This is particularly important, as the power dynamics of actors have shifted within these new media landscapes. Whereas traditionally, the gatekeeping function and thus the agenda-setting process were exclusively reserved for journalists, digital media opens up the opportunity for grassroots media productions, which are, at least in theory, accessible to everyone (Mattoni & Ceccobelli, 2018). In addition, the various forms of interaction reveal the network logic of communication, as they facilitate the analysis of both the sender and the receiver (Klinger & Svensson, 2015).

While forest-related studies have considered the role of digital media in sectoral communication (Korhonen et al., 2016; Słupińska et al., 2022), forest debates (Caetano, 2021; Daume & Füldner, 2016) or risk communication (Unal Colak & Yllmaz, 2021), hybrid analyses could contribute to a more differentiated picture of how forests are communicated in the public sphere. In an attempt to conceptualise discursive power in Hybrid Media Systems, based on a forest debate in Germany, Mack et al. (2025) highlight the differences in dominance among actors. Whereas forestry actors and frames dominate the debate in traditional forms of communication, forestry critical actors are more prominent in the X (previously Twitter) debate. However, considering the hybridity of media communication, the results show little outreach and relevance of the X debate on German forests in policy processes. Further, the results show that not the actors with the highest number of followers achieved discursive power in the X debate. Instead, actors that have a general standing in forest debates show high centrality within the X network.

Despite social media offering manifold opportunities for public participation in communication, scholars have also highlighted its negative impact. For example, regarding climate change debates, social media shows signs of increasing polarisation that often result in value-aligned echo chambers that reproduce the power relations of previously dominant actors (Boyd, 2023; Brüggemann et al., 2020; Elgesem & Brüggemann, 2023; Falkenberg et al., 2022; Heiberger et al., 2022; Kreiss & McGregor, 2024; Pearce et al., 2019). Additionally, out-group hostility is further exacerbated by the anonymity of communication in digital spaces (Boyd, 2023). A lack of scientific analyses of these forms of deliberation and participation persists. As a highlighted example, the European Forest Institute's (EFI) project "Out of the Flames" mapped online engagement and public narratives surrounding the 2019 Amazon rainforest fires.

7.5.1.1 Lightbox: Potentials and pitfalls of conducting social network analysis (SNA) with social media data

Social network analysis (SNA) can potentially be conducted using data collected from social media platforms, for example, by examining author-mention networks to understand "who engaged with whom" on social media. When examining the data related to forest biomass debates at the European level, collected through the academic X API (the platform at the time of data collection) between January 2020 and April 2023, a notable increase in interest in this theme was observed in 2021. When zooming in on the most significant spike in July 2021, one can observe concerted, large-scale civil society campaigns, namely #StopFakeRenewables and #ForestsAreNotFuel, calling against the classification of forest biomass as renewable energy in the EU Renewable Energy Directive. These campaigns coincided with another hashtag, #FitFor55, referring to the legislative package introduced as part of the EU Green Deal in the same month.

As shown in Fig. 7.1, three distinct clusters can be observed in the author-mention network, namely, the civil society cluster (right), where Greta Thunberg and Greenpeace, were heavily promoted by activists, a EU-related cluster (middle), where EU officials and institutional accounts appeared, and the forestry cluster (left), where forest industry and forest science accounts emerged with almost no interactions with civil society accounts. When zooming in on the popular posts by the most mentioned accounts, posts by Thunberg and Greenpeace were not only promoted by activists but also further shared by Green MEPs, evidenced by the overlap between the civil society and EU clusters.

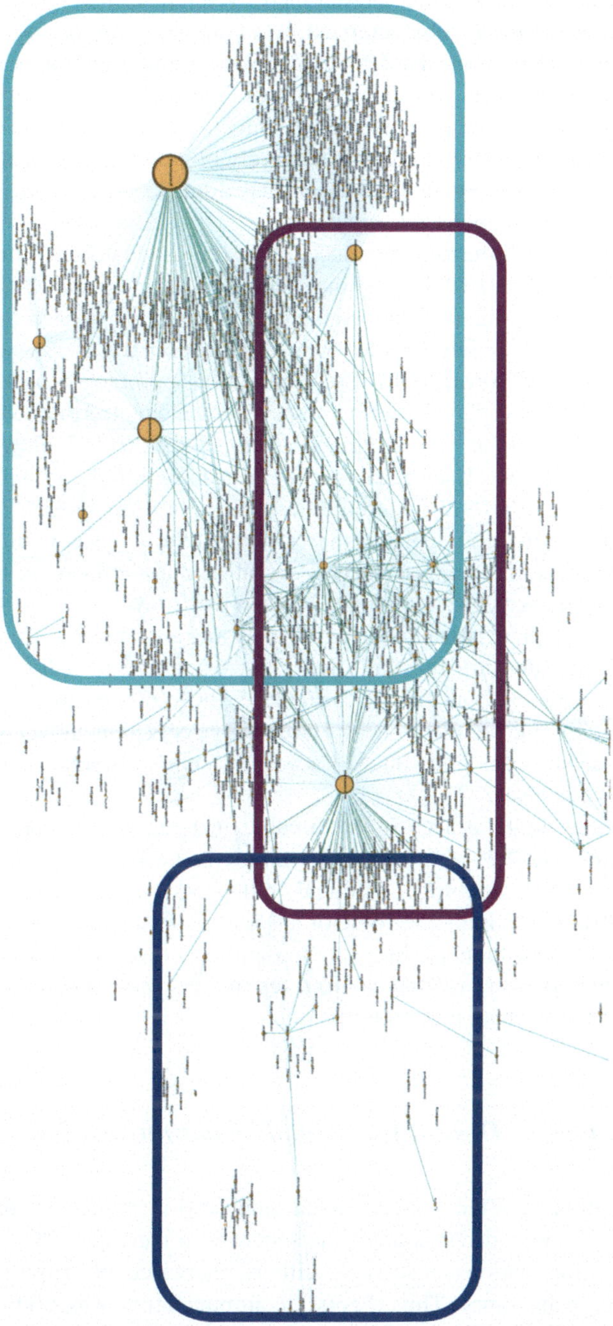

Fig. 7.1 Author-mention network of European forest biomass Twitter debates during the period from 01.07.2021 to 31.07.2021 (visualized with ForceAtlas2 using Gephi)

However, several methodological and interpretative risks must be acknowledged when using social media data in SNA. Digital media are inherently "natively relational" (Venturini et al., 2019: 4), characterised by a complex array of connections between digital objects. This means that SNA may risk oversimplification or misrepresentation of the data by obscuring the layered nature of these networks, for example, by failing to account for how hashtags link diverse content. The illustrative case above is based on a relatively small dataset composed largely of identical retweets from a few prominent actors. Scaling up to larger datasets may dilute key discursive nuances if qualitative interpretation cannot be conducted.

Further limitations include the failing to account for the varied usage of functions like "mentions" (Boyd et al., 2010). Depending on the dataset and context, mentions have multiple meanings, ranging from endorsements to attributions, or even criticism. Moreover, analyses that focus solely on "mentions" risk overlooking users who do not utilise this feature (Moats & Borra, 2018). These limitations point to a broader issue: the need to understand the interplay between user practices and platform affordances. This is where the concept of *device culture* (Rogers et al., 2013; Weltevrede & Borra, 2016) becomes relevant. User behaviours significantly influence on platform architectures, and in turn, platform functionalities shape platform-specific cultures (Weltevrede & Borra, 2016). Recognising these technocultural systems highlights the importance of comparative analyses across social media platforms in order to better understand how device culture emerges on each platform.

Temporality poses another critical challenge. Analysing social interaction within a fixed time frame may obscure the fluid and evolving nature of actor relationships. Additionally, technical constraints, particularly those related to data access and structure, further complicate the process. Here, API-accessed metadata is essential for SNA. Yet, recent restrictions on API access present substantial barriers to longitudinal network analyses.

Despite limitations, digital traces offer valuable insights into social interactions (Rogers & Marres, 2002). Combining digital methods (Rogers, 2013, 2019) with qualitative approaches enables a practical framework for forest governance research. Rogers (2018) highlights the value of analysing' social issue networking' by mapping key actors and their positions, alignment and divergence (ibid: 455–456). As Weltevrede and Borra (2016) caution, researchers must remain aware of how platform dynamics mediate these interactions.

7.5.2 Case Study: UX Design for "Good" Communication

"Good" communication is often defined as communication processes that have successfully lead to the intended results of the sender of a message. These results may include policy implications, resolved conflicts, increased participation, and changed attitudes or behaviours. Thus, "good" communication was often understood as the process of conveying specific content—such as facts, ideas, or meanings—to the target audience, who then adopt this content and behave accordingly.

Related research questions often address which kind of content has been successfully conveyed by whom, through which media, and with the support of which mediators or arguments.

Unarguably, the content itself (e.g., simple facts, complex meanings, or entire ideologies) has a significant impact on the successful delivery of a specific message. Considering theoretical approaches such as selective perception (Hansen et al., 2019; Merten, 1999) and cognitive dissonance and its avoidance (Festinger, 1957), the course content influences the recipient's selection and development of messages.

However, there are other very important elements of communication besides the content (see Fig. 7.1), such as syntax, lexis, and performance (Bucher, 2014). These elements should also be acknowledged in media analysis, as they have a significant influence on which messages resonate with the recipients. Thus, communication analysis in the field of forest policy should not only focus on the contents further. Diction and performance must also be considered to investigate why some messages are delivered more effectively than others.

Performance can be considered the design or format of a presentation. However, performance is not only the visible packaging of a message, but also a conscious way of emphasising the function of the message for recipients (Bucher, 2014). Thus, performance is the deliberate arrangement of data to align the message with the needs of the target audience. As with most other forms of communication, dozens of specific target groups exist for forest-related messaging. This again raises the claim for research on target groups, their socio-psychological characteristics, needs, and behaviours. Furthermore, a conceptual framework is needed to investigate the performance of multimodal messages, since the performance of communication has been widely neglected in forest policy and forest-related communication research. In multimodal messages, data with different semiotic resources (e.g., texts, figures, colours, sounds) or different content types (e.g., texts, pictures, audio- or film sequences) are combined (Bucher, 2012). For example, a message on a website typically contains text in a specific font, size, colour, and spacing, as well as accompanying images, design elements (e.g., frames, lines, background colours) or even embedded videos or animations.

In a recent study about the performance of websites provided by institutionalised, forest-related actors in Germany (e.g., NGOs, forest-related ministries, state forestry enterprises, state offices, universities offering higher forestry education), the user experience (UX) was used as a conceptual framework. At this point, it is essential to note that there has been a significant shift in the perspectives used for media analysis from a sender's to a recipient's perspective in recent years (Bucher & Schumacher, 2012). In line with this, the UX concept adopts a user-centred approach to communication in digital media. This approach covers not only information and communication technologies and human resource management, but also includes organisational and psychological aspects, such as human–computer interactions. The holistic UX concept, therefore, includes multiple features such as:

– the characteristics of the recipients or users, including their needs, attitudes, motivations, habits and behaviours.

- the communication goals of the communication sender in connection with the function(s) of the sent message.
- the information architecture, which is determined by the content and its characteristics (e.g., visual design, amount of data, accessibility, objectivity, linguistic features).
- the interaction design, which describes the linkages between the communication sender, user, and the type of media or device used for communication.

Corey Stern has developed the most comprehensive conceptualisation of user experience so far by combining various UX models known to date. His CUBI-UX model includes the four typical factors of UX (C—content, U—user goals, B—business goals (which refers to the communication sender) and I—interaction) and relates them to each other. An illustration of the CUBI-UX model has been published by Stern (2014).

However, the CUBI-UX model is insufficient for describing the performance of a message, mainly, as the technical description of its content and arrangement in an information architecture overlooks the highly important aspects of information quality. In line with Wang and Strong (1996: 6), information quality can be defined as "data that are fit for use by data consumers". It ensures that the content of a message is presented in a way that attracts the recipient's attention and encourages them to engage with information in more detail. These aspects are visualised in Fig. 7.2 as thresholds of attention, perception and comprehension that need to be overcome. Wang and Strong (1996) identified 15 dimensions of information quality, which are widely accepted so far: accuracy, credibility, reputation, objectivity, timeliness, appropriate amount of data, completeness, relevancy, added value, representational consistency, interpretability, ease of understanding, concise representation, access security, and accessibility. The better these dimensions align with the needs of the target audience, the better. Thus, the higher the information quality, the more likely it is that your message will attract attention and be understood.

Comprehensive conceptual frameworks now exist to investigate the communication performance. These frameworks ensure that messages are tailored to the goals and needs of the target audiences in terms of visual aesthetics and language, adjusted to the media used and its logic, and of high informational quality in terms of accuracy, timeliness, and completeness. In short, it is assumed that performance involves consciously arranging the (different) data in an appropriate was that aligns with the needs of users and meets the requirements of high-quality information.

Based on this conceptual framework of performance, a suitable method for assessing the performance of communication on websites was sought. The Modular User Experience Questionnaire Plus (UEQ+), developed by Schrepp and Thomaschewski (2019), was used in this case. This modular user experience questionnaire offers 16 assessment dimensions: attractiveness, efficiency, perspicuity, dependability, stimulation, originality, trust, visual aesthetics, adaptability, usefulness, intuitive operation, value, content credibility, content quality, haptics, and acoustics. All UX dimensions are evaluated with four items each on a seven-point Likert scale. Depending on the object of evaluation, the questionnaire can

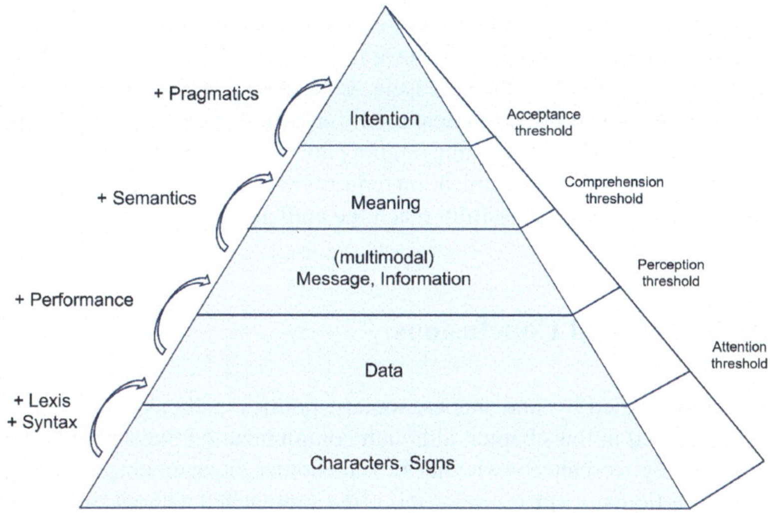

Fig. 7.2 Multilayer-model of communication by Liebal et al. (2025)

be composed of the appropriate dimensions. Based on the insights gained from a previous cognitive walkthrough (for methodology, see Thielsch & Hirschfeld, 2018) of 30 websites, the following eight dimensions were included in the performance analysis: attractiveness, stimulation, originality, visual aesthetics, adaptability, usefulness, intuitive operation, and content quality. Originally, the UX questionnaire was developed to be directly applied by users. However, this study showed that it can also be applied from an expert perspective, whereby the "experts involved put themselves in the role of the potential users" (Thielsch & Hirschfeld, 2018: 36).

The results provided a clear insights into why certain content—in this study, all related to the topics of forest conditions and restoration—is or is not successfully conveyed to the public. On the one hand, the websites of forestry-related authorities provide a much more comprehensive range of evidence-based information on forest conditions and restoration than most actors in the interest mediation system (e.g., forestry-related or environmental NGOs, game and hunting associations). On the other hand, websites operated by actors in the interest mediation system, especially environmental and nature conservation NGOs, provide a better user experience in each UX dimension than websites operated by forest authorities. Thus, there is an obvious difference in the performance of online communication on these types of websites. As discussed in this chapter, a lack of performance can prevent audiences from paying attention to certain content (e.g., facts, arguments, opinions), thereby hindering their perception, elaboration, and comprehension of this content. Although having a rather poor website performance, with comprehensive content, as observed on the websites of several forest authorities, will likely result in less attention to the content provided by forest-related authorities. In this regard, the study sheds a new light on the debate as to why, in Germany, ecosystem services related to the protection of nature, climate and biodiversity have increasingly been rated higher than the wood-supplying function of forests (Eisele & Juschka, 2022; Lorenz & Elsasser, 2018), despite the intensive communication of forest authorities and forestry-related NGOs. Consequently, when discussing ongoing and necessary transitions in communication practices, "good" communication requires not only high-quality content, but also high performance to successfully reach the audience.

7.6 Discussion and Conclusions

Communication shaped by, and shapes, society, politics, and environmental governance. As discussed in this chapter, although communication plays a pivotal role in forest policy and governance—whether in stakeholder engagement, science-policy interaction or participatory processes—it is often approached without sufficient theoretical grounding or conceptual clarity. The review reveals that communication is frequently considered a technical or instrumental tool, rather than as a dynamic, context-dependent, and relational process that is deeply embedded in broader power structures and social meanings.

One of the central issues raised is the ongoing reliance on the information-deficit model, which posits that better-informed citizens will make better decisions. This assumption has been criticised (e.g., Takala et al., 2022). Contemporary research emphasises that more information is not necessarily better. What is needed is better communication: communication that is attuned to context, responsive to values and emotions, and grounded in mutual trust and understanding. The exact nature of "better" communication depends on the chosen research perspective. This could involve frequent in-person communication that emphasises shared values and fosters trust, target-group-specific public communication, emotive storytelling, mass media outreach, or a user-friendly website experience. Each of these approaches provides valuable insights and adds another piece to the puzzle. The aim of this chapter was to highlight the complexity inherent in communication processes. There is no one-size-fits-all template for effective communication, and what constitutes "good communication" varies significantly depending on the context. For example, mass media communication differs fundamentally from dialogical, interpersonal settings. The effectiveness of communication is always context- and case-specific. Nevertheless, communication models and insights from communication psychology and related research fields can provide valuable guidance in navigating these complexities.

This chapter uses a variety of theoretical approaches, such as communication psychology, media studies, social network analysis, and deliberative theory, to demonstrate the intricacy of communication. Rather than offering prescriptive solutions, the intention is to demonstrate the value of these approaches in informing more reflective, inclusive, and effective communication strategies. Effective communication is never one-size-fits-all solution; it is always context-specific, actor-dependent, and value-laden.

These implications are particularly relevant for science communication, especially when engaging with policymakers, the public, or specific stakeholder groups, such as forest owners. As Hetemäki (2019) notes, scientific knowledge is necessary but not sufficient for policy formulation. Participatory communication models emphasising the co-production of knowledge and recognising non-scientific forms of expertise (e.g., local, indigenous) have the potential to enhance the legitimacy, relevance, and democratic engagement of forest governance (Metcalfe, 2019). Future research could productively explore how different media and formats could support co-production and facilitate meaningful dialogue across epistemic and cultural divides.

At the same time, researchers must critically reflect on their own communicative roles. Scientific knowledge is never neutral; it can perpetuate existing power structures and reproduce dominant narratives. This raises important questions: Which stories are being told, and which are being left out? Who gets to speak, and who is silenced? Fundamentally, narratives provide the reference points that shape the stories told about the world and the kind of future that can be imagined. Narratives are particularly relevant in the context of transformation research, where they shape what futures are considered possible, desirable, or conceivable. As Veland et al., (2018: 42) write: "This implies scientists, journalists, and other actors alike perform 'storytelling', using written word, images, and figures. Each narrative constrains and enables what is thinkable and sayable about the past, present, and future". Hajer and Versteeg (2019) therefore argue that scholars should take an active role in collaborating with societal stakeholders to develop alternative narratives and visions for the future.

The deliberative potential of new media, especially X, and hybrid interactions has attracted the attention of scholars (e.g., Milleville et al., 2023; Pellicer-Chenoll et al., 2023; Sanchez & Lobo, 2023), yet this remains under-explored, particularly in forest science. In political science and media studies, digital media are viewed not only as communication tools, but also as active spaces for public discourse and democratic engagement. Crozier's (2007) concept of recursive governance highlights this presenting policymaking as an ongoing communicative process in which public input, expressed via the media, continually influences governance structures. This perspective positions digital media as crucial sites of discursive participation, where actors shape and contest policy narratives. However, these deliberative dynamics remain under-explored in forest science and environmental policy.

In recent years, forest governance has opened up to participatory processes (Maier et al., 2014), and knowledge of communication may be important in facilitating these processes. However, Arts (2014) has noted that participation processes may contribute to the reinforcement of existing power structures. Therefore, when analysing and designing participation processes, it is important to consider the political context. Further exploration is needed to determine whether and how new forms of communication either reinforce or disrupt power structures.

According to Habermas, communication should aim to facilitate rational debate, guided by formal expertise, which establishes legitimacy. By opening spaces for deliberation in which participants recognise their respective arguments, the model highlights the potential for reaching a consensus. Although the deliberative potential has been discussed in relation to forest policy (Giessen et al., 2009; Kleinschmit, 2012), the model of deliberative politics has been criticised for ignoring underlying power relations. The critiques argue that Habermas' model cannot recognise the link between legitimacy and power because his concept of legitimacy is based on an understanding of pure rationality which obscures the social constitution of objectivity as an act of power (Mouffe, 1999; Wahl-Jorgensen, 2019). In response to the "idea that public debate should be rational, impartial, disembodied, dispassionate, and objective" (Wahl-Jorgensen, 2019: 2), the emotional turn in media studies has demonstrated the importance of considering subjectivity in debates and rethinking what is normatively desirable. This is particularly relevant given that social and political actions are considered to be fundamentally driven by emotions articulated and provoked by the media (Wahl-Jorgensen, 2019; Wettergren, 2005). Rather than striving for dispassionate neutrality, communication research must acknowledge the affective and performative dimensions of discourse, particularly in a networked society where emotion-driven "affective publics" can mobilise across ideological boundaries (Papacharissi, 2015).

Finally, as Van Woerkum et al. (2011) argue, change is fundamentally social, and communication plays a key role in how it emerges, becomes visible, and is acted upon. Communication not only reflects transformation, it enables it. However, transformation is rarely linear or predictable. Embracing this unpredictability—acknowledging uncertainty and opening up space for discussion about different possible futures—may be essential for fostering genuine change, rather than simply reproducing existing paradigms.

In conclusion, this chapter has called for communication science to be more systematically and theoretically integrated into forest policy and governance research. Rather than being viewed as an add-on, communication must be recognised as a core component of governance processes—a space where power is negotiated, knowledge is co-produced, identities are constructed, and futures are envisioned. Addressing complex forest governance challenges requires an understanding of communication that goes beyond its technical or instrumental aspects to encompass its relational, emotional, political, and transformative dimensions.

References

Aasetre, J. (2006). Perceptions of communication in Norwegian forest management. *Forest Policy and Economics, 8*(1), 81–92.

Aicher, C. (2003). Discourse analysis: A fruitful method in the forestry science? Taking the story lines of Venezuelan forest policy as an illustration. *Allgemeine Forst- und Jagdzeitung, 174*(4), 61–66.

Ajzen, I., & Fishbein, M. (2011). Attitudes and the attitude-behavior relation: Reasoned and automatic processes. *European Review of Social Psychology, 11*(1), 1–33.

Ambroise, L., Abildtrup, J., & Pülzl, H. (2023). Analysis of French forest policy in the context of ecosystem services: From the national to the regional level. *Environemental Policy and Governance, 33*(2), 153–163.

Andersson, M., Bostedt, G., & Sandström, C. (2022). The role of Swedish forests in climate change mitigation—A frame analysis of conflicting interests. *Forest Policy and Economics, 144*, Article 102842.

André, K., Baird, J., Gerger Swartling, Å., Vulturius, G., & Plummer, R. (2017). Analysis of Swedish Forest Owners' information and knowledge-sharing networks for decision-making: Insights for climate change communication and adaptation. *Environmental Management, 59*(6), 885–897.

Ardoin, N. M., DiGiano, M. L., O'Connor, K., & Podkul, T. E. (2017). The development of trust in residential environmental education programs. *Environmental Education Research, 23*(9), 1335–1355.

Ardoin, N. M., Heimlich, J., Braus, J., & Merrick, C. (2013). *Influencing conservation action: What research says about environmental literacy, behavior, and conservation results*. National Audubon Society.

Arnould, M., Morel, L., & Fournier, M. (2021). Developing the persona method to increase the commitment of non-industrial private forest owners in French forest policy priorities. *Forest Policy and Economics, 126*, 1–13.

Arts, B. (2021). *Forest governance: Hydra or chloris?* Cambridge University Press.

Arts, B., & Buizer, M. (2009). Forests, discourses, institutions. A discursive-institutional analysis of global forest governance. *Forest Policy and Economics, 11*(5–6), 340–347.

Arts, B., Appelstrand, M., Kleinschmit, D., Pülzl, H., & Visseren-Hamakers, I. (2010). Discourses, actors and instruments in international forest governance. In J. Rayner, A. Buck, & P. Katila (Eds.), *Embracing complexity: Meeting the challenges of International Forest Governance. A global assessment report*. Prepared by the Global Forest Expert Panel on the International Forest Regime (Vol. 28, pp. 57–73). International Union of Forest Research Organizations.

Asselin, J., Asselin, G., & Egli, F. (2022). The discursive context of forest in land use documents an Irish case study. *Nature + Culture, 17*(2), 170–190.

Atkin, C. K., & Rice, R. E. (2013). Theory and principles of public communication campaigns. In R. E. Rice & C. K. Atkin (Eds.), *Public communication campaigns* (pp. 16–50). Sage.

Baten, M. A. (2024). Revisiting the success of co-management in forest protected areas of Bangladesh: Untangling the interplay of power, participation, and community empowerment. *Environmental Policy and Governance, 34*(6), 649–662.

Beck, S. (2012). From truth to trust: Lessons learned form 'Climategate.' In K. Hogl, E. Kvarda, R. Nordbeck, & M. Pregernig (Eds.), *Environmental governance: The challenge of legitimacy and effectiveness* (pp. 220–241). Edward Elgar.

Behagel, J., & Turnhout, E. (2011). Legitimacy in the implementation of the water framework directive in the Netherlands: Towards participatory and deliberative norms? *Journal of Environmental Policy and Planning, 13*(3), 297–316.

Bentele, G., & Seidenglanz, R. (2008). *Public relations research: European and international perspectives and innovations*. Springer Fachmedien.

Bentele, G., Brosius, H. P., & Jarren, O. (2003). *Öffentliche Kommunikation: Handbuch Kommunikations- und Medienwissenschaft*. VS Verlag für Sozialwissenschaft.

Berger, A. A. (1995). *Essentials of mass communication theory*. Sage Publications.

Berger, P. L., & Luckmann, T. (1967). *The social construction of reality*. Penguin Books.

Bethmann, S., Simminger, E., Baldy, J., & Schraml, U. (2018). Forestry in interaction. Shedding light on dynamics of public opinion with a praxeological methodology. *Forest Policy and Economics, 96*, 93–101.

Beymer-Farris, B. A., & Bassett, T. J. (2012). The REDD menace: Resurgent protectionism in Tanzania's mangrove forests. *Global Environmental Change, 22*(2), 332–341.

Bhatt, P., & Pickering, C. M. (2021). Public perceptions about Nepalese National Parks: A global Twitter discourse analysis. *Society and Natural Resources, 34*(6), 685–702.

Bidone, F. (2022). Driving governance beyond ecological modernization: REDD plus and the Amazon Fund. *Environmental Policy and Governance, 32*(2), 110–121.

Blicharska, M., Aggestam, P., Giessen, L., Hilszczański, J., Hermanowicz, E., Holeksa, J., Jacobsen, J. B., et al. (2020). Between biodiversity conservation and sustainable forest management: A multidisciplinary assessment of the emblematic Białowieża Forest case. *Biological Conservation, 248*, Article 108614.

Blondet, M., de Koning, J., Borrass, L., Ferranti, F., Geitzenauer, M., Weiss, G., Turnhout, E., & Winkel, G. (2017). Participation in the implementation of Natura 2000: A comparative study of six EU member states. *Land Use Policy, 66*, 346–355.

Bogen, K. W., Bleiweiss, K. K., Leach, N. R., & Orchowski, L. M. (2021). #MeToo: Disclosure and response to sexual victimization on Twitter. *Journal of Interpersonal Violence, 36*(17–18), 8257–8288.

Bolay, E., & Reichle, B. (2015). *Waldpädagogik: Handbuch der waldbezogenen Umweltbildung*. Schneider Verlag Hohengehren GmbH.

Bonfadelli, H., & Friemel, T. N. (2015). *Medienwirkungsforschung*. UKV Verlagsgesellschaft mbH.

Boyd, K. (2023). Group epistemology and structural factors in online group polarization. *Episteme, 20*(1), 57–72.

Boyd, D., Golder, S., & Lotan, G. (2010). Tweet, tweet, retweet: Conversational aspects of retweeting on Twitter. In *43rd Hawaii International Conference on System Sciences System Sciences (HICSS)*, 5 January 2010 (pp. 1–10). IEEE.

Brämer, R. (2006). *Natur obskur: Wie Jugendliche heute Natur erfahren*. Oekom.

Brito, B., Micol, L., Davis, C., Nakhooda, S., Daviet, F., & Thuault, A. (2009). *The governance of forests toolkit: A draft framework of indicators for assessing governance of the forest sector*. World Resource Institute. Retrieved July 4, 2025, from http://pdf.wri.org/working_papers/gfi_tenure_indicators_sep09.pdf

Brüggemann, M., Elgesem, D., Bienzeisler, N., Gertz, H. D., & Walter, S. (2020). Mutual group polarization in the blogosphere: Tracking the hoax discourse on climate change. *International Journal of Communication, 14*, 1025–1048.

Bruns, A. (2023). From "the" public sphere to a network of publics: Towards an empirically founded model of contemporary public communication spaces. *Communication Theory, 33*(2–3), 70–81.

Bruzzese, S., Ahmed, W., Blanc, S., & Brun, F. (2022). Ecosystem services: A social and semantic network analysis of public opinion on Twitter. *International Journal of Environmental Research and Public Health, 19*(22), 15012.

Bubela, T., Nisbet, M. C., Borchelt, R., Brunger, F., Critchley, C., Einsiedel, E., et al. (2009). Science communication reconsidered. *Nature Biotechnology, 27*(6), 514–518.

Bucher, H. J. (2012). Multimodalität - Ein universelles Merkmal der Medienkommunikation: Zum Verhältnis von Medienangebot und Medienrezeption. In H. J. Bucher & P. Schumacher (Eds.), *Interaktionale Rezeptionsforschung: Theorie und Methode der Blickaufzeichnung in der Medienforschung* (pp. 51–82). Springer.

Bucher, H. J. (2014). Sprach- und Diskursanalyse in der Medienforschung. In M. Karmasin, M. Rath, & B. Thomaß (Eds.), *Kommunikationswissenschaft als Integrationsdisziplin* (pp. 271–298). Springer Fachmedien.

Bucher, H. J., & Schumacher, P. (Eds.). (2012). *Interaktionale Rezeptionsforschung: Theorie und Methode der Blickaufzeichnung in der Medienforschung*. Springer.

Bujis, A., & Lawrence, A. (2013). Emotional conflicts in rational forestry: Towards a research agenda for understanding emotions in environmental conflicts. *Forest Policy and Economics, 33*, 104–111.

Butler, B. J., Tyrrell, M., Feinberg, G., VanManen, S., Wiseman, L., & Wallinger, S. (2007). Understanding and reaching family forest owners: Lessons from social marketing research. *Journal of Forestry, 105*(7), 348–357.

Burkart, R. (1995). *Kommunikationswissenschaft: Grundlagen und Problemfelder*. Böhlau.

Burke, K. (1969). *A grammar of motives*. University of California Press.

Caetano, M. A. L. (2021). Political activity in social media induces forest fires in the Brazilian Amazon. *Technological Forecasting and Social Change, 167*, Article 120676.

Calibeo, D. L., & Hindmarsh, R. (2022). From fake news to echo-chambers: On the limitations of new media for environmental activism in Australia, and "Activist-Responsive Adaptation." *Environmental Communication, 16*(4), 490–504.

Cashore, B. (2002). Legitimacy in global environmental governance: How non-state market driven (NSDM) governance systems gain rule-making authority. *Governance, 15*(4), 503–529.

Castells, M. (2009). *Communication power*. Oxford University Press.

Chadwick, A. (2013). *The hybrid media system: Politics and power*. Oxford University Press.

Chadwick, A. (2017). *The hybrid media system: Politics and power* (2nd ed.). Oxford University Press.

Chung, J. H., Sarmiento, I. G., Van Swol, L. M., Shaw, B. R., Koshollek, A., & Hangsan Ahn, P. (2020). Promoting reforestation to landowners: The role of advice-giving through information, efficacy, narratives, and identification in storytelling. *Journal of Forestry, 118*(5), 474–486.

Crozier, M. (2007). Recursive governance: Contemporary political communication and public policy. *Political Communication and Public Policy, 24*(1), 1–18.

Curran, J. (2002). *Media and power*. Routledge.

Cutlip, S. M., Center, A. H., & Broom, G. M. (2000). *Effective public relations*. Prentice Hall.

Daniels, S. E., & Walker, G. B. (2001). *Working through environmental conflict: The collaborative learning approach*. Praeger Publishers.

Daume, S., & Füldner, K. (2016). "Forest Tweets": Informal digital coverage of the oak processionary moth or why foresters should care about social media. *Allgemeine Forst- und Jagdzeitung, 187*(9/10), 185–197.

De Benedictis, S., Orgad, S., & Rottenberg, C. (2019). #MeToo, popular feminism and the news: A content analysis of UK newspaper coverage. *European Journal of Cultural Studies, 22*(5–6), 718–738.

De Groot, M., Pocock, M. J., Bonte, J., Fernandez-Conradi, P., & Valdés-Correcher, E. (2022). Citizen science and monitoring forest pests: A beneficial alliance? *Current Forestry Reports, 9*, 15–32.

De Koning, J., Winkel, G., Sotirov, M., Blondet, M., Borras, L., Ferranti, F., & Geitzenauer, M. (2014). Natura 2000 and climate change: Polarisation, uncertainty, and pragmatism in discourses on forest conservation and management in Europe. *Environmental Science and Policy, 39*, 129–138.

Deutsche Forstchefkonferenz. (2007). Waldpädagogik-Zertifikat: Gemeinsame Rahmenregelungen und Mindest-Standards des bundesländerübergreifend von den Forstverwaltungen getragenen Waldpädagogik-Zertifikats, Anlage 2 zum FCK-Beschluss vom 26./27.04.2007.

Dobler, G. (2013). Im Wald Nachhaltigkeit lernen. *LWF Wissen, 72*, 86–91.

Dragoi, M., Popa, B., & Blujdea, V. (2011). Improving communication among stakeholders through ex-post transactional analysis—Case study on Romanian forestry. *Forest Policy and Economics, 13*(1), 16–23.

Druckmann, J. N. (2021). A framework for the study of persuasion. *Annual Review of Political Science, 25*(1), 65–88.

Dunwoody, S. (2014). Science journalism prospects in the digital age. In M. Bucchi & B. Trench (Eds.), *Handbook of public communication of science and technology* (pp. 43–55). Routledge.

Duvall, S. S., & Heckemeyer, N. (2018). #BlackLivesMatter: Black celebrity hashtag activism and the discursive formation of a social movement. *Celebrity Studies, 9*(3), 391–408.

Edelman, M. (1993). Contestable categories and public opinion. *Political Communication, 10*(3), 231–242.

Eisele, H., & Juschka, J. (2022). Wald bewirtschaften? Ja? Nein? – Und wie? *LWF aktuell, 1*, 4–7.

Ekayani, M., Nurrochmat, D. R., & Darusman, D. (2016). The role of scientists in forest fire media discourse and its potential influence for policy-agenda setting in Indonesia. *Forest Policy and Economics, 68*, 22–29.

Elgesem, D., & Brüggemann, M. (2023). Polarisation or just differences in opinion: How and why Facebook users disagree about Greta Thunberg. *European Journal of Communication, 38*(3), 237–254.

Elomina, J., & Pülzl, H. (2021). How are forests framed? An analysis of EU forest policy. *Forest Policy and Economics, 127*, Article 102448.

Entman, R. M. (1993). Framing: Toward clarification of a fractured paradigm. *Journal of Communication, 43*(4), 51–58.

Esselin, A., & Ljung, M. (2006). Bridging the gap: The Mountain Mistra Programme as an arena for collaborative learning. *International Journal of Biodiversity Science and Management, 2*(4), 315–325.

Fabra-Crespo, M., & Rojas-Briales, E. (2015). Analysis of mass media news on forest issues: A case study of Spain. *Forest Systems, 24*(2), Article e029.

Fabricius, C., Koch, E., & Magome, H. (2001). Towards strengthening collaborative ecosystem management: Lessons from environmental conflict and political change in southern Africa. *Journal of the Royal Society of New Zealand, 31*(4), 831–844.

Fähnrich, B., Weitkamp, E., & Kupper, J. F. (2023). Exploring 'quality' in science communication online: Expert thoughts on how to assess and promote science communication quality in digital media contexts. *Public Understanding of Science, 32*(5), 605–621.

Falkenberg, M., Galeazzi, A., Torricelli, M., Marco, N., Di Larosa, F., Sas, M., Mekacher, A., Pearce, W., Zollo, F., Quattrociocchi, W., & Baronchelli, A. (2022). Growing polarization around climate change on social media. *Nature Climate Change, 12*, 1114–1121.

FAO. (2010). *Developing effective forest policy: A guide*. Food and Agriculture Organization of the United Nations.

FAO-PROFOR. (2011). *Framework for assessing and monitoring forest governance*. Food and Agriculture Organization of the United Nations.

Feindt, P. H., & Kleinschmit, D. (2011). The BSE crisis in German newspapers: Reframing responsibility. *Science as Culture, 20*(2), 183–208.

Feindt, P. H., & Oels, A. (2005). Does discourse matter? Discourse analysis in environmental policy making. *Journal of Environmental Policy and Planning, 7*(3), 161–174.

Festinger, L. (1957). *A theory of cognitive dissonance*. Row, Peterson.

Ficko, A., Lidestav, G., Dhubhain, A. N., Karppinen, H., Zivojinovic, I., & Westin, K. (2019). European private forest owner typologies: A review of methods and use. *Forest Policy and Economics, 99*, 21–31.

Fienitz, M., Busse, M., Fienitz, M., & Heiland, S. (2022). Analysing the impact of communication and public participation on the acceptability of Germany's Black Forest National Park. *Journal for Nature Conservation, 67*, Article 126155.

Finley, A. O., & Kittredge, D. B., Jr. (2006). Thoreau, Muir, and Jane Doe: Different types of private forest owners need different kinds of forest management. *Northern Journal of Applied Forestry, 23*(1), 27–34.

Fischer, F., & Forester, J. (1993). *The argumentative turn in policy analysis and planning*. Duke University Press.

Fischer, F., & Gottweis, H. (2013). The argumentative turn in public policy revisited: Twenty years later. *Critical Policy Studies, 7*(4), 425–433.

Fischer, K., Stenius, T., & Holmgren, S. (2020). Swedish Forests in the bioeconomy: Stories from the national forest program. *Society & Natural Resources, 33*(7), 896–913.

Ford, J. K., Riley, S. J., Lauricella, T. K., & Van Fossen, J. A. (2020). Factors affecting trust among natural resources stakeholders, partners, and strategic alliance members: A meta-analytic investigation. *Frontiers in Communication, 5*.

Fox, N., Chamberlain, B., Lindquist, M., & Van Berkel, D. (2022). Understanding landscape aesthetics using a novel viewshed assessment of social media locations within the Troodos UNESCO Global Geopark, Cyprus. *Frontiers in Environmental Science, 10*, Article 884115.

Frei, T., Edou, K., Fernandez-Blanco, C. R., & Winkel, G. (2022). Governing abandoned land: Storylines on natural forest regrowth in France and Spain. *Environmental Science and Policy, 135*, 58–66.

Gerhards, J. (1994). Politische Öffentlichkeit: Ein system- und akteurstheoretischer Bestimmungsversuchin. In F. Neidhardt (Ed.), *Öffentlichkeit, öffentliche Meinung, soziale Bewegungen,*

Kölner Zeitschrift fur Soziologie und Sozialpsychologie, Sonderheft 34/1994 (pp. 77–105). Westdeutscher Verlag.

Gerhard, J., & Schäfer, M. S. (2007). Demokratische Internet-Öffentlichkeit? Ein Vergleich der öffentlichen Kommunikation im Internet und in den Printmedien am Beispiel der Humangenomforschung. *Publizistik, 52*(5), 210–228.

Gerosideris, C., & Ferra, I. (2020). Digital media and environmental protests in Greece: #Skouries. In A. Veneti & A. Karatzogianni (Eds.), *The emerald handbook of digital media in Greece: Digital activism and society* (pp. 399–418). Emerald Publishing Limited.

Ghorbani, M., & Azadi, H. (2021). A social-relational approach for analyzing trust and collaboration networks as preconditions for rangeland comanagement. *Rangeland Ecology and Management, 75*(1), 170–184.

Giessen, L., & Buttoud, G. (2014). Defining and assessing forest governance. *Forest Policy and Economics, 49*, 1–3.

Giessen, L., Kleinschmit, D., & Böcher, M. (2009). Between power and legitimacy—Discourse and expertise in forest and environmental governance. *Forest Policy and Economics, 11*, 452–453.

Gisselquist, R. M. (2012). What does good governance mean? UNU WIDER. Retrieved December 26, 2024, from https://www.wider.unu.edu/publication/what-does-good-governance-mean

Gorris, P., & Koch, L. (2024). Building trust in environmental co-management: Social embeddedness in a contested German biodiversity conservation governance process. *Environmental Science and Policy, 154*, Article 103695.

Graves, R. A., Nielsen-Pincus, M., Haugo, R. D., & Holz, A. (2022). Forest carbon incentive programs for non-industrial private forests in Oregon (USA): Impacts of program design on willingness to enroll and landscape-scale program outcomes. *Forest Policy and Economics, 141*, Article 102778.

Grundel, I., Christenson, N., & Dahlström, M. (2022). Identifying interests and values in forest areas through collaborative processes and landscape resource analysis. *Forest Policy and Economics, 142*, Article 102801.

Grunig, J. E., & Grunig, L. S. (1984). *Managing public relations*. Holt.

Grunig, J. E., & Grunig, L. S. (1989). Toward a theory of the public relations behavior of organizations: Review of a program of research. *Public Relations Research Annual, 1*(1–4), 27–63.

Habermas, J. (2006). Political communication in media society: Does democracy still enjoy an epistemic dimension? The impact of normative theory on empirical research. *Communication Theory, 16*(4), 411–426.

Hajdu, M., & Simoneau, C. (2020). Communicating science in a policy context to a broader audience. In V. Šucha & M. Sienkiewicz (Eds.), *Science for policy handbook* (pp. 166–179). Elsevier.

Hajer, M., & Versteeg, W. (2005). A decade of discourse analysis of environmental politics: Achievements, challenges, perspectives. *Environmental Policy and Planning, 7*(3), 175–184.

Hajer, M., & Versteeg, W. (2019). Imagining the post-fossil city: Why is it so difficult to think of new possible worlds? *Territority, Politics, Governance, 7*(2), 122–134.

Halla, T., & Laine, J. (2022). To cut or not to cut: Emotions and forest conflicts in digital media. *Journal of Rural Studies, 94*, 439–453.

Hallberg-Sramek, I., Bjärstig, T., & Nordin, A. (2020). Framing woodland key habitats in the Swedish media—How has the framing changed over time? *Scandinavian Journal of Forest Research, 35*(3–4), 198–209.

Halonen, M., Näyhä, A., & Kuhmonen, I. (2022). Regional sustainability transition through forest-based bioeconomy? Development actors' perspectives on related policies, power, and justice. *Forest Policy and Economics, 142*, Article 102775.

Hamunen, K., Virkkula, O., Hujala, T., Hiedanpaa, J., & Kurttila, M. (2015). Enhancing informal interaction and knowledge co-construction among forest owners. *Silva Fennica, 49*(1).

Hansen, A., Herrmann, M., Ehlers, J. P., Mondritzki, T., Hensel, K. O., Truebel, H., & Boehme, P. (2019). Perception of the progressing digitization and transformation of the German health care system among experts and the public: Mixed methods study. *JMIR Public Health & Surveillance, 5*(4), Article e14689.

Häublein, S., & Granzow, T. (2024). The forest-water-nexus: A narrative construction of the (ir-) relevance of cooperation. *Forest Policy and Economics, 168*, Article 103318.

Heiberger, R., Majó-Vázquez, S., Castro Herrero, L., Nielsen, R. K., & Esser, F. (2022). Do Not Blame the Media! The role of politicians and parties in fragmenting online political debate. *International Journal of Press/Politics, 27*(4), 910–941.

Hetemäki, L. (2019). The role of science in forest policy: Experiences by EFI. *Forest Policy and Economics, 105*, 10–16.

Hogl, K., Kvarda, E., Nordbeck, R., & Pregernig, M. (2012). *Environmental governance: The challenge of legitimacy and effectiveness*. Edward Elgar.

Hujala, T., & Tikkanen, J. (2008). Boosters of and barriers to smooth communication in family forest owners' decision making. *Scandinavian Journal of Forest Research, 23*(5), 466–477.

Hulbert, J. M., Hallett, R. A., Roy, H. E., & Cleary, M. (2023). Citizen science can enhance strategies to detect and manage invasive forest pests and pathogens. *Frontiers in Ecology and Evolution, 11*.

Ince, J., Rojas, F., & Davis, C. A. (2017). The social media response to Black Lives Matter: How Twitter users interact with Black Lives Matter through hashtag use. *Ethnic and Racial Studies, 40*(11), 1814–1830.

Jäckel, M. (2019). *Medienwirkungen kompakt: Einführung in ein dynamisches Forschungsfeld*. Springer Fachmedien.

Jakobsson, R., Olofsson, E., & Ambrose-Oji, B. (2021). Stakeholder perceptions, management and impacts of forestry conflicts in southern Sweden. *Scandinavian Journal of Forest Research, 36*(1), 68–82.

Janse, G. (2005). European co-operation and in forest communication. European Forestry Institute Technical Report. EFI.

Janse, G., & Schifferdecker, G. (2024). Successful communication of forest science. In IUFRO (Ed.), *Communicating Forest Science 2024*. Retrieved December 26, 2024, from https://www.iufro.org/science/divisions/division-9/90000/90100/90102/publications/communicating-forest-science-2/

Jardine, N., Secord, J., & Spary, E. (1996). *Cultures of natural history*. Cambridge University Press.

Joa, B., Paulus, A., Mikoleit, R., & Winkel, G. (2020). Decision making in tree selection—Contemplating conflicting goals via Marteloscope exercises. *Rural Landscapes, 7*(1), 3.

John, M., Wirth, K., Kaufmann, A., Ertelt, H., & Frei, T. (2024). Forest deliberations: Marteloscopes as sites of encounter between climate activists and forest managers. *Forest Policy and Economics, 169*, Article 103356.

Jones, M. D., Shanahan, E. A., & McBeth, M. K. (2014). Chapter 1: Introducing the narrative policy framework. In *The science of stories* (pp. 1–27). Palgrave Macmillan.

Jungherr, A., Posegga, O., & An, J. (2019). Discursive power in contemporary media systems: A comparative framework. *International Journal of Press/Politics, 24*(4), 404–425.

Kalaba, F. K. (2016). Barriers to policy implementation and implications for Zambia's forest ecosystems. *Forest Policy and Economics, 69*, 40–44.

Kangas, A. (2010). Value of forest information. *European Journal of Forest Research, 129*, 863–874.

Kansanga, M., Atuoye, K., & Luginaah, I. (2019). Same problem, conflicting 'truths': Rethinking the missing links in forest degradation narrativization in Ghana. *African Geographical Review, 38*(4), 283–295.

Keib, K., Himelboim, I., & Han, J. Y. (2018). Important tweets matter: Predicting retweets in the #BlackLivesMatter talk on twitter. *Computers in Human Behavior, 85*, 106–115.

Kelly, E. C., & Crandall, M. S. (2022). State-level forestry policies across the US: Discourses reflecting the tension between private property rights and public trust resources. *Forest Policy and Economics, 141*, Article 102757.

Kepplinger, H. M. (2002). Mediatization of politics: Theory and data. *Journal of Communication, 52*(4), 972–986.

Kingdon, J. W. (2003). *Agendas, alternatives and public policies.* Addison-Wesley Educational Publishers Inc.

Klapper, J. T. (1960). *The effects of mass communication.* Free Press.

Kleinschmit, D. (2012). Confronting the demands of a deliberative public sphere with media constraints. *Forest Policy and Economics, 16*, 71–80.

Kleinschmit, D., & Krott, M. (2008). The media in forestry: Government, governance and social visibility. In T. Sikor (Ed.), *Public and private in natural resource governance: A false dichotomy?* Earthscan.

Kleinschmit, D., & Sjöstedt, V. (2014). Between science and politics: Swedish newspaper reporting on forests in a changing climate. *Environmental Science and Policy, 35*, 117–127.

Kleinschmit, D., Arts, B., Giurca, A., Mustalahti, I., Sergent, A., & Pülzl, H. (2017). Environmental concerns in political bioeconomy discourses. *International Forestry Review, 19*, 41–55.

Kleinschmit, D., Böcher, M., & Giessen, L. (2009). Discourse and expertise in forest and environmental governance—An overview. *Forest Policy and Economics, 11*, 309–312.

Kleinschmit, D., Pülzl, H., Secco, L., Sergent, A., & Wallin, I. (2018). Orchestration in political processes: Involvement of experts, citizens, and participatory professionals in forest policy making. *Forest Policy and Economics, 89*, 4–15.

Kleinschmit, D., Wildburger, C., Grima, N., & Fisher, B. (2024). *International forest governance: A critical review of trends, drawbacks, and new approaches, IUFRO world series* (Vol. 43). IUFRO.

Klinger, U., & Svensson, J. (2015). The emergence of network media logic in political communication: A theoretical approach. *New Media and Society, 17*(8), 1241–1257.

Koch, L. (2023). "Us versus them" mentalities in co-managing a Natura 2000 forest: Narratives, identities, and a culture of conflict. *Environmental Policy and Governance, 34*(6), 582–597.

Koch, L., Gorris, P., Prell, C., & Pahl-Wostl, C. (2023). Communication, trust and leadership in co-managing biodiversity: A network analysis to understand social drivers shaping a common narrative. *Journal of Environmental Management, 336*, Article 117551.

Kohler, B. (2000). Im Spannungsfeld von Öffentlichkeitsarbeit und Umweltbildung – Was genau ist Waldpädagogik. *Afz/der Wald, 55*(1), 28–30.

Kollmuss, A., & Agyeman, J. (2002). Mind the gap: Why do people act environmentally and what are the barriers to pro-environmental behavior? *Environmental Education Research, 8*(3), 239–260.

Kreiss, D., & McGregor, S. C. (2024). A review and provocation: On polarization and platforms. *New Media & Society, 26*(1), 556–579.

Krott, M. (2005). *Forest policy analysis.* Springer.

Krumland, D., & Krott, M. (2004). Forestry: A star in the media, The image of forestry actors in the media public and public opinion. *Allgemeine Forst- und Jagdzeitung, 175*(1–2), 34–38.

Kubo, H., & Supriyanto, B. (2010). From fence-and-fine to participatory conservation: Mechanisms of transformation in conservation governance at the Gunung Halimun-Salak National Park, Indonesia. *Biodiversity and Conservation, 19*(6), 1785–1803.

Kuipers, B. T., Shivan, G. C., & Potter-Witter, K. (2013). Identifying appropriate communication means for reaching nonindustrial private forest landowners. *Journal of Forestry, 111*(1), 34–41.

Lähtinen, K., Toppinen, A., Suojanen, H., Stern, T., Ranacher, L., Burnard, M., & Kuzman, M. K. (2017). Forest sector sustainability communication in Europe: A systematic literature review on the contents and gaps. *Current Forestry Reports, 3*, 173–187.

Leipold, S. (2014). Creating forests with words—A review of forest-related discourse studies. *Forst Policy and Economics, 40*, 12–20.

Lemos, M. C., & Agrawal, A. (2006). Environmental governance. *Annual Review of Environment and Resources, 31*, 325.

Leskinen, L. A. (2004). Purposes and challenges of public participation in regional and local forestry in Finland. *Forest Policy and Economics, 6*(6), 605–618.

Liebal, S. (2011). *Waldpädagogik. Theoretische Grundlagen, Bildungskonzept für die Schulpraxis, Wirkungen.* Verlag Kessel.

Liebal, S. (2013). *Wald zwischen energetischer Nutzung und Wildnis – Handlungsempfehlungen für Bildungsanbieter und ausgewählte Stakeholder*. Universität Rostock.
Liebal, S., Köhler, J., & Weber, N. (2025). Forest condition and forest restoration as topics of online communication—An analysis of website performance of selected forest actors in Germany. *Forest Policy and Economics, 172*, Article 103454.
Lindahl, K. B., & Westholm, E. (2012). Future forests: Perceptions and strategies of key actors. *Scandinavian Journal of Forest Research, 27*(2), 154–163.
Lindahl, K. B., Baker, S., & Waldenström, C. (2013). Place perceptions and controversies over forest management: Exploring a Swedish example. *Journal of Environmental Policy & Planning, 15*(2), 201–223.
Lindqvist, L., & Lindgren, S. (2023). Mapping an emerging hashtag ecosystem: Connective action and interpretive frames in the Swedish #MeToo movement. *Feminist Media Studies, 23*(8), 4089–4106.
Lingua, F., Coops, N. C., Lafond, V., Gaston, C., & Griess, V. C. (2022). Characterizing, mapping and valuing the demand for forest recreation using crowdsourced social media data. *PLoS One, 17*(8), Article e0272406.
Lorenz, M., & Elsasser, P. (2018). Ansichten und Einstellungen zu Wald und Forstwirtschaft in Deutschland. *Allgemeine Forst- und Jagdzeitung, 189*(1/2), 1–15.
Louv, R. (2011). *Das letzte Kind im Wald? Geben wir unseren Kindern die Natur zurück*. Verlagsgruppe Beltz.
Littlejohn, S. W., & Foss, K. A. (2011). *Theories of human communication*. Waveland Press.
Ma, Z., Kittredge, D. B., & Catanzaro, P. (2012). Challenging the traditional forestry extension model: Insights from the Woods Forum Program in Massachusetts. *Small-Scale Forestry, 11*(1), 87–100.
Mack, P., Kremer, J., & Kleinschmit, D. (2023). Forest dieback reframed and revisited? Forests (re)negotiated in the German media between forestry and nature conservation. *Forest Policy and Economics, 147*, Article 102833.
Mack, P., Wallin, I., Zwickel, M. S., Pfistner, J., König, L., & Kleinschmit, D. (2025). Calling into the void? German forest dieback 2.0 debate on Twitter. A case study to operationalize the analysis of discursive power in hybrid media systems. *Forest Policy and Economics, 172*, 103447.
Mäder, A., Bachinger, M., Ziermann, A., Harprecht, P., Kromrey, V., & Schlemmer, F. (2025). Social media, deliberative communication and conflict management: Reflections based on the example case of Freiburg Im Breisgau (Germany). *Forest Policy and Economics, 170*, Article 103390.
Maier, C., Lindner, T., & Winkel, G. (2014). Stakeholders' perceptions of participation in forest policy: A case study from Baden-Württemberg. *Land Use Policy, 39*, 166–176.
Mattoni, A., & Ceccobelli, D. (2018). Comparing hybrid media systems in the digital age: A theoretical framework for analysis. *European Journal of Communication, 33*(5), 540–557.
Mayer, P. (2006). Biodiversity: The appreciation of different thought styles and values helps to clarify the term. *Restoration Ecology, 14*, 105–111.
Mazzoleni, G., & Schulz, W. (1999). Mediatization of politics: A challenge for democracy? *Political Communication, 16*(3), 247–261.
McNair, B. (2011). *An introduction to political communication*. Routledge.
McQuail, D. (2010 [1983]). *McQuail's mass communication theory*. Sage Publications
Mehrabian, A., & Wiener, M. (1967). Nonverbal communication in verbal and nonverbal studies. *Journal of Consulting Psychology, 31*, 253–258.
Metcalfe, J. (2019). Comparing science communication theory with practice: An assessment and critique using Australian data. *Public Understanding of Science, 28*(4), 382–400.
Merten, K. (1999). *Einführung in die Kommunikationswissenschaft*. LIT Verlag.
Milleville, K., Van Ackere, S., Verdoodt, J., Verstockt, S., De Maeyer, P., & Van de Weghe, N. (2023). Exploring the potential of social media to study environmental topics and natural disasters. *Journal of Location Based Services, 17*(4), 355–369.

Mirza, I. M., & Kusumasari, B. (Dec 31, 2022). Reality construction of disaster discourses on Twitter: Analysis of corpus-assisted discourse study on forest fires in Indonesia from 2014–2019. *Annals of Burns and Fire Disasters, 35*(4), 334–345.

Moats, D., & Borra, E. (2018). Quali-quantitative methods beyond networks: Studying information diffusion on Twitter with the Modulation Sequencer. *Big Data & Society, 5*(1).

Moormann, A., Lude, A., & Möller, A. (2021). Wirkungen von Naturerfahrungen auf Umwelteinstellungen und Umwelthandeln. In U. Gebhard, A. Lude, A. Möller, & A. Moormann (Eds.), *Naturerfahrung und Bildung* (pp. 57–78). Springer Fachmedien.

Moscardo, G., Pearce, P., & Morrison, A. (2001). Evaluating different bases for market segmentation. *Journal of Travel & Tourism Marketing, 10*(1), 29–49.

Mouffe, C. (1999). Deliberative democracy or agonistic pluralism? *Social Research, 66*(3), 745–758.

Müller, W., & Kruse, S. (2021). Modes of drought climatization: A frame analysis of drought problematization in Germany across policy fields. *Environmental Policy and Governance, 31*(5), 546–559.

Mylek, M. R., & Schirmer, J. (2020). Understanding acceptability of fuel management to reduce wildfire risk: Informing communication through understanding complexity of thinking. *Forest Policy and Economics, 113*, Article 102120.

Neuman, R. W., Just, M. R., & Crigler, A. N. (1992). *Common knowledge. News and the construction of political meaning*. University of Chicago Press.

Nhem, S., Jin Lee, Y., & Phin, S. (2017). Sustainable management of forest in view of media attention to REDD + policy, opportunity and impact in Cambodia. *Forest Policy and Economics, 85*(1), 10–21.

Nieborg, D. B., & Poell, T. (2018). The platformization of cultural production: Theorizing the contingent cultural commodity. *New Media & Society, 20*(11), 4275–4292.

Niederer, S. M. C. (2019). *Networked content analysis: The case of climate change*. Universiteit van Amsterdam.

Nisbet, M. C., & Scheufele, D. A. (2009). What's next for science communication? Promising directions and lingering distractions. *American Journal of Botany, 96*(10), 1767–1778.

Nousiainen, D., & Mola-Yudego, B. (2022). Characteristics and emerging patterns of forest conflicts in Europe—What can they tell us? *Forest Policy and Economics, 136*, Article 102671.

Novak, K., McCaffrey, S., & Schultz, C. A. (2023). Comparing land manager and community perceptions of a Colorado prescribed fire outreach program. *Journal of Forestry, 121*(1), 27–36.

Ofoegbu, C., & Speranza, C. I. (2021). Discourses on sustainable forest management and their integration into climate policies in South Africa. *International Forestry Review, 23*(2), 168–181.

O'Keefe, D. J. (2016). *Persuasion: Theory and research*. SAGE Publications.

Palier, B., & Surel, Y. (2005). The "Three I's" and the analysis of the state in action. *Revue Française de Science Politique, 55*(1), 7–32.

Papacharissi, Z. (2015). *Affective publics: Sentiment, technology, and politics* (p. 2015). Oxford University Press.

Papaspyropoulos, K. G. G., Liakou, H., & Dimopoulos, P. (2023). Climate change in the biodiversity and forest strategies in Greece using discourse analysis and text mining: Is an integration into a cost-efficient natural resources policy feasible? *Sustainability, 15*(7), 6127.

Park, M. S., & Kleinschmit, D. (2016). Framing forest conservation in the global media: An interest-based approach. *Forest Policy and Economics, 68*(3), 7–15.

Parkins, J. R., & Sinclair, A. J. (2023). Public participation at a crossroads: Manipulation or meaningful engagement in the Boreal Region. In M. M. Girona, H. Morin, S. Gauthier, & Y. Bergeron (Eds.), *Boreal forests in the face of climate change: Advances in global change research* (pp. 575–588). Springer.

Pearce, W., Niederer, S., Özkula, S. M., & Sánchez Querubín, N. (2019). The social media life of climate change: Platforms, publics, and future imaginaries. *WIREs Climate Change, 10*(2), 1–13.

Pearson, J. C., Nelson, P. E., Titsworth, S., & Harter, L. (2011). *Human communication*. Macgrow.

Pellicer-Chenoll, M. T., Taczanowska, K., Serra-Añó, P., Toca-Herrera, J. L., & González, L. M. (2023) National Parks in Twitter: A German-speaking perspective. *eco.mont—Journal on Protected Mountain Areas Research, 15*(1), 25–37.

Porter, L. (2007). Producing forests: A colonial genealogy of environmental planning in Victoria, Australia. *Journal of Planning Education and Research, 26*, 466–477.

Pregernig, M. (2000). Putting science into practice: The diffusion of scientific knowledge exemplified by the Austrian 'Research Initiative Against Forest Decline.' *Forest Policy and Economics, 1*(2), 165–176.

Pregernig, M. (2002). Perceptions, not facts: How forestry professionals decide on the restoration of degraded forest ecosystems. *Journal of Environmental Management and Planning, 45*(1), 25–38.

Prochaska, J. O., Redding, C. A., & Evers, K. E. (2015). The transtheoretical model and stages of change. In K. Glanz, B. K. Rimer, & K. Viswanath (Eds.), *Health behavior and health education: Theory, research, and practice San Francisco* (pp. 97–121). Jossey-Bass.

Reese, S. D. (1991). Setting the media's agenda: A power balance perspective. *Annals of the International Communication Association, 14*(1), 309–340.

Rein, M., & Schön, D. (1993). Reframing policy discourse. In F. Fischer & J. Forester (Eds.), *The argumentative turn in policy analysis and planning* (pp. 145–166). Duke University Press.

Rogers, R. (2013). *Digital methods*. MIT Press.

Rogers, R. (2018). Otherwise engaged: Social media from vanity metrics to critical analytics. *International Journal of Communication, 12*, 450–472.

Rogers, R. (2019). *Doing digital methods*. Sage.

Rogers, R., & Marres, N. (2002). French scandals on the Web, and on the streets: A small experiment in stretching the limits of reported reality. *Asian Journal of Social Science, 30*(2), 339–353.

Rogers, R., Weltevrede, E., Borra, E., & Niederer, S. (2013). National web studies: The case of Iran online. In J. Hartley, J. Burgess, & A. Bruns (Eds.), *A companion to new media dynamics* (pp. 142–166). Wiley-Blackwell.

Riemann, L., Giurca, A., & Kleinschmit, D. (2022). Contesting the framing of bioeconomy policy in Germany: The NGO perspective. *Journal of Environmental Policy and Planning, 24*(6), 822–838.

Rientjes, S. (2002). Making nature conservation modern: An analysis of developments in nature conservation policy in relation to macro-social changes, The Netherlands as a case study. *Journal of Environmental Policy and Planning, 4*, 1–21.

Rollhansen, N. (1994). Science, politics and the mass media: On biased communication of environmental issues. *Science, Technology & Human Values, 19*(3), 324–341.

Romina, R. (2014). Social learning, natural resource management, and participatory activities: A reflection on construct development and testing. *NJAS—Wageningen Journal of Life Sciences, 69*, 15–22.

Rosengren, K. E. (2000). *Communication: An introduction*. Sage.

Rubino, E. C., Tian, N., & Pelkki, M. H. (2022). Improving communications to increase Nonindustrial Private Forest Landowner (NIPF) participation in forest certification programs: A case study in Arkansas, USA. *Forests, 13*(1), 86–99.

Ruppert-Winkel, C., & Winkel, G. (2011). Hidden in the woods? Meaning, determining, and practicing of 'common welfare' in the case of the German public forests. *European Journal of Forest Research, 130*, 421–434.

Saarikoski, H., Raitio, K., & Barry, J. (2013). Understanding 'successful' conflict resolution: Policy regime changes and new interactive arenas in the Great Bear Rainforest. *Land Use Policy, 32*, 271–280.

Sanchez, D. M., & Lobo, N. G. (2023). Urban Forest tweeting: Social media as more-than-human communication in Tokyo's Rinshinomori Park. *Land, 12*(4), 727–742.

Schoeneborn, D., & Tittin, H. (2013). Transcending transmission: Towards a constitutive perspective on CSR communication. *Corporate Communications, 18*(2), 193–211.

Schrepp, M., & Thomaschewski, J. (2019). *Eine modulare Erweiterung des User Experience Questionnaire. Mensch und Computer 2019—Usability Professionals.* Gesellschaft für Informatik e.V.

Schulz, W. (2004). Reconstructing mediatization as an analytical concept. *European Journal of Communication, 19*(1), 87–101.

Secco, L., Da Re, R., Gatto, P., & Tassa, T. (2011). How to measure governance in forestry: Key dimensions and indicators from emerging economic mechanisms. *Allgemeine Forst Und Jagd Zeitung, 182*(5/6).

Sergent, A., Arts, B., & Edwards, P. (2018). Governance arrangements in the European forest sector: Shifts towards 'new governance' or maintenance of state authority? *Land Use Policy, 79*, 968–976.

Setiono, D. (2024). Enhancing forest fire communication: A case study of Oregon Government social media. *Jurnal Manajemen Hutan Tropika, 30*(2), 202–2012.

Singer, J. B. (2005). The marketplace of ideas: With a vengeance. *Media Ethics, 16*(2), 14–16.

Skill, K., Passero, S., & Francisco, M. (2021). Assembling Amazon fires through English hashtags. Materializing environmental activism within Twitter networks. Computer Supported Cooperative Work (CSCW). *The Journal of Collaborative Computing and Work Practice, 30*(5–6), 715–732.

Smallwood, J. M., Delabre, I., Vergara, S. P., & Rowhani, P. (2023). The governmentality of tropical forests and sustainable food systems, and possibilities for post-2020 sustainability governance. *Journal of Environmental Policy and Planning, 25*(1), 103–117.

Steffek, J. (2009). Discursive legitimation in environmental governance. *Forest Policy and Economics, 11*(5–6), 313–318.

Stern, C. (2014). CUBI: A user experience model for project success. UX Magazine, September 25, 2014. Retrieved December 26, 2024, from https://uxmag.com/articles/cubi-a-user-experience-model-for-project-success

Stockmann, J., Franz, K., Seintsch, B., & Neitzel, C. (2024). Factors explaining the willingness of small-scale private forest owners to engage in forestry: A German case study. *Forests, 15*(2), 319–349.

Stoeber, R. (2008). *Kommunikations- und Medienwissenschaften: Eine Einführung*. C. H. Beck.

Stone, D. (1988). *Policy paradox and political reason.* Scott Foresman & Co.

Strömbäck, J. (2008). Four phases of mediatization: An analysis of the mediatization of politics. *The International Journal of Press & Politics, 13*(3), 228–246.

Stroud, N. J. (2008). Media use and political predispositions: Revisiting the concept of selective exposure. *Political Behavior, 30*(3), 341–366.

Suk, J., Sun, Y., Sun, L., Li, M., Farías, C., Kwon, H., Ghosh, S., Borah, P., Mini, D. S., Correa, T., Garlough, C., & Shah, D. V. (2024). 'Think global, act local': How #MeToo hybridized across borders and platforms for contextual relevance. *Information, Communication & Society, 27*(3), 498–519.

Takacs, D. (1996). *The idea of biodiversity: Philosophies of paradise.* John Hopkins University Press.

Takala, T., Brockhaus, M., Hujala, T., Tanskanen, M., Lethinen, A., Tikkanen, J., & Toppinen, A. (2022). Discursive barriers to voluntary biodiversity conservation: The case of Finnish forest owners. *Forest Policy and Economics, 136*, 102681.

Takala, T., Lehtinen, A., Tanskanen, M., Hujala, T., & Tikkanen, J. (2019). The rise of multi-objective forestry paradigm in the Finnish print media. *Forest Policy and Economics, 106*, Article 101973.

Tarrant, A. M., Overdevest, C., Bright, A. D., Cordell, H. K., & English, D. B. (1997). The effect of persuasive communication strategies on rural resident attitudes toward ecosystem management. *Society & Natural Resources, 10*(6), 537–550.

Thi, K. P. D., Turnhout, E., & Arts, B. (2012). Changing forestry discourses in Vietnam in the past 20 years. *Forest Policy and Economics, 25*, 31–41.

Thielsch, M. T., & Hirschfeld, G. (2018). Woran erkenne ich einen guten User Experience Fragebogen?. Mensch und Computer 2018 - Workshopband. Bonn: Gesellschaft für Informatik e.V.

Tiebel, M., Mölder, A., Bieling, C., Hansen, P., & Plieninger, T. (2024). Understanding small-scale private forest owners is a basis for transformative change towards integrative conservation. *People & Nature, 6*(1), 337–353.

Tietze, N., Gerhold, L., & Ibisch, P. L. (2023). Policy narratives in forest fire management. *Environmental Research Letters, 18*, Article 114023.

Trudgill, S. (2001). Psychobiogeography: Meanings of nature and motivations for a democratized conservation ethic. *Journal of Biogeography, 28*, 677–698.

Ullah, A., Bavorova, M., Shah, A. A., & Kandel, G. P. (2023). Community participation in development programs: Key lessons from the billion trees afforestation project (BTAP). *Environmental Science & Policy, 150*, Article 103581.

Unal Colak, F., & Yllmaz, S. (2021). Social media use during August 2021 forest fire disaster in Turkey. *Disaster Medicine and Public Health Preparedness, 15*(6), 679–681.

Vainio, A., & Paloniemi, R. (2012). Forest owners and power: A Foucauldian study on Finnish forest policy. *Forest Policy and Economics, 21*, 118–125.

Van Dijk, J. (2019). Governing digital societies: Private platforms, public values. *Computer Law & Security Review, 36*, Article 105377.

Van Ruler, B. (2004). The communication grid: An introduction of a model of four communication strategies. *Public Relations Review, 30*, 123–143.

van Ruler, B. (2018). Communication theory: An underrated pillar on which strategic communication rests. *International Journal of Strategic Communication, 12*(4), 367–381.

Van Woerkum, C. M. J. (1982). *Voorlichtingskunde en massacommunicatie, het werkplan van de massamediale voorlichting [Extension science and mass communication: Action plan of mass media enlightenment]*. Wageningen.

Van Woerkum, C., Aarts, N., & Van Herzele, A. (2011). Changed planning for planned and unplanned change. *Planning Theory, 10*(2), 144–160.

Veland, S., Scoville-Simonds, M., Gram-Hanssen, I., Schorre, A. K., El Khoury, A., Nordbø, M. J., et al. (2018). Narrative matters for sustainability: The transformative role of storytelling in realizing 1.5°C futures. *Current Opinion in Environmental Sustainability, 31*, 41–47.

Venturini, T., Munk, A. K., & Jacomy, M. (2019). Actor-network VS network analysis VS digital networks. In J. Vertesi & D. Ribers (Eds.), *DigitalSTS: A handbook and fieldguide* (pp. 510–523). Princeton University Press.

Virkkula, O., & Hujala, T. (2014). Potentials of forestry extension encounters: A conversation analysis approach. *Small-Scale Forestry, 13*, 407–423.

Weingart, P., & Guenther, L. (2016). Science communication and the issue of trust. *Journal of Science Communication, 15*(5), C01.

Wang, R. Y., & Strong, D. M. (1996). Beyond accuracy: What data quality means to data consumers. *Journal of Management Information Systems, 12*(4), 5–33.

Wells, W., Burnett, J., & Moriarty, S. E. (1989). *Advertising: Principles and practice*. Prentice Hall.

Weltevrede, E., & Borra, E. (2016). Platform affordances and data practices: The value of dispute on Wikipedia. *Big Data & Society, 3*(1).

Wettergren, Å. (2005). Mobilization and the moral shock: Adbusters Media Foundation. In H. Flam & D. King (Eds.), *Emotions and social movements* (pp. 99–118). Routledge.

Wilber, K. (1998). *Marriage of sense and soul: Integrating science and religion*. Random House.

Wilkes-Allemann, J., Deuffic, P., Jandl, R., Westin, K., Lieberherr, E., Foldal, C., Lidestav, G., et al. (2021). Communication campaigns to engage (non-traditional) forest owners: A European perspective. *Forest Policy and Economics, 113*, Article 102621.

Wilson, R. S. (2008). Balancing emotion and cognition: A case for decision aiding in conservation efforts. *Conservation Biology, 22*(6), 1452–1460.

Winkel, G. (2012). Foucault in the forests: A review of the use of 'Foucauldian' concepts in forest policy analysis. *Foest Policy and Economics, 16*, 81–92.

Yasmi, Y., Colfer, C. J. P., Yuliani, L., Indriatmoko, Y., & Heri, V. (2007). Conflict management approaches under unclear boundaries of the commons: Experiences from Danau Sentarum National Park, Indonesia. *International Forestry Review, 9*(2), 597–609.

Chapter 8
Forest Governance 4.0: The Good, the Bad, and the Ugly

Alexandru Giurca and Tanya Baycheva-Merger

Abstract This chapter aims to explore the intersection between digital technologies and forest governance. This will advance our theoretical understanding of how information can be a transformative resource for current and future forest governance research, management, and policy. It focuses on the key concepts that underpin the relationship between digital technologies, information, and environmental and forest governance. This includes an overview of core principles of forest and environmental governance, including effectiveness, efficiency, accountability, transparency, equity/fairness, and participation. The chapter begins by introducing the Information Age, its emerging trends, key information and communication technologies, as well as their potential to enhance forest monitoring, information collection, sharing, access, and use of information. Drawing on theories of old and new forms of governance, as well as insights from public administration, digital government and the study of society and technology, the chapter then proposes a theoretical framework: **Forest Governance 4.0 (FG4.0)**. This framework helps to make sense of the growing body of research in this area and identifies issues that require increased research and policy attention. Through this lens, the subsequent sections explore and discuss critical emerging issues related to, but not limited to, data privacy, security, ownership, digital literacy, and equitable access to digital resources and information among others. Finally, the chapter proposes a research agenda to further our understanding of critical developments in forest governance 4.0.

A. Giurca (✉)
European Forest Institute, Platz der Vereinten Nationen 7, 53113 Bonn, Germany
e-mail: alex.giurca@efi.int

T. Baycheva-Merger
ICLEI European Secretariat GmbH, Leopoldring 3, 79098 Freiburg Im Breisgau, Germany

F. Aggestam (ed.), *Governing Forests Through Transformative Change*, Advances in Global Change Research 70, https://doi.org/10.1007/978-3-032-19072-7_8

8.1 Forest Governance in a World of Information

We live in an information-driven world. For centuries, forest information has been fundamental to forest governance. It was initially used as a mechanism for state oversight and resource management through national forest inventories (NFIs), which focused on timber management and economic planning (McRoberts et al., 2010). These inventories enabled governments to monitor, control, and utilise forest resources more efficiently, thereby ensuring a stable supply of timber and other resources to meet economic demands. However, the role of forest information has expanded significantly over time, particularly since the 1992 UN Conference on Environment and Development recognised the critical role of forests in achieving global environmental sustainability (Vidal et al., 2016).

Today, information is a multifaceted resource in forest governance. It is essential for promoting sustainable management, enhancing accountability and transparency, and fostering community engagement and inclusion in decision-making processes. Reliable data on forest health, biodiversity, carbon stocks, and changes in land-use is now considered vital for both short-term planning and long-term conservation efforts. It plays a critical role in forest governance, helping to ensure sustainable management, accountability, transparency, and community involvement. Thus, forest information is not merely a management tool, but a strategic asset that enables adaptive, responsive, and inclusive governance practices to meet environmental, economic, and social objectives.

The evolution of the role of information's in forest governance is closely tied to broader shifts within the Information (or Digital) Age, which has reshaped society, the economy, and the forest sector alike. This historical period began in the mid-twentieth century with the Third Industrial Revolution, which marked a transition from mechanical and analogue electronic technology to digital electronics. The development of the computer symbolised the dawn of this new era. However, it was the integration of computers with telecommunications (particularly the internet) that truly launched the Information Age by enabling the global exchanges and analysis of information in real-time (Mol, 2008). By 2000, the beginning of the Fourth Industrial Revolution was evident, characterised by an explosion of data and information driven by increasingly sophisticated information and communication technologies (ICTs). These technological advancements introduced new opportunities for rapid, data-driven decision-making and resource management across all sectors, including forestry. The Fourth Industrial Revolution is already impacting the forest sector, giving rise to what is referred to in this chapter as "forest governance 4.0."

Forest Governance 4.0 is a new approach (or paradigm) to managing and regulating forest resources, driven by the transformative advancements in the Fourth Industrial Revolution. This chapter explores the application of cutting-edge digital technologies, such as big data analytics, the Internet of Things (IoT), artificial intelligence (AI), and blockchain, is referred to as it is applied to forest governance frameworks. Essentially, Forest Governance 4.0 uses these technologies to provide access to vast amounts of forest data, enabling real-time monitoring and supporting data-driven decision-making processes. For example, the IoT allows for the deployment of interconnected devices, such as sensors and drones, thereby enhancing large-scale and real-time forest monitoring and surveillance. Big data analytics can process vast amounts of forest-related information from sources such as satellite imagery and sensors to track real-time harvesting activities, fire hazards, forest health, and changes in biodiversity (Gabrys et al., 2022; Nitoslawski et al., 2021). AI algorithms can analyse complex datasets to identify patterns and predict future trends. They can also automate tasks such as species identification and forest mapping (Ibid). Like the IoT and AI, blockchain technology can ensure transparency and traceability in wood supply chains. In sustainable forest management (SFM), for example, blockchain technology can verify the origin of wood and forest products, ensuring they come from certified sources and reducing illegal logging (Stopfer et al., 2024). Social media platforms and open forest data platforms increase public and stakeholder participation in forest and environmental governance (Riemann et al. in Colombo et al., 2023). However, technological solutionism carries certain risks, such as addressing symptoms rather than causes, leading to unintended consequences, and potentially compromising equity and justice (Banwell & Roelens, 2024; Kloppenburg et al., 2022).

Forest Governance 4.0 also raises a series of critical governance-related questions concerning effectiveness, efficiency, accountability, transparency, equity/fairness, and participation. For example, experience has shown that access to more information or "better evidence" does not necessarily guarantee better decision-making (Hetemäki, 2019). Policy decisions are, in the end, "political" (Ibid.), and the interests of policymakers play an important role, reflecting tensions between transitioning towards greater openness and diversity of values in the forest sector (Baycheva-Merger & Sotirov, 2020; Baycheva-Merger et al., 2018; Rantala et al., 2020). Furthermore, with their historical conversion into administrative entities through territorialisation processes, forests have become political entities where forest management operates as a technology of state power (Gabrys et al., 2022; Peluso & Vandergeest, 2020). As a growing number of studies argue (Gabrys et al., 2022; Goldstein, 2020; Goldstein & Faxon, 2020), digital technologies also generate distinct "political forests". Although the rapidly evolving infrastructure of technologies and information sharing offers promise in terms of efficiency, optimisation, and economic growth (Scholz et al., 2018), a conceptualisation of governance in the digital era that is dominated by technology is insufficient. Better conceptualisations are needed to address the many ways in which social and political environments, governance processes and the engagement and interaction of people and societal groups with advancing technologies are affected (Dawes, 2009; Rantala et al., 2020). Just as "classical" forest governance issues relate to accountability, transparency, equity/fairness, and participation (Arts, 2021), Forest Governance 4.0 presents a number of known and unknown challenges. These require attention to public policies, legitimacy, trust, power relationships, stakeholder engagement, and balancing competing and conflicting goals (Dawes, 2009).

This chapter primarily aims to explore the intersection of digital technologies, information, and forest governance. It seeks to enhance the theoretical understanding of how information can transform the practices of public and private actors in the forest sector. To this end, the chapter introduces the Forest Governance 4.0 (FG4.0) framework, which is designed to critically evaluate and guide this transformation, taking into account the complex interactions among technological, social, political, and ecological factors. The framework is intended to assist researchers and policy-makers in evaluating the opportunities and challenges of digital forest governance in the Information Age. The metaphor in the title therefore suggests that the effects of the Information Age on traditional forest governance do not only bring opportunities, nor can they be directed towards the desired "good" outcomes that align with the principles of good forest governance, such as accountability, effectiveness, efficiency, equity, participation, and transparency (Arts, 2021; FAO-PROFOR, 2011). However, challenges ("the bad"), which oppose these principles, as well as unforeseen issues or "ugly" surprises, may ultimately undermine the forest governance transformation.

The chapter is structured as follows: the next section presents the different theories for understanding Forest Governance 4.0. It begins by revisiting established, contemporary and emerging conceptions of forest governance, such as the transition from top-down, state-centred models to multi-level and networked governance approaches, and the more recent discussions surrounding algorithmic and data-driven governance. Drawing on seminal contributions from public administration and digital government, the chapter reflects on the application of various theoretical concepts and frameworks in the limited field of digital forest governance. Building on these insights, it proposes an integrative framework for understanding research and policy trends. Through the lens of this framework, the emerging issues in forest and environmental governance in the Information Age are reviewed and a research agenda is proposed. In the final section, the chapter sets out a way forward, reflecting on the transformational potential of information for current and future forest governance research, management, and policy.

8.2 Theories for Understanding Forest Governance in the Information Age

8.2.1 Forest Governance: From Old to New to Newer Forms of Governance

Over the years, "old" conceptions of governance evolved from state-centric, e.g., the capacity of government to "steer society" (Pierre & Peters, 2000) to society-centric or "new" governance approaches referring to the coordination of collective action by public and private actors (Arts, 2021). This coordination is achieved through various policy instruments (Howlett, 2004) or different institutional arrangements and societal interactions. For example, taking a broader definition of forest governance, "all formal and informal, public and private institutional arrangements, the social interactions therein, and the effects of these on forests" can be understood as forest governance (Giessen & Buttoud, 2014). "New" governance thus implies that more horizontal modes of policymaking emerge (Arts, 2021).

In the forest sector, new modes of governance, such as decentralisation and privatisation, are believed to gradually enable the transition from top-down regulatory forest policies (Agrawal et al., 2008). This view is rooted in the idea that a shift from government to governance is underway, and that state bureaucracies are transitioning to a form of governance without government (Rhodes, 1996; Rosenau & Czempiel, 1992). Arguments in favour of this shift include the withdrawal or reduced influence of national governments as a direct result of globalisation as well as the emergence of neoliberalism and New Public Management (Arts, 2021; Van Tatenhove et al., 2000).

However, some scholars are sceptical about the idea of governance without government (Giessen et al., 2016; Howlett et al., 2009). A compelling counterexample is made by Giessen and colleagues (2016) who argue that forest certification is far from being a "non-state market-driven governance system" (Cashore, 2002) but rather a form of de facto co-governance between private and public bureaucracies (Giessen et al., 2016). Along the same lines, Howlett et al. (2009) point out that what makes "new governance" something really new and different is not always clear in the case of the forest-based sector. Sergent et al. (2018) reveal that, despite the emergence of network governance and new policy instruments in the European forest sector, state forestry still dominates. These scholarly debates demonstrate that, in practice, the lines between modes of governance become blurred, and often neither old nor new types of governance exclude each other, rather, they coexist (Zwitter & Hazenberg, 2020). Gradual changes in governance arrangements in the forest sector are driven by macro-political trends, such as global and European discourses and guidelines on sustainability and good governance (Sergent et al., 2018). Digitalisation is among these macro-trends.

The changes in environmental governance (and forests) are largely a product of technological developments that advanced information capabilities, particularly at the beginning of the 2000s. From satellite imagery and GIS to blockchain and open data platforms, these technologies have expanded the capacity of non-state actors to engage in governance, advocate for sustainable practices, and hold governments and corporations accountable. While macro-political trends, such as global sustainability discourses and good governance frameworks, provide the context, technological advancements have enabled these actors to shape and participate in contemporary governance processes actively (Duberry, 2019; Mol, 2008). The discussion around new or newer forms of governance is relevant as it may help to uncover ongoing debates and theories about the evolving role of governance in the Information Age.

8.2.2 Perspectives on Governance 4.0

8.2.2.1 Public Administration and Digital Government

Discussions about new governance, governance without government, and decentralised forms have been a focus of public management and digital government literature for some time (Dunleavy & Margetts, 2023; Gil-Garcia et al., 2018). As a broad research field, digital government emerged from multiple disciplines, including public administration, information science, management information systems, computer science, communication, and political science (Gil-Garcia et al., 2018). Various models and conceptual frameworks have gained prominence within this broad field (Dunleavy & Margetts, 2023). For example, Fountain (2004) introduced the concept of the "virtual state" to describe how governments utilise information technology to operate beyond traditional bureaucratic boundaries. Clift (2004) is often credited with early work on e-governance, as evidenced by his writings on "e-democracy" and the use of digital tools to enhance democratic engagement. Dunleavy and Margetts (2023) note that since then, other concepts, such as collaborative government (Gil-Garcia et al., 2018) and agile government (Mergel et al., 2021) have also gained prominence.

Following new technological developments, such as blockchain technology (Sethuramalingam, 2020; Zwitter & Hazenberg, 2020) or social media (Criado et al., 2013; Mergel, 2013), the discussion on the role of government and governance in the digital age has been reinvigorated. For example, in their analysis of blockchain technology, Zwitter and Hazenberg (2020) introduce the concept of "digital network governance", maintaining that current dominant modes of governance, whether "old" or "new", are inadequate in understanding governance in the digital realm. This is because the emerging new roles and power relationships in the digital domain are neither hierarchical nor horizontal. Instead, they are fluid, with different roles and power relationships often residing in a single, anonymous actor, as is the case with blockchain technology (Ibid). Following the introduction of social media in the mid-2000s, new frameworks have emerged to help us understand and measure the impact of these new forms of bidirectional, networked interactions between private actors and the government (Mergel, 2013) or to propose different dimensions (e. g., tools, goals, and topics) through which to understand the role of social media in government (Criado et al., 2013). During the COVID-19 pandemic, the limited capacity of governments to govern data and digital platforms became obvious (Mazzucato & Kattel, 2020) but eventually led to the realisation that such external pressures push governments and organisations to digitalise (Moser-Plautz & Schmidthuber, 2023). However, this raises new concerns and governance questions related to data privacy, ethics, and human rights (Alanoca et al., 2021; Beduschi, 2021).

One widely cited theory covering the long-running relationship between digital change and the administrative state is the "digital-era governance theory" (Dunleavy et al., 2006). It has emerged as a response to the limitations of New Public Management, emphasising reintegration, needs-based holism, and digitisation of administrative processes. This theory posits that developments in digital technology shape three foundational elements of public governance and state operations (digitalisation, vertical integration, and reorganisation of government structures to create integrated services that cut across traditional departmental boundaries). These changes requires a rethink of past public administration and public governance models of public bureaucracies (Dunleavy & Margetts, 2023; Dunleavy et al., 2006; Lytras et al., 2022). However, scholars point to the approach's over-emphasis on technological solutions without adequately addressing the socio-political dimensions of governance (Bellanova, 2016; Dawes, 2009). This includes the involvement of non-state actors (such as NGOs, civil society, private companies, or international organisations) in governance processes, which have become increasingly significant in digitalised administrative contexts. For example, Bellanova (2016) and Dawes (2009) argue that governance, especially in the digital era, involves complex interactions between state and non-state actors, focused not only on technological efficiencies but also on power dynamics, accountability, and participation in decision-making.

Along these lines of argument, Mol (2008) introduces the concept of informational governance, which challenges traditional, state-centric governance models. He suggests that environmental governance is no longer the exclusive domain of governments. Instead, it is shaped by a multi-actor system where NGOs, private companies, and civil society organisations use digital technologies to reconfigure power dynamics in favour of greater accountability and inclusivity. This shift in governance is evident in how NGOs now collaborate across borders, leveraging ICTs to campaign for environmental stewardship and encourage governments and corporations to adopt more sustainable practices. Complementing Mol's perspective is Paolo Gerbaudo's concept of digital activism (2012), which highlights the power of ICTs in mobilising civil society. Gerbaudo argues that digital technologies have transformed social movements, enabling civil society groups, including those advocating for environmental causes, to organise large-scale campaigns and mobilise supporters through online platforms. This digital activism aligns with Mol's informational governance but places more emphasis on the activist power of NGOs and civil society to influence corporate and governmental policies through digital advocacy. A relevant example is the *Fridays for Future* movement, where social media and digital platforms played a key role in spreading environmental awareness and mobilising millions of young activists worldwide to advocate for stronger climate policies.

Furthermore, Florini's theory of accountability networks (2007) examines how ICTs enable transparency and accountability in governance by creating global networks of actors—corporations, civil society, and governments—that monitor and report on environmental practices. Manuel Castells' network society theory (2010) further elaborates on this shift towards networked governance, where ICTs facilitate new forms of governance that diffuse power across multiple actors, including NGOs, civil society, and corporations. In Castells' view, governance is no longer dominated by traditional state hierarchies; instead, global networks of non-state actors collaborate through digital platforms to influence environmental policies and shape corporate behaviour. Jerome Duberry's work on global environmental governance in the digital age (2019) extends these ideas by introducing the concept of a digital civil society, where NGOs and grassroots movements leverage ICTs—such as social media, websites, and online petitions—to engage in environmental advocacy. Duberry highlights how digital tools enable smaller, grassroots organisations to mobilise public opinion and challenge larger corporations and governments on critical environmental issues, such as deforestation and biodiversity loss. The conceptual model of "digital citizen empowerment" (Sharma et al., 2022) complements this view, showing how ICTs enhance citizen engagement through digital activism, multi channel service delivery, and participatory governance. Citizens now have more opportunities to directly influence environmental governance by utilising digital platforms to voice their concerns, participate in policymaking, and hold governments and corporations accountable, for example, through participatory and deliberative online platforms.

Despite the different disciplines and schools of thought underlying these conceptual frameworks on governance in the Information Age, they all point to a shift from traditional state-centric models to more decentralised, multi-actor systems where technology plays a pivotal role. Different theoretical perspectives, such as informational governance, digital network governance, or network society theory, emphasise how digital tools—ranging from ICTs and blockchain to social media—have empowered non-state actors to participate actively in environmental governance. An overview of the various conceptual frameworks and their interrelationships is presented in Table 8.1.

8.2.2.2 Digital Governance in Forest Governance Research

Although empirical examples of different digital technologies and information management in the forest-based sector have increased in the literature (Nitoslawski et al., 2021; Scholz et al., 2018), surprisingly few studies address complex Forest Governance 4.0-related issues. Studies commonly mention concerns related to data governance and regulations (Nitoslawski et al., 2021) but often neglect to address more specific governance-related concerns. To date, the few existing forest governance contributions have focused primarily on information management, actor interests, and power struggles in the "political-" or "digital" forest (Gabrys et al., 2022).

An important conceptual contribution was made by Rantala and colleagues (2020), who examined the increasingly inseparable links between the governance of forests and the governance of information. By adopting an actor-centred institutionalist approach, they start with the premise that institutional structures influence the views considered in data policy and legal reforms. Rantala's proposed framework helps conceptualise how data and information link the governance of tangible natural resources, such as forests, with knowledge governance, uncovering different interests and power relations among actors and institutions that preside over data management

Table 8.1 Multiple conceptual strands in digital government and public administration are relevant for forest and environmental governance

Thematic focus	Conceptual strand	Key Ideas	Key references
Technological transformation of public administration	Digital-era governance (DEG)	Focuses on reintegration, needs-based holism, and digital transformation of service delivery	Dunleavy et al. (2006), Dunleavy and Margetts (2023), Lytras et al. (2022)
	Virtual state	Focuses on reshaping bureaucratic boundaries, technology integration and transforming public services	Fountain (2004)
	Agile government	Emphasises adaptability, iterative design, and co-creation in digital service delivery	Mergel et al. (2021)
Transparency and accountability	Accountability networks	Focuses on ICTs enabling transnational networks of monitoring actors (CSOs, corporations, states)	Florini (2007)
	Open government	Suggests a shift towards transparency and access to government data and processes	Gil-Garcia et al. (2018)
	E-Democracy	Early vision for ICTs improving democratic participation	Clift (2004)
Civil society empowerment and participation	Informational governance, digital civil society and digital citizen empowerment	Postulate power reshaped by access to information and describe cases of NGOs and civil society shaping environmental governance using ICTs, through, for example, digital activism and digital tools (budgeting, petitions)	Mol (2008), Gerbaudo (2012), Duberry (2019), Sharma et al. (2022)

(continued)

Table 8.1 (continued)

Thematic focus	Conceptual strand	Key Ideas	Key references
Networked and decentralised governance	Network society	Governance becomes decentralised and multi-actor, facilitated by ICTs	Castells (2010)
	Digital network governance	Blockchain and digital actors create anonymous, networked, non-hierarchical power structures	Zwitter and Hazenberg (2020)
Crisis-driven digitalisation and institutional change	Push to digitalisation under external pressures	Crises like COVID-19 accelerate digital governance, exposing institutional gaps	Moser-Plautz and Schmidthuber (2023), Mazzucato and Kattel (2020)

and forest management. Building on previous work, Rantala et al. (2022) adopt a rational choice and historical institutionalist approach to analyse the interlinkages between data governance and natural resource governance in Finland and Canada. This work lead the authors to suggest that data governance can reinforce the distribution of benefits according to path-dependent patterns in natural resource governance. However, it was also suggested that data governance may change the balance of power and redirect institutional pathways (Ibid.).

Building on Scharpf's (1997) actor-centred institutionalist approach, Baycheva-Merger and colleagues (2018) provide an overview of the most salient institutional and actor-oriented factors that constrain forest information exchanges at the national level across European countries. They found that expert-based forest information exchange is hindered by several closely interlinked institutional (e.g., data protection policies, inter-organisational information arrangements, organisational cultures, and a lack of incentives) and actor-oriented factors (e.g., distrust, diverging preferences and perceptions, intellectual property rights, and technical capabilities). Focusing on the EU forest information system, Baycheva-Merger and Sotirov (2020) expand the actor-centred institutionalist approach with the analytical concept of positive coordination, which highlights how actors work together to achieve collective goals (Scharpf, 1994). This angle enables them to uncover conflicts related to political authority, competence, and powerful interests, leading them to conclude that information is more than a neutral and objective form of scientific and technical expertise, it is an asset that can be strategically used for political gain (Baycheva-Merger & Sotirov, 2020).

A similar argument relating to the politicisation of forests through technology and information is made by Gabrys et al. (2022). In their critical analysis of the "political forest," referring to the actors, institutions, systems, and practices that operate within the context of neoliberal power, and influence over forests. They adopt a more holistic, "cosmopolitical approach", expanding the political to a broader array of (more-than-human) entities and considering different environmental and political relations and subjects that multiply ways of knowing, inhabiting, and cultivating forests (Ibid.). By focusing on digital technologies, approaches, and processes central to forest research, conservation, and management, Gabrys and colleagues identify observational technologies, datafication processes, participatory approaches, techniques for automation and optimisation, as well as regulation and transformation processes that call into question the "political" in "political forests" (see also Gabrys, 2020).

These contributions have significantly advanced the understanding of the emerging issues in digital forest governance. Despite the various epistemological and theoretical approached adopted in these studies, they all highlight similar concerns regarding the social and political impact of information technology on forests and forest governance worldwide. These include issues related to the values of different stakeholders and power struggles in an increasingly politicised forest sector, where access to technology and information can lead to power imbalances (Table 8.2). A more holistic framework is needed to consider technological change and political interests while also being aware of broader societal trends and human and institutional elements, given the complexity of the rapidly evolving digital landscape and its significant impact on how forests are constructed and governed.

8.2.3 FG4.0: A Framework for Understanding Research and Policy Trends in Forest Governance in the Information Age

Drawing on lessons from the broader literature on the intersection between public management and digital (environmental) governance, as well as existing contributions from forest governance in the Information Age-from actor-centred institutionalist perspectives and science, technology, and society scholarship-this section proposes a holistic framework for understanding forest governance in the Information Age, the "FG4.0 framework". This approach draws inspiration from Dawes's (2009) research and action framework for addressing complex and dynamic challenges in digital governance (see also Maxwell & Dawes, 2009). This approach distinguishes between six dimensions:

(i) "The purpose and role of government" refers to elements such as the need for a clear legal framework, distribution of power and responsibilities (like accountability, transparency and stewardship), and the impact of external influences. It also considers civil service capabilities, process efficiency, and organisational diversity.

(ii) "Societal trends" refers to demographic trends such as age distribution, birth rates, and migration, combined with globalisation, multiculturalism, privatisation, and institutional change. This category encompasses factors that influence

Table 8.2 Theoretical perspectives and key ideas in the forest governance 4.0 literature

Thematic focus	Conceptual strand	Key ideas	References
Data governance and institutional power	Actor-centered institutionalism	Forest and data governance are intertwined; institutional structures shape which actors' views are considered in policy and legal reforms. Actor preferences and institutional contexts mediate data use, influencing the outcomes of forest governance	Rantala et al. (2020, 2022), Scharpf (1997)
	Historical institutionalism	Data governance practices reinforce path dependencies in forest governance. Information flows can perpetuate or shift power balances among actors	Rantala et al. (2022)
Data governance and institutional power	Information as a political asset	Forest information is not neutral; it is shaped and strategically deployed by actors for political advantage. Institutional constraints and actor-level factors (e.g., trust, IP rights, technical capabilities) hinder exchange	Baycheva-Merger et al. (2018), Baycheva-Merger and Sotirov (2020)
	Positive coordination	Actors attempt to cooperate despite conflicting interests, authority, and institutional arrangements, highlighting tensions in EU forest information systems	Baycheva-Merger and Sotirov (2020), Scharpf (1994)
Techno-political critiques of digital forests	Political forest/ cosmopolitical approach	Digital tools reshape forest governance in neoliberal contexts. "Political forests" encompass a range of actors, institutions, technologies, and more-than-human entities. Digital technologies are transforming environmental governance and challenging dominant epistemologies	Gabrys et al. (2020, 2022)

political participation, including the digital divide rooted in economic conditions, social tensions, geographic mobility, global wealth, job availability, and labour distribution.

(iii) "Changing technologies" refers to all ICTs significantly impacting societies and governments, ranging from electronic devices to ubiquitous sensors capable of automatic data gathering, monitoring, and information exchange.

(iv) "Information management" comprises concerns related to the handling of metadata, ensuring the accuracy, quality, and integrity of information content in repositories like digital libraries and archives.
(v) "Human elements" refers to the factors associated with individuals, such as integrity of self, identity, autonomy, personal choice, privacy, trust, adjustment, learning, and acceptance of change.
(vi) "Cross-boundary interaction and complexity" cluster includes elements that highlight the challenges of crossing technical, organisational, institutional, and personal boundaries, such as distributed and multi-agent systems, cooperation, and competition.

The dimensions are interconnected and cover the dynamic interaction between human and technical developments as well as the values and institutions inherent to democracy (Dawes, 2009). However, an important supplement to Dawe's proposed conception of interpreting digital governance as a "dynamic, open socio-technical system" is necessary before delving deeper into each dimension. While this perspective is certainly valuable in guiding the understanding of societal trends and changing technology, it overlooks a crucial component for these endeavours, namely, the natural environment. Neglecting the environment or reducing it to a "technical system" risks perpetuating the technical optimism that digital technologies can maintain a separation from nature, attempting to establish environmental facts, govern land use, and preserve, conserve, or manage natural resources (Gabrys et al., 2022). The conversion of forest environments into state or private ownership generates political entities that act as an extension of state power (Ibid.). Therefore, it is crucial to maintain awareness of the fact that forest environments are entities in their own right, as well as political entities that extend state power, when conceptualising the governance of socio-technical interactions.

A more suitable inter-disciplinary approach strengthening the FG4.0 framework is a "Socio-Technical-Ecological Systems Approach (STES)", combining two fields: social-ecological systems (SES) and socio-technical systems (STS) (Ahlborg et al., 2019). According to Ahlborg and colleagues (2019), the primary rationale is that an accurate understanding of interlinked phenomena at the interface of human society and ecosystems is only possible if there is an understanding of the technological mediation of human–environment relationships. The commonalities and potential controversies of merging the two fields are beyond the scope of this discussion. However, it is sufficient to borrow some key ideas from the STES approach to gain a better understanding of the complex interlinkages between society, technology, and the environment. Most importantly, Ahlborg and colleagues' approach helps frame digital technology as a mediator between humans and the natural world. Similar to previous conceptualisations, it raises awareness that technology can produce ambivalent outcomes. However, it also warns that technology can transform power relations, give rise to systemic pressures, and change interactions with and impacts on the natural world (Ibid). The FG4.0 framework describes and expands on these dimensions by adapting them to specific challenges related to digital forest governance, technology, and information management in forestry. Figure 8.1 brings these perspectives together, expanding on the dimensions and their co-construction of the dynamic, open, socio-technical-ecological system.

Against this conceptual backdrop, the following six dimensions emerge as particularly relevant digital forest governance:

1. **Purpose and role of government**

This dimension relates to the role of government in the Information or Digital Age, whether governance without government (Rhodes, 1996; Rosenau & Czempiel, 1992) truly exists in forest governance, and whether other types of government,

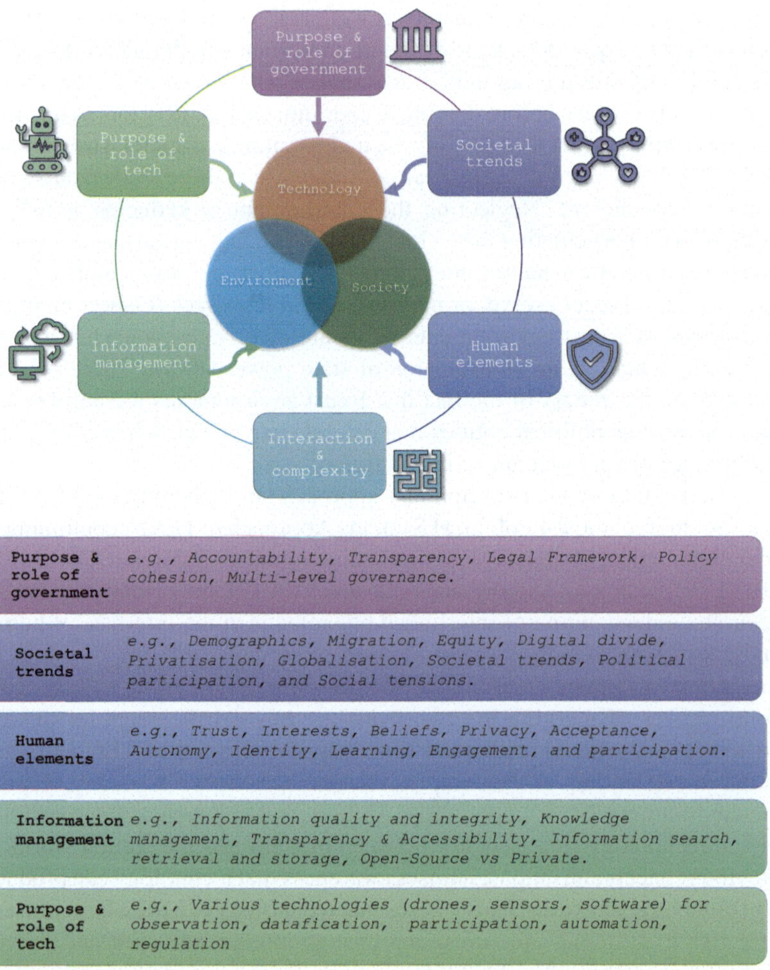

Fig. 8.1 Forest governance 4.0 as a dynamic, open, socio-technical-ecological system. Adapted after Dawes (2009)

i.e., open, collaborative, or agile (Dunleavy & Margetts, 2023; Gil-Garcia et al., 2018; Mergel et al., 2021) can emerge in the Information Age. It relates to the different modes of governance (new and old) and to the existing legal, international, and national frameworks, as well as coherent and integrated cross-sectoral policies that directly or indirectly affect forests. Additionally, this dimension emphasise the distribution and exercise of governmental power (e.g., in the case of data management for state-owned or privately-owned forests) as well as raises questions related to accountability and transparency in collecting, using, and sharing forest-related information at local, national or international levels (Baycheva-Merger & Sotirov, 2020). Lastly, it also considers the efficiency and effectiveness of processes and practices that directly impact forest governance.

2. **Purpose and role of technologies**

As the pace of digitalisation advances at a breathtaking speed, new technologies emerge that significantly impact forests governance. Instead of listing existing and upcoming technologies, the chapter borrows from Gabrys et al. (2022) to differentiate between the purpose and roles of these technologies, which may range between:

- observation (e.g., satellite observation and remote sensing, technologies for monitoring and surveillance, and pattern detection).
- datafication (e.g., software technologies for data processing, and mathematical and statistical models for environmental phenomena).
- participation in decision making and monitoring (e.g., mobile technologies, online platforms, and Geographical Information Systems (GIS).
- other data collection platforms enabling data collection and citizen science, or social media platforms expanding the actors, networks, and practices of forest engagement).
- automation (e.g., digitalised environmental data collection, processing, and validation for fast and cost-effective procedures meant to decrease dependency on human labour, e.g., wireless sensor networks, machine learning, airborne and terrestrial UAVs used to quantify and analyse forest dynamics).
- regulation (e.g., technologies and software oriented toward forms of detection and prediction, such as forest loss, land use changes, carbon storage, and illegal logging tracking). Examples include AI and machine learning systems processing satellite imagery, composing renderings of changing land use, and triggering alerts while informing policy (Gabrys et al., 2022).

3. **Information management**

Related to what type of information is produced by digital technologies and how it is managed, this dimension covers a wide range of tools and practices for open/private access, data storage, and retrieval. It also connects to concerns about the quality and integrity of generated data and how this knowledge is being managed. Furthermore, it relates to broader concerns about the transparency, sharing, and accessibility of data. The EU Forest Information System (FISE) is a good example of an information management initiative, bringing together data, information, and knowledge gathered or derived through key forest-related policy drivers (Baycheva-Merger et al., 2018; Baycheva-Merger & Sotirov, 2020). It is an integrated platform designed to collect, harmonise, and disseminate comprehensive forest-related data across the EU. It facilitates data-driven policymaking by providing insights into forest conditions, biodiversity, carbon stocks, and forest management practices. However, as highlighted by Baycheva-Merger and Sotirov (2020), it also faces significant challenges related to data sharing and sovereignty issues among member states. Many EU countries exhibit reluctance to fully share their national forest data with the European Commission due to concerns over data control, national forest policy autonomy, and perceived risks to national interests

4. **Societal trends**

Societal trends (e.g., demography, age structures, migration, changing values and digital literacy) influence technological adoption, interaction, and their impact on forest governance. Demographic trends, such as changing age structures and rural-to-urban migration, play a crucial role. In many regions, land abandonment and migration from rural areas to cities are on the rise, altering how forest lands are managed (Giurca et al., 2023; Zgłobicki et al., 2020). Urban populations, with differing environmental values and concerns, shape new demands for forest governance, often as a result of globalisation and privatisation. These trends influence the way technology is utilised in forest management, ranging from passive ownership to recreational engagement (Nordlund et al., 2017; Racevskis & Lupi, 2006). While urban dwellers tend to have higher digital literacy, rural communities may face digital divides, which can hinder equitable political participation in (forest) governance (Roberts et al., 2017; Young, 2019).

5. **Human elements**

This dimension covers human elements such as trust, interests, beliefs, privacy, acceptance, autonomy, learning, participation and engagement. For instance, research on forest governance in the Information Age highlights conflicts between the right to information and the right to privacy, alongside the interests of stakeholders with established roles (and power) in forest governance (Rantala et al., 2020). The role of trust-building and collective action in data governance is central to these endeavours (Ibid.). Different public and private forest values are present in forest governance approaches in different countries (Rantala et al., 2022). In some cases, actors exchange forest information primarily within a trusted group of actors, such as partners, member organisations, or organisations with shared interests or belief systems (Baycheva-Merger et al., 2018).

6. **Interaction and complexity**

This dimension encompasses the complexity of crossing personal, institutional and organisational boundaries. More broadly, it refers to the increasingly inextricable interactions between the human, technological, and natural worlds. Different lenses, such as the Socio-Technical-Ecological system (Ahlborg et al., 2019), may help make sense of this complexity. Other approaches, such as viewing forests as "cosmopolitical" worlds, also contribute to this complexity (Gabrys et al., 2022). "More-than-human" entities—such as beings, forces, and phenomena that go beyond the traditional Western concept of "human" but have agency, consciousness, or spiritual importance in environmental cosmologies (O'Gorman & Gaynor, 2020)—and diverse relationships with the land, as perceived by local and Indigenous communities, can also be considered integral to the complexity surrounding digital technologies, land tenure and intangible or immeasurable forest values and services (Westerlaken et al., 2023).

8.3 Implications for Research and Policy Through the Lens of FG4.0: A Research Agenda

The integration of digital technologies into forest governance raises several questions for research and policy, requiring a deeper understanding of the complex and dynamic factors at play. Each of the six dimensions previously outlined, ranging from technological adoption to human elements, offers unique entry points for future research. The complexity inherent in these dimensions presents a rich field of inquiry, potentially offering insights that can shape both academic understanding and practical governance models. Table 8.3 provides an overview of these dimensions, highlighting emerging topics, conceptual approaches, and research questions that can guide scholars and policymakers in exploring the good, the bad, and the ugly aspects of Forest Governance 4.0.

The integration of ICTs into environmental governance has revolutionised how data is collected, shared and utilised (Kloppenburg et al., 2022; Mol, 2008). Advanced digital technologies have significantly improved transparency, accountability, and real-time monitoring of environmental practices. However, these advancements come with significant challenges; some are already known, and some are still unknown or emerging. Concerns over data quality, power asymmetries, algorithmic bias, and surveillance have emerged, shaping the impact of ICTs on governance. As ICTs continue to evolve, understanding these challenges is crucial for ensuring that these tools foster equitable and effective governance. By examining existing literature on forest and environmental governance through the lens of the FG4.0 framework, this section explores both the good, ugly, and bad aspects of forest governance in the Information Age (see Table 8.3).

Table 8.3 Overview of the FG4.0, emerging topics, conceptual approaches, and examples of research questions (adapted from Dawes, 2009)

Dimension	Emerging topic(s)	Examples of conceptual approach(es)	Examples of research questions
Purpose and role of government	• Digital forest governance with or without government • Efficiency and effectiveness of processes and practices that directly impact the governance of forests	*(Digital) Network Governance*Zwitter and Hazenberg (2020), Multi-level governance Bache et al. (2016) *Institutional change*Kingston and Caballero (2009) *Digital activism*Gerbaudo (2012) Accountability networks Florini (2007)	• How do agile governance models facilitate the integration of new technologies in forest management across different governance levels? • Why do some governments adopt open and collaborative forest governance models while others resist these changes? • Who are the key actors and institutions driving policy innovation in forest governance, and how do they interact with digital platforms for public participation? • What are the risks of depoliticisation in forest governance due to the reliance on blockchain, AI and algorithmic decision-making? • What role should regulation play in ensuring transparency and accountability in the use of digital technologies for forest governance?

(continued)

Table 8.3 (continued)

Dimension	Emerging topic(s)	Examples of conceptual approach(es)	Examples of research questions
Purpose and role of technologies	• Purpose and role of different existing and emerging technologies concerning observation, datafication, automation, participation and regulation	Socio-technical-ecological systems Ahlborg et al. (2019), Cosmopolitical forests Biemann (2015), More-than-human entities Gabrys et al. (2022) Informational governance Mol (2008) Network society theory Castells (2010) Digital civil society Duberry (2019) Digital citizen empowerment Sharma et al. (2022)	• How do diverse observation practices reshape forest governance? • How do new data practices contribute to or support existing forest governance arrangements while reducing or omitting other non-computable forest experiences? • How can digital technologies facilitate engagement and participation of different stakeholder groups? • What are the potential pitfalls of increased digital observation, automation and regulation on forest governance? • Are there any patterns of coloniality? How does the "colonisation of private life" through digital technologies affect public trust and engagement in environmental governance? • What are the ethical implications of the panoptic nature of digital forest governance? • How and why are certain technological solutions featured and mainstreamed, while others are marginalised?
Information management	• Information quality, reliability and integrity • Transparency and accessibility in forest data • Open versus private data access	Data governance framework Abraham et al. (2019) *Knowledge commons* Frischmann et al. (2014)	• Why do certain stakeholders prioritise proprietary data management over open data approaches, and what are the implications for management and governance? • How do decentralised data platforms (e.g., blockchain) impact the transparency and accountability of forest management practices, and how do these affect forest governance?

(continued)

Table 8.3 (continued)

Dimension	Emerging topic(s)	Examples of conceptual approach(es)	Examples of research questions
Societal trends	• Digital divides equitable participation in rural vs. urban forest governance • Political participation, participatory practices and engagement	Socio-technical-ecological systems Ahlborg et al. (2019), Digital civil society Duberry (2019) Digital citizen empowerment Sharma et al. (2022)	• What are the potential consequences of digital divides for equitable participation in forest governance? • How do urbanisation and migration patterns influence public attitudes towards forest conservation and land management? • What are the most effective ways to support smaller organisations or those with limited financial resources in adopting digital technologies for environmental governance? • How can the digital divide between well-funded organisations and smaller entities be narrowed, ensuring that resource constraints do not hinder their participation in digital governance? • What different types of participatory practices and other forms of engagement result from the use of digital technologies and information sharing? • What strategies can be implemented to address the skills divide and enhance digital literacy for underrepresented communities? • What role can international organisations, the private sector and governments play in closing the digital divide and ensuring that digital tools are used to promote equitable, participatory forest governance?

(continued)

Table 8.3 (continued)

Dimension	Emerging topic(s)	Examples of conceptual approach(es)	Examples of research questions
Human elements	• Trust, privacy, and autonomy in digital forest governance • Beliefs, identity, ideas, discourses and interests	Forest information governance Rantala et al. (2020) Actor-centred institutionalist approaches Baycheva-Merger et al. (2018)	• How do concerns about privacy and trust affect stakeholder participation in digital platforms for forest management? • Who decides which types of digital tools or platforms are implemented for public engagement in forest governance, and how do these decisions affect autonomy and equity among stakeholders? • How can the social and cultural resistance to digital technologies, particularly the neo-Luddite attitudes that see them as dehumanising or harmful to societal values, be addressed?
Interaction and complexity	• Cross-sectoral and cross-boundary collaboration is enabled through digital forest governance • More-than-human entities • Plural knowledge practices	Socio-technical-ecological systems Ahlborg et al. (2019) Forests as political entities, or the cosmopolitics of socio-technical worlds Gabrys et al. (2022)	• How does the complexity of cross-boundary collaboration influence the effectiveness of digital tools in forest governance, especially in regions with diverse ecological and social conditions? • What are the opportunities and pitfalls of digital technologies when it comes to integrating other forest cosmologies, indigenous or plural knowledge practices?

8.3.1 Data Governance

The integration of digitalisation in forest and environmental governance offers significant potential to enhance data-driven decision-making, but it also brings forth critical challenges related to data governance. While there is no single definition of data governance, it can be understood as the frameworks, policies, roles, responsibilities, procedures, standards or rules that govern how digital data is managed, but also how it is shared, accessed, and used across organisations (Abraham et al., 2019; Davidson et al., 2023). As advanced digital tools facilitate more open and transparent data sharing and access, they also introduce several pressing challenges. Linked to multiple interrelated dimensions described in the FG4.0 framework, e.g., information management or human elements, several concerns over data privacy, security, and ownership and management of data arise with digital governance (Dhagarra et al., 2020; Gebremeskel et al., 2023; Quach et al., 2022; Westerlund et al., 2021). In the environmental context, vast amounts of data collected through digital technologies, such as satellite imagery and IoT sensors, can expose sensitive information about natural resources and land ownership, thereby raising risks of misuse and privacy breaches (Dillon et al., 2019). Moreover, the question of who owns and controls this data creates tensions, as environmental data often involves intersecting public, private, and community interests, making it challenging to balance transparency with proprietary claims (Reyes-García et al., 2022).

Additionally, inconsistent data quality and the lack of standardisation hinder effective cross-sector collaboration and decision-making (Dillon et al., 2019; Gaffoor et al., 2020). These issues often lead to policy inefficiencies due to the use of inaccurate or incomplete datasets. System interoperability, where different platforms struggle to communicate across jurisdictions and sectors, further complicates the integration of environmental data. Ethical concerns also arise with the increasing use of AI and machine learning, which can embed biases into decision-making processes, disproportionately impacting marginalised communities and exacerbating existing environmental injustices (see the following section).

8.3.2 Accountability and Transparency

Digital technologies and information-sharing mechanisms have undoubtedly enhanced transparency in environmental governance (Gupta et al., 2020; Kloppenburg et al., 2022), providing real-time data on critical issues like deforestation, carbon emissions, and biodiversity loss. The integration of digital transparency in environmental governance significantly empowers both state and non-state actors by providing timely and accurate data on critical issues, such as deforestation and emissions. This supports accountability, effective decision-making, and collaboration. For state actors, such as government agencies, real-time access to environmental data enables proactive monitoring and enforcement. Tools such as satellite imagery and

IoT sensors allow authorities and civil society to address illegal logging or pollution problems swiftly. For example, an electronic monitoring of wood traceability system, SUMAL, is used in Romania to prevent illegal logging and deforestation (Dogaru, 2024). NGOs use open platforms like Global Forest Watch to monitor deforestation, mobilise public opinion, and drive policy change. In the private sector, transparent data enables sustainable supply chain management, as companies track raw materials to ensure compliance with green standards, thereby building trust with consumers and investors. Local communities and indigenous groups gain a vital tool for protecting their lands and resources. Access to environmental data empowers them to document illegal activities and present evidence for defending their land rights. The inclusion of local knowledge into broader governance frameworks enhances equity and promotes collaborative environmental management (Mol, 2008; Duberry, 2019).

However, while transparency is often viewed as a tool to increase accountability and democratise decision-making, it presents significant challenges related to power dynamics, governance, and equity. One major issue is that transparency does not automatically lead to better outcomes (Gupta, 2008; Gupta et al., 2020). The availability of vast amounts of environmental data often leads to conflicting interpretations, as stakeholders such as governments, NGOs, and corporations utilise data to support their respective agendas. This can stall governance processes and prevent effective environmental interventions. Additionally, the fragmented nature of environmental data complicates consensus-building efforts, further preventing effective governance.

Another central issue in data transparency is that it is rarely neutral. Control over data generation and dissemination often rests with powerful state or corporate actors, who decide what information is collected and made public. This "selective transparency" (Birchall, 2011) allows them to curate data disclosures, projecting accountability while potentially concealing environmental mismanagement or misconduct. Gupta et al. (2020) further highlight transparency issues in digital sustainability governance, noting that selective data disclosure, proprietary algorithms, and restricted data ownership by private companies create barriers to understanding, limit public insight, and obstruct independent verification. The diffusion of accountability across multiple actors also complicates the process of holding anyone responsible, weakening public trust. Additionally, digital inequalities prevent equitable access to data, with under-resourced communities disproportionately affected, thereby undermining inclusivity in environmental governance.

The lack of clarity in algorithmic decision-making presents additional risks. As AI becomes central to environmental governance—whether in predicting climate risks, managing resources, or tracking emissions—responsibility for poor decisions grows unclear (Scoville et al., 2021; Tironi & Rivera Lisboa, 2023). When algorithms result in adverse outcomes, it can be challenging to assign accountability among technology developers, data providers, or policymakers. Algorithmic bias is a significant concern, as these systems often mirror the biases of their creators and reinforce existing inequalities (Ziaja, 2021). This is particularly problematic in forest and environmental governance, where marginalised communities, especially in the Global South, may bear the brunt of policies shaped by biased algorithms. Several dimensions of the FG4.0 framework can critically address these issues, including the government's purpose and role, societal trends, and information management.

8.3.3 The Adoption and Use of Technologies

The digital transformation of environmental governance is far more than a straightforward technical procedure. As suggested by the interaction and complexity dimension of the FG4.0 framework, digital transformation is a complex, multi-dimensional process that involves significant organisational, social, and economic shifts (Hinings et al., 2018). While digital technologies offer tremendous potential for improving decision-making, enhancing transparency, and addressing environmental challenges, their adoption and use are often met with reluctance by organisations and stakeholders alike. This hesitancy stems from the profound changes that digital transformation induces in organisational structures, job roles, and societal norms (Dąbrowska et al., 2022; Hinings et al., 2018).

One of the main barriers to adopting digital technologies in environmental governance is the significant transformation they require within organisations. Implementing these technologies is not just about upgrading tools or systems, it often requires a fundamental reorganisation of workflows, communication channels, and decision-making processes (Brous et al., 2020; Baycheva-Merger et al., 2024). For example, the integration of AI or big data analytics may centralise decision-making in technical departments, diminishing the influence of traditional departments that rely on manual processes or human expertise. This shift can cause resistance from employees who may fear losing their jobs or seeing their roles diminished in the face of automation. Moreover, organisations need substantial financial resources to undertake this transformation. Investing in advanced digital infrastructures, training staff, and maintaining up-to-date technology is costly. Smaller organisations or those with limited budgets may struggle to keep pace, widening the gap between well-funded entities and those lacking the means to adopt these technologies effectively. Additionally, the ongoing need for technical expertise presents another hurdle, as organisations must recruit or retrain employees to manage and optimise digital systems (Baycheva-Merger et al., 2024).

Beyond organisational barriers, the digital transformation of forest and environmental governance faces challenges related to broader social acceptance (Kumar et al., 2022; Papagiannidis & Marikyan, 2022; Westerlund et al., 2021). As suggested by the human elements dimension of the FG4.0 framework, society is not homogeneous. Different actors hold varied interests, values, and beliefs about the use of technology. This divergence often manifests in what has been described as a "neo-Luddite" attitude (Mol, 2008), referring to a scepticism or outright opposition to the use of technology. Some people view digital technologies, especially those involving automation and AI, as dehumanising or potentially harmful to societal values. Moreover, there are competing values and beliefs about the role of technology in addressing environmental issues. For example, environmental activists might distrust corporate-controlled digital platforms, fearing that these tools could prioritise profit over genuine ecological sustainability. Others may be concerned that data-driven technologies oversimplify complex environmental issues by reducing them to numbers and algorithms, potentially overlooking important social or ethical dimensions (Duberry, 2019).

8.3.4 Participation and Engagement

Digital technologies allow local communities, NGOs, and individuals to contribute to and influence decision-making processes that impact environmental or management outcomes. However, while digital tools can bridge gaps, they also introduce new challenges, including disparities in access, digital literacy, and the practical benefits derived from their use. As indicated by the two dimensions of FG4.0 on societal trends and human elements, these issues may lead to a "digital divide" (Massimo & Muschert, 2017), which manifests most fundamentally in unequal physical access to digital tools and infrastructure. Well-funded organisations, can fully leverage sophisticated tools like satellite imagery, environmental sensors, and real-time data platforms. In contrast, smaller organisations—particularly those in rural areas often lack digital literacy, and basic digital infrastructure, such as internet access. In this context, small-holders, local NGOs and indigenous communities frequently struggle to access technology essential for meaningful participation in forest and environmental governance, resulting in their exclusion from decision-making processes that directly affect their lives and environments. Without the necessary tools, these groups are often unable to contribute to or question the data that informs forest and environmental policy (Gain & Fritsch, 2024).

Even when physical access to digital tools is achieved, a persistent skills gap remains. This divide highlights differences in digital literacy (Denfanapapol et al., 2024; Ferreira et al., 2021)—the ability to effectively use technology and interpret the data it produces. Many communities, particularly those in underserved regions, lack the technical expertise to interpret complex environmental data, leading to tokenistic participation. In such cases, marginalised groups may be included in governance processes in name only, as they lack the capacity to engage with the digital tools available to them fully. Consequently, decision-making remains dominated by actors with the expertise to fully exploit digital technologies, further reinforcing existing power imbalances.

The third level of the digital divide concerns the unequal benefits derived from digital engagement. Even when marginalised groups have access to digital tools and possess the necessary skills, they may not experience the same tangible advantages as larger, more powerful organisations. This disparity in benefits reinforces global inequalities in environmental governance, highlighting the need for more inclusive digital strategies that support meaningful participation from all sectors of society (Denfanapapol et al., 2024).

8.3.5 Surveillance, Privacy, and Ethical Concerns

Digital technologies are often viewed as neutral, objective tools that deliver critical data for decision-making. Governments, corporations, and environmental organisations utilise remote sensing, AI, and big data analytics to gather insights that inform policy, optimise resources, and curb environmental degradation. This technocratic framing presents digital technologies as enhancing governance by enabling efficient data collection and dissemination (Bakker & Ritts, 2018; Dillon et al., 2019). However, while these tools enhance the ability to monitor and address environmental issues, they also raise concerns about surveillance, control, and encroachments on personal privacy (Kloppenburg et al., 2022, Mol, 2008).

Mol (2008) explores the concept of *panopticism*, arguing that digital tools, though intended to increase transparency, can serve as *instruments of surveillance*. Panopticism, a concept developed by Michel Foucault, refers to a system of power in which people modify their behaviour due to the awareness of being observed, even if they cannot see their observers. In forest and environmental governance, digital tools create this "panoptic" structure, where governments, corporations, and even citizens are continuously monitored by those controlling the data (Mol, 2008). Here, digital technologies do more than offer data; they monitor and regulate behaviour. Technologies such as satellite tracking, sensor networks, and data platforms, which aim to increase transparency in forest management, can produce panoptic effects, causing forest-dependent communities, indigenous peoples, and local stakeholders to modify their traditional practices due to ongoing digital surveillance. These technologies could result in indigenous and local communities losing control over information about their territories and practices (Kukutai & Taylor, 2016).

Additionally, forest and environmental governance increasingly rely on data collected from individuals, often through personal devices such as smartphones, smart meters, or home sensors. Although intended to encourage sustainable behaviours, these technologies collect extensive personal data, tracking energy consumption, travel habits, water use, and land management activities. Shoshana Zuboff (2019) coins the economic system centered on the pervasive tracking, collection, and commodification of personal data by corporations to predict and influence human behavior as "*surveillance capitalism*". Mol describes this as the "*colonisation*" of private life, where personal choices are subject to external oversight. For example, drones are increasingly used to monitor forest-related activities, such as logging, grazing, or forest clearing, in real-time. While beneficial for overseeing conservation efforts, these drones may inadvertently capture data on individuals, such as their presence in specific forest areas, the types of resources they gather, or their land use patterns, creating an external record of personal activities tied to forest resources (Buchelt et al., 2024). This extension of surveillance into private spaces raises ethical questions: Who has access to this data, and how is it used? The commodification of personal data, especially when handled by private companies, blurs the line between governance and surveillance, with implications for individual autonomy and freedom.

Table 8.4 Key issue areas in forest governance seen through the FG4.0 framework

Issue area	Main issues to be considered	Forest governance 4.0 dimensions
Data governance	• Ambiguity over data ownership and control • Risks of data misuse and privacy breaches • Lack of standardisation and interoperability	• Information management • Human elements • Purpose and role of technologies
Accountability and transparency	• Enhanced monitoring via ICTs empowers state and non-state actors • Risk of "selective transparency" by powerful actors • Data used to support conflicting agendas • Algorithmic opacity and inequity	• Purpose and role of government • Information management • Societal trends
Technology adoption and organisational change	• Organisational resistance to digital transformation • Disruption of internal power dynamics • Cost and capacity barriers • Varied societal values toward digital tech	• Interaction and complexity • Human elements • Purpose and role of technologies
Participation and digital divide	• Unequal access to digital tools • Gaps in digital literacy • Unequal benefits from digital participation • Risk of tokenism and exclusion of marginalised communities	• Societal trends • Human elements • Information management
Surveillance and Ethics	• Expansion of monitoring into private life • Commodification of personal environmental data • Ethical concerns over consent, autonomy, and oversight	• Purpose and role of government • Purpose and role of technologies • Human elements

Finally, Mol also discusses *synopticism*, where digital tools enable citizens to monitor powerful entities. Platforms like Global Forest Watch enable real-time public monitoring of environmental offenders. While this "bottom-up" surveillance empowers citizens to hold actors accountable, it also introduces complexities. Public reporting of environmental violations can foster a culture of surveillance, potentially pitting communities against each other. Furthermore, synopticism does not entirely shift power dynamics. Although citizens can monitor environmental harm, they often lack the resources to effect change, as the power to act primarily rests with governments and corporations (Mol, 2008) (Table 8.4).

8.4 The Way Forward

Exploring the intersection of digital technologies and forest governance is vital, as it offers new ways to understand and address the complex challenges forests face in a rapidly changing world. By examining how information enables unprecedented precision in tracking forest health, biodiversity, carbon storage, and illegal activities, scholars can deepen their understanding of information's role in shaping governance mechanisms that are both adaptive and resilient. Understanding these dynamics is crucial to designing future forest governance that is both inclusive and technologically equipped to respond to global environmental challenges.

The previous paragraph was generated with AI. In contrast, the rest of this chapter was not. It contains thoughts and ideas, distilled into our own words, informed by critical thinking, which is shaped by our disciplinary backgrounds and theoretical schools of thought. This must not be taken for granted. Soon, critical thinking may become the exception rather than the norm. Information is increasingly being collected, produced, framed, and reframed within algorithmic "black boxes" that are beyond our access, comprehension, and control. Bearing in mind, the purpose of this chapter is to raise awareness of the importance of critical, interdisciplinary thinking when analysing the role of information technology in how people understand, manage, and interact with the natural environment.

In the rapidly changing Information Age, forest and environmental governance is moving beyond conventional "old" and "new" models, instead adopting a more horizontal, open and interconnected approach. Thus, Information Age governance involves complex interactions between state and non-state actors, focusing not only on technological efficiencies, but also on power dynamics, accountability and participation in decision-making. A holistic, interdisciplinary approach is necessary to better understand the complex relationships between society, technology, information and the natural environment. The FG4.0 framework proposed in this chapter achieves this. It introduces a comprehensive, interdisciplinary approach to understanding the interconnected elements at play. This contribution can be understood as a guide to comprehensiveness, helping to make sense of the growing body of research in this area, as well as identifying concerns that require increased research and policy attention. It is hoped that this will serve as a starting point for academic and policy debates on the future of good forest governance in the context of the Information Age.

Against this background, it is time to reconsider the assumption in the title of this chapter that Forest Governance 4.0 could have positive, negative and unknown implications. Each of the five topics discussed under the implications for research and policy has its own opportunities and pitfalls. In terms of data governance, this chapter argues that digitalisation in forest and environmental governance can improve data-driven decision-making. However, data governance also raises concerns relating to misuse, privacy breaches, and data ownership (Dillon et al., 2019). Similarly, while transparency certainly increases accountability and democratises decision-making, it also poses significant challenges regarding power dynamics and equity (Gupta,

2008; Gupta et al., 2020). Although technologies offer opportunities to improve transparency and support decision-making, they are often met with reluctance by stakeholders. Indeed, one of the main challenges to adopting digital technologies in forest and environmental governance is the substantial organisational change they entail (Baycheva-Merger et al., 2024). Participation and engagement are crucial for enabling this change, and digital technologies certainly facilitate stakeholders' contributions to and influence on decision-making processes. However, digital technologies present a series of new challenges, such as access barriers and limited digital literacy, perpetuating the growing "digital divide" (Massimo & Muschert, 2017). While access to digital tools and data can facilitate decision-making, concerns are raised about surveillance, control, and encroachment on personal privacy (Kloppenburg et al., 2022; Mol, 2008).

These are just a few of the issues that emerged from this brief analysis of the literature. Every opportunity that digital technologies and information bring to governance is met with a challenge. Some of these challenges can be anticipated. However, the unknown unknowns, or 'ugly' challenges, remain. While it is beyond the scope of this contribution to conduct foresight exercises to predict the future of technology, areas that require further critical analysis can be identified. One underlying assumption observed throughout the literature is the techno-optimistic view that greater access to technology and information leads to greater sustainability, particularly in the forestry sector (Damaševičius et al., 2024). This way of thinking is rooted in the neoliberalisation of nature (Castree, 2008), involving various processes, some of which are contradictory, such as land privatisation, marketisation, deregulation, reregulation, liberalisation of resource trade, and competitiveness in resource markets (Birch et al., 2010). Such rationales are particularly prevalent in techno-scientific visions for expanding the bioeconomy, driven by global market demands that rely on the increased consumption of natural resources (Birch, 2006; Giurca & Befort, 2023; Holmgren et al., 2020). The literature is still scant on the potential negative impacts of ICTs on environmental sustainability is still scant. For instance, studies on urbanisation indicate that economic growth driven by technological advancements frequently results in heightened environmental degradation due to increased resource consumption and pollution (Ahmad et al., 2022; Ahmed et al., 2021). The warnings outlined here, alongside those of other scholars, emphasise the substantial sociopolitical risks associated with digitised forest environments. Notably, there is a risk of exacerbating existing environmental injustices and introducing new ones (Gabrys et al., 2022). Despite technology and ICTs being increasingly integrated into forest management, the potentially adverse impacts of this integration on the sustainability of forest management remain largely unexplored in the literature.

As this book announces the transformative potential of forest governance, it is worth reflecting on the potential of information, particularly in relation to forests, as a transformative resource. The transformational potential of technologies and ICTs for forest and environmental governance must be questioned. Digital transformation refers to the unprecedented disruption to society, industry and organisations caused by advances in digital technology (Feroz et al., 2021). However, this transformation encompasses more than technological change; it has profound implications for power dynamics, social equity, and environmental justice. Specifically, sustainability transformations specifically refer to fundamental, system-wide reorganisations across technological, economic, and social factors (Díaz et al., 2019). They require more than the adoption of technology. They demand shifts in values, institutional arrangements and inclusivity frameworks in order to accommodate the diversity of forest-dependent stakeholders, including indigenous communities and local populations. Both types of transformation are based on the principles of inclusivity, equity, and fairness towards the environment and the stakeholders whose livelihoods depend on it. However, inclusivity in forest governance must extend beyond mere access to data or digital platforms, empowering marginalised voices to influence decision-making processes. Transformational forest governance should go beyond digitalisation and information governance to be based on guiding principles such as participation, representation, empowerment, transparency, accountability, justice, equality, and adaptive learning (Larson et al., 2021). For instance, it is important to recognise that, although they are promising, digital tools can sometimes exacerbate existing disparities in access to resources, training and infrastructure. Approaches to forest governance must be critical and inclusive, evolving in tandem with digital tools rather than following them. The FG4.0 framework supports this goal.

References

Abraham, R., Schneider, J., & vom Brocke, J. (2019). Data governance: A conceptual framework, structured review, and research agenda. *International Journal of Information Management, 49*, 424–438. https://doi.org/10.1016/J.IJINFOMGT.2019.07.008

Agrawal, A., Chhatre, A., & Hardin, R. (2008). Changing governance of the world's forests. *Science, 320*(5882), 1460–1462. https://doi.org/10.1126/SCIENCE.1155369/SUPPL_FILE/AGRAWAL_SOM.PDF

Ahlborg, H., Ruiz-Mercado, I., Molander, S., & Masera, O. (2019). Bringing technology into social-ecological systems research—Motivations for a socio-technical-ecological systems approach. *Sustainability, 11*(7). https://doi.org/10.3390/SU11072009

Ahmad, M., Ahmed, Z., Gavurova, B., & Oláh, J. (2022). Financial risk, renewable energy technology budgets, and environmental sustainability: Is going green possible? *Frontiers in Environmental Science, 10*, Article 909190. https://doi.org/10.3389/FENVS.2022.909190/BIBTEX

Ahmed, Z., Nathaniel, S. P., & Shahbaz, M. (2021). The criticality of information and communication technology and human capital in environmental sustainability: Evidence from Latin American and Caribbean countries. *Journal of Cleaner Production, 286*, Article 125529. https://doi.org/10.1016/J.JCLEPRO.2020.125529

Alanoca, S., Guetta-Jeanrenaud, N., Ferrari, I., Weinberg, N., Buse, R., & Miailhe, N. (2021). Digital contact tracing against COVID-19: a governance framework to build trust Global context. *International Data Privacy Law, 11*. https://edition.cnn.com/2020/07/01/americas/latin-

Arts, B. (2012). Forests policy analysis and theory use: Overview and trends. *Forest Policy and Economics, 16*, 7–13. https://doi.org/10.1016/j.forpol.2010.12.003

Arts, B. (2021). Forest governance: Hydra or Chloris? In F. Biermann & A. Gupta (Eds.), *Elements in earth system governance*. Cambridge University Press. https://doi.org/10.1017/9781108863551

Bache, I., Bartle, I., & Flinders, M. (2016). Multi-level governance. In A. Christopher & T. Jacob (Eds.), *Handbook on theories of governance*. Edward Elgar Publishing. https://doi.org/10.4337/9781782548508.00052

Bakker, K., & Ritts, M. (2018). Smart Earth: A meta-review and implications for environmental governance. Global environmental change, 52, 201–211. https://doi.org/10.1016/j.gloenvcha.2018.07.011

Banwell, N., & Roelens, C. (2024). Challenging techno-solutionism in engineering education. Ethical and pedagogical reflections for sustainability. *GAIA—Ecological Perspectives for Science and Society, 33*(4), 357–362. https://doi.org/10.14512/GAIA.33.4.6

Baycheva-Merger, T., & Sotirov, M. (2020). The politics of an EU forest information system: Unpacking distributive conflicts associated with the use of forest information. *Forest Policy and Economics, 112*, Article 102110. https://doi.org/10.1016/J.FORPOL.2020.102110

Baycheva-Merger, T., Selter, A., Seijger, C., & Häublein, S. (2024). *Digital opportunity or a threat? Adoption of Internet of Things (IoT) monitoring systems for natural resources in Germany.*

Baycheva-Merger, T., Sotirov, M., Holmgren, S., & Selter, A. (2018). Institutional and actor-oriented factors constraining expert-based forest information exchange in Europe: A policy analysis from an actor-centred institutionalist approach. *Forests, 9*(3), 129. https://doi.org/10.3390/F9030129

Beduschi, A. (2021). Rethinking digital identity for post-COVID-19 societies: Data privacy and human rights considerations. *Data & Policy, 3*, Article e15. https://doi.org/10.1017/dap.2021.15

Bellanova, R. (2016). Digital, politics, and algorithms. *European Journal of Social Theory, 20*(3), 329–347. https://doi.org/10.1177/1368431016679167

Biemann, U. (2015). The cosmo-political forest: A theoretical and aesthetic discussion of the video forest law. *GeoHumanities, 1*(1), 157–170. https://doi.org/10.1080/2373566X.2015.1075359

Birch, K. (2006). The neoliberal underpinnings of the bioeconomy: The ideological discourses and practices of economic competitiveness. *Genomics, Society and Policy, 2*(3), 1–15. http://strathprints.strath.ac.uk/25814/1/neoliberal_bioeconomy_GSP.pdf

Birch, K., Levidow, L., & Papaioannou, T. (2010). Sustainable capital? The neoliberalization of nature and knowledge in the European "Knowledge-based bio-economy." *Sustainability, 2*(9), 2898–2918. https://doi.org/10.3390/su2092898

Birchall, C. (2011). Introduction to 'Secrecy and Transparency' The Politics of Opacity and Openness. Theory, Culture & Society, 28(7–8), 7–25.

Brous, P., Janssen, M., & Herder, P. (2020). The dual effects of the Internet of Things (IoT): A systematic review of the benefits and risks of IoT adoption by organizations. *International Journal of Information Management, 51*, Article 101952. https://doi.org/10.1016/j.ijinfomgt.2019.05.008

Buchelt, A., Adrowitzer, A., Kieseberg, P., Gollob, C., Nothdurft, A., Eresheim, S., Tschiatschek, S., Stampfer, K., & Holzinger, A. (2024). Exploring artificial intelligence for applications of drones in forest ecology and management. *Forest Ecology and Management, 551*, Article 121530. https://doi.org/10.1016/J.FORECO.2023.121530

Cashore, B. (2002). Legitimacy and the privatization of environmental governance: How Non-State Market–Driven (NSMD) governance systems gain rule-making authority. *Governance, 15*(4), 503–529. https://doi.org/10.1111/1468-0491.00199

Castells, M. (2010). *The rise of the network society* (2nd ed., pp. 271–282). Wiley-Blackwell.21. https://doi.org/10.53227/101943

Castree, N. (2008). Neoliberalising nature: The logics of deregulation and reregulation. *40*(1), 131–152. https://doi.org/10.1068/A3999

Clift, S. (2004). *E-democracy, E-governance and public net-work.*

Colombo, G., Gray, J., & Bounegru, L. (2023). Visual models for social media image analysis: groupings, engagement, trends, and rankings. *International Journal of Communication, 17*, 1956–1983. http://ijoc.org

Criado, J. I., Sandoval-Almazan, R., & Gil-Garcia, J. R. (2013). Government innovation through social media. *Government Information Quarterly, 30*(4), 319–326. https://doi.org/10.1016/J.GIQ.2013.10.003

Dąbrowska, J., Almpanopoulou, A., Brem, A., Chesbrough, H., Cucino, V., Di Minin, A., Giones, F., Hakala, H., Marullo, C., Mention, A., Mortara, L., Nørskov, S., Nylund, P. A., Oddo, C. M., Radziwon, A., & Ritala, P. (2022). Digital transformation, for better or worse: A critical multi-level research agenda. *R&D Management, 52*(5), 930–954. https://doi.org/10.1111/radm.12531

Damaševičius, R., Mozgeris, G., Kurti, A., & Maskeliūnas, R. (2024). Digital transformation of the future of forestry: an exploration of key concepts in the principles behind Forest 4.0. *Frontiers in Forests and Global Change, 7*, 1424327. https://doi.org/10.3389/FFGC.2024.1424327/BIBTEX

Davidson, E., Wessel, L., Winter, J. S., & Winter, S. (2023). Future directions for scholarship on data governance, digital innovation, and grand challenges. *Information and Organization, 33*(1), Article 100454. https://doi.org/10.1016/j.infoandorg.2023.100454

Dawes, S. S. (2009). Governance in the digital age: A research and action framework for an uncertain future. *Government Information Quarterly, 26*(2), 257–264. https://doi.org/10.1016/j.giq.2008.12.003

Denfanapapol, S., Setthasuravich, P., Rattanakul, S., Pukdeewut, A., & Kato, H. (2024). The digital divide, wealth, and inequality: An examination of socio-economic determinants of collaborative environmental governance in Thailand through provincial-level panel data analysis. *Sustainability, 16*(11), 4658. https://doi.org/10.3390/su16114658

Dhagarra, D., Goswami, M., & Kumar, G. (2020). Impact of trust and privacy concerns on technology acceptance in healthcare: An Indian perspective. *International Journal of Medical Informatics, 141*, Article 104164. https://doi.org/10.1016/j.ijmedinf.2020.104164

Díaz, S., Settele, J., Brondízio, E. S., Ngo, H. T., Guèze, M., Agard, J., Arneth, A., Balvanera, P., Brauma, K. A., Butchart, S. H. M., Chan, K. M. A., Garibaldi, L. A., Ichii, K., Liu, J., Subramanian, S. M., Midgley, G. F., Miloslavich, P., Molnár, Z., Obura, D., & Zayas, C. N. (2019). *The global assessment report on biodiversity and ecosystem services summary for policymakers.* www.ipbes.net

Dillon, L., Lave, R., Mansfield, B., Wylie, S., Shapiro, N., Chan, A. S., & Murphy, M. (2019). Situating data in a Trumpian era: The environmental data and governance initiative. *Annals of the American Association of Geographers, 109*(2), 545–555. https://doi.org/10.1080/24694452.2018.1511410

Dogaru, L. (2024). Timber tracking information system, a necessity in forest protection and sustainability. In *SUMAL information system, an initiative in Romania. Lecture notes in networks and systems, 928 LNNS* (pp. 63–71). https://doi.org/10.1007/978-3-031-54671-6_5

Duberry, J. (2019). *Global environmental governance in the information age: Civil society organizations and digital media.*

Dunleavy, P., & Margetts, H. (2023). Data science, artificial intelligence and the third wave of digital era governance. *Public Policy and Administration.* https://doi.org/10.1177/09520767231198737/ASSET/IMAGES/10.1177_09520767231198737-IMG4.PNG

Dunleavy, P., Margetts, H., Bastow, S., & Tinkler, J. (2006). New public management is dead—long live digital-era governance. *Journal of Public Administration Research and Theory, 16*(3), 467–494. https://doi.org/10.1093/JOPART/MUI057

FAO-PROFOR. (2011). *Framework for assessing and monitoring forest governance.*

Feroz, A. K., Zo, H., & Chiravuri, A. (2021). Digital transformation and environmental sustainability: A review and research agenda. *Sustainability, 13*(3), 1530. https://doi.org/10.3390/SU13031530

Ferreira, D., Vale, M., Miguel Carmo, R., Encalada-Abarca, L., & Marcolin, C. (2021). The three levels of the urban digital divide: Bridging issues of coverage, usage and its outcomes in VGI platforms. *Geoforum, 124*, 195–206. https://doi.org/10.1016/j.geoforum.2021.05.002

Florini, A. (2007). *The right to know: Transparency for an open world*. Columbia University Press.

Fountain, J. E. (2004). *Building the virtual state : Information technology and institutional change.* Brookings Institution Press.

Frischmann, B. M., Madison, M. J., & Strandburg, K. J. (2014). *Governing knowledge commons* (p. 499).

Gabrys, J. (2020). Smart forests and data practices: From the Internet of Trees to planetary governance. *Big Data and Society, 7*(1). https://doi.org/10.1177/2053951720904871/ASSET/IMAGES/LARGE/10.1177_2053951720904871-FIG2.JPEG

Gabrys, J., Westerlaken, M., Urzedo, D., Ritts, M., & Simlai, T. (2022). Reworking the political in digital forests: The cosmopolitics of socio-technical worlds. *Progress in Environmental Geography, 1*(1–4), 58–83. https://doi.org/10.1177/27539687221117836

Gaffoor, Z., Pietersen, K., Jovanovic, N., Bagula, A., & Kanyerere, T. (2020). Big data analytics and its role to support groundwater management in the Southern African development community. *Water, 12*(10), 2796. https://doi.org/10.3390/w12102796

Gain, A. K., & Fritsch, O. (2024). Bridging the digital divide for effective environmental management. *Australasian Journal of Environmental Management, 31*(2), 129–135. https://doi.org/10.1080/14486563.2024.2373629

Gebremeskel, B. K., Jonathan, G. M., & Yalew, S. D. (2023). Information security challenges during digital transformation. *Procedia Computer Science, 219*, 44–51. https://doi.org/10.1016/j.procs.2023.01.262

Gerbaudo, P. (2012). *Tweets and the streets social media and contemporary activism.* London Pluto Press.

Gerbaudo, P. (2021). Social media activism and the funnelling of participation. *Rivista di Digital Politics*, 2/20

Giessen, L., & Buttoud, G. (2014). Defining and assessing forest governance. *Forest Policy and Economics, 49*, 1–3. https://doi.org/10.1016/J.FORPOL.2014.11.009

Giessen, L., Burns, S., Sahide, M. A. K., & Wibowo, A. (2016). From governance to government: The strengthened role of state bureaucracies in forest and agricultural certification. *Policy and Society, 35*(1), 71–89. https://doi.org/10.1016/J.POLSOC.2016.02.001

Gil-Garcia, J. R., Dawes, S. S., & Pardo, T. A. (2018). Digital government and public management research: Finding the crossroads. *Public Management Review, 20*(5), 633–646. https://doi.org/10.1080/14719037.2017.1327181

Giurca, A., & Befort, N. (2023). Deconstructing substitution narratives: The case of bioeconomy innovations from the forest-based sector. *Ecological Economics, 207*, Article 107753. https://doi.org/10.1016/J.ECOLECON.2023.107753

Giurca, A., Nichiforel, L., Halici, M., Pasarel, A., & Htenas, A. (2023). *Unleashing opportunities for the bioeconomy in the mountain regions of Romania.* www.poca.ro

Goldstein, J. E. (2020). The volumetric political forest: territory, satellite fire mapping, and Indonesia's burning Peatland. *Antipode, 52*(4), 1060–1082. https://doi.org/10.1111/ANTI.12576

Goldstein, J. E., & Faxon, H. O. (2020). New data infrastructures for environmental monitoring in Myanmar: Is digital transparency good for governance? *5*(1), 39–59. https://doi.org/10.1177/2514848620943892

Gupta, A. (2008). Transparency under scrutiny: Information disclosure in global environmental governance. *Global Environmental Politics, 8*(2), 1–7. https://doi.org/10.1162/glep.2008.8.2.1

Gupta, A., Boas, I., & Oosterveer, P. (2020). Transparency in global sustainability governance: to what effect? *Journal of Environmental Policy & Planning, 22*(1), 84–97. https://doi.org/10.1080/1523908X.2020.1709281

Gurin, J. (2014). Open governments, open data: A new lever for transparency, citizen engagement, and economic growth. *SAIS Review of International Affairs, 34*(1), 71–82. https://doi.org/10.1353/sais.2014.0009

Hetemäki, L. (2019). The role of science in forest policy–Experiences by EFI. *Forest Policy and Economics, 105*, 10–16. https://doi.org/10.1016/J.FORPOL.2019.05.014

Hinings, B., Gegenhuber, T., & Greenwood, R. (2018). Digital innovation and transformation: An institutional perspective. *Information and Organization, 28*(1), 52–61. https://doi.org/10.1016/j.infoandorg.2018.02.004

Holmgren, S., D'Amato, D., & Giurca, A. (2020). Bioeconomy imaginaries: A review of forest-related social science literature. *Ambio, 49*(12), 1860–1877. https://doi.org/10.1007/s13280-020-01398-6

Howlett, M. (2004). Beyond good and evil in policy implementation: Instrument mixes, implementation styles, and second generation theories of policy instrument choice. *Policy and Society, 23*(2), 1–17. https://doi.org/10.1016/S1449-4035(04)70030-2

Howlett, M., Rayner, J., & Tollefson, C. (2009). From government to governance in forest planning? Lessons from the case of the British Columbia Great Bear Rainforest initiative. *Forest Policy and Economics, 11*(5–6), 383–391. https://doi.org/10.1016/J.FORPOL.2009.01.003

Kelly, S., Kaye, S.-A., & Oviedo-Trespalacios, O. (2023). What factors contribute to the acceptance of artificial intelligence? A systematic review. *Telematics and Informatics,77*, 101925. https://doi.org/10.1016/j.tele.2022.101925

Kingston, C., & Caballero, G. (2009). Comparing theories of institutional change. *Journal of Institutional Economics, 5*(2), 151–180. https://doi.org/10.1017/S1744137409001283

Kloppenburg, S., Gupta, A., Kruk, S. R. L., Makris, S., Bergsvik, R., Korenhof, P., Solman, H., & Toonen, H. M. (2022). Scrutinizing environmental governance in a digital age: New ways of seeing, participating, and intervening. *One Earth, 5*(3), 232–241. https://doi.org/10.1016/j.oneear.2022.02.004

Kukutai, T., & Taylor, J. (2016). Indigenous data sovereignty. In *Indigenous data sovereignty*. ANU Press. https://doi.org/10.22459/CAEPR38.11.2016

Kumar, A., Agrawal, R., Wankhede, V. A., Sharma, M., & Mulat-weldemeskel, E. (2022). A framework for assessing social acceptability of industry 4.0 technologies for the development of digital manufacturing. *Technological Forecasting and Social Change, 174*, 121217. https://doi.org/10.1016/j.techfore.2021.121217

Larson, A. M., Mausch, K., Bourne, M., Luttrell, C., Schoneveld, G., Cronkleton, P., Locatelli, B., Catacutan, D., Cerutti, P., Chomba, S., Djoudi, H., Ihalainen, M., Lawry, S., Minang, P., Monterroso, I., Myers, R., Naito, D., Pham, T. T., Reed, J., Stoian, D., et al. (2021). Hot topics in governance for forests and trees: Towards a (just) transformative research agenda. *Forest Policy and Economics, 131*, 102567. https://doi.org/10.1016/J.FORPOL.2021.102567

Linkov, I., Trump, B. D., Poinsatte-Jones, K., & Florin, M. V. (2018). Governance strategies for a sustainable digital world. *Sustainability, 10*(2), 440. https://doi.org/10.3390/SU10020440

Lytras, D., Visvizi, A., Nikiforova, A., Ravšelj, D., Umek, L., Todorovski, L., & Aristovnik, A. (2022). A review of digital era governance research in the first two decades: A bibliometric study. *Future Internet, 14*(5), 126. https://doi.org/10.3390/FI14050126

Massimo, R., & Muschert, G. W. (Eds). (2017). Theorizing digital divides. 1st ed. Routledge. https://doi.org/10.4324/9781315455334

Maxwell T., & Dawes, S. S. (2009). Public governance as a socio-technical system: from concept to application. In *Proceedings of the 10th annual international conference on digital government research social networks: Making connections between citizens, data and government*.

Mazzucato, M., & Kattel, R. (2020). COVID-19 and public-sector capacity. *Oxford Review of Economic Policy, 36*(Supplement_1), S256–S269. https://doi.org/10.1093/OXREP/GRAA031

McRoberts, R. E., Tomppo, E. O., & Næsset, E. (2010). Advances and emerging issues in national forest inventories. *Scandinavian Journal of Forest Research, 25*(4), 368–381. https://doi.org/10.1080/02827581.2010.49673

Mergel, I. (2013). A framework for interpreting social media interactions in the public sector. *Government Information Quarterly, 30*(4), 327–334. https://doi.org/10.1016/J.GIQ.2013.05.015

Mergel, I., Ganapati, S., & Whitford, A. B. (2021). Agile: A new way of governing. *Public Administration Review, 81*(1), 161–165. https://doi.org/10.1111/PUAR.13202

Mol, A. P. J. (2008). *Environmental reform in the information age*. Cambridge University Press.

Moser-Plautz, B., & Schmidthuber, L. (2023). Digital government transformation as an organizational response to the COVID-19 pandemic. *Government Information Quarterly, 40*(3), Article 101815. https://doi.org/10.1016/J.GIQ.2023.101815

Nitoslawski, S. A., Wong-Stevens, K., Steenberg, J. W. N., Witherspoon, K., Nesbitt, L., & Konijnendijk van den Bosch, C. C. (2021). The digital forest: Mapping a decade of knowledge on technological applications for forest ecosystems. *Earth's Future, 9*(8), e2021EF002123. https://doi.org/10.1029/2021EF002123

Nordlund, A., Schenk, T., & Westin, K. (2017). Forest beliefs in an urbanizing world: Views on and usage of forest areas among persons with and without a migration biography in Germany and Sweden. *Society & Natural Resources, 30*(2), 160–176. https://doi.org/10.1080/08941920.2016.1238988

O'Gorman, E., & Gaynor, A. (2020). More-than-human histories. *25*(4), 711–735. https://doi.org/10.1093/ENVHIS/EMAA027

Papagiannidis, S., & Marikyan, D. (2022). Environmental sustainability: A technology acceptance perspective. *International Journal of Information Management, 63*, Article 102445. https://doi.org/10.1016/j.ijinfomgt.2021.102445

Peluso, N. L., & Vandergeest, P. (2020). Writing political forests. *Antipode, 52*(4), 1083–1103. https://doi.org/10.1111/ANTI.12636

Pierre, J., & Peters, G. (2000). *Governance.* Macmillan.

Quach, S., Thaichon, P., Martin, K. D., Weaven, S., & Palmatier, R. W. (2022). Digital technologies: Tensions in privacy and data. *Journal of the Academy of Marketing Science, 50*(6), 1299–1323. https://doi.org/10.1007/s11747-022-00845-y

Racevskis, L. A., & Lupi, F. (2006). Comparing urban and rural perceptions of and familiarity with the management of forest ecosystems. *Society and Natural Resources, 19*(6), 479–495. https://doi.org/10.1080/08941920600663862

Rantala, S., Swallow, B., Lähteenmäki-Uutela, A., & Paloniemi, R. (2022). Forest data governance as a reflection of forest governance: Institutional change and endurance in Finland and Canada. *Environmental Science & Policy, 136*, 751–760. https://doi.org/10.1016/J.ENVSCI.2022.07.031

Rantala, S., Swallow, B., Paloniemi, R., & Raitanen, E. (2020). Governance of forests and governance of forest information: Interlinkages in the age of open and digital data. *Forest Policy and Economics, 113*, Article 102123. https://doi.org/10.1016/J.FORPOL.2020.102123

Reyes-García, V., Tofighi-Niaki, A., Austin, B. J., Benyei, P., Danielsen, F., Fernández-Llamazares, Á., Sharma, A., Soleymani-Fard, R., & Tengö, M. (2022). Data sovereignty in community-based environmental monitoring: Toward equitable environmental data governance. *BioScience, 72*(8), 714–717. https://doi.org/10.1093/biosci/biac048

Rhodes, R. A. W. (1996). The new governance: Governing without government. *44*(4), 652–667. https://doi.org/10.1111/J.1467-9248.1996.TB01747.X

Roberts, E., Anderson, B. A., Skerratt, S., & Farrington, J. (2017). A review of the rural-digital policy agenda from a community resilience perspective. *Journal of Rural Studies, 54*, 372–385. https://doi.org/10.1016/J.JRURSTUD.2016.03.001

Rosenau, J. N., & Czempiel, E.-O. (1992). *Governance without government: Order and change in world politics.* Cambridge University Press. https://doi.org/10.1017/CBO9780511521775

Salam, A. (2020). Internet of things for sustainable forestry. In *Internet of Things for sustainable community development, Internet of Things* (pp. 147–181). Springer International Publishing, https://doi.org/10.1007/978-3-030-35291-2_5

Scharpf, F. W. (1994). Games real actors could play. *6*(1), 27–53. https://doi.org/10.1177/0951692894006001002

Scharpf, F. W. (1997). *Games real actors play : actor-centered institutionalism in policy research.* Taylor & Francis Group. https://www.routledge.com/Games-Real-Actors-Play-Actor-centered-Institutionalism-In-Policy-Research/Scharpf/p/book/9780813399683

Scholz, J., De Meyer, A., Marques, A. S., Pinho, T. M., Boaventura-Cunha, J., Van Orshoven, J., Rosset, C., Künzi, J., Kaarle, J., & Nummila, K. (2018). Digital technologies for forest supply chain optimization: Existing solutions and future trends. *Environmental Management, 62*(6), 1108–1133. https://doi.org/10.1007/S00267-018-1095-5/METRICS

Scoville, C., Chapman, M., Amironesei, R., & Boettiger, C. (2021). Algorithmic conservation in a changing climate. *Current Opinion in Environmental Sustainability, 51*, 30–35. https://doi.org/10.1016/j.cosust.2021.01.009

Sergent, A., Arts, B., & Edwards, P. (2018). Governance arrangements in the European forest sector: Shifts towards 'new governance' or maintenance of state authority? *Land Use Policy, 79*, 968–976. https://doi.org/10.1016/J.LANDUSEPOL.2016.08.036

Sethuramalingam, D. (2020). Digital governance: Theory, policy and practice. *SSRN Electronic Journal*. https://doi.org/10.2139/SSRN.3679639

Sharma, S., Kar, A. K., Gupta, M. P., Dwivedi, Y. K., & Janssen, M. (2022). Digital citizen empowerment: A systematic literature review of theories and development models. Information Technology for Development, 28(4), 660-687. https://doi.org/10.1080/02681102.2022.2046533

Singh, R., Gehlot, A., Vaseem Akram, S., Kumar Thakur, A., Buddhi, D., & Kumar Das, P. (2021). Forest 4.0: Digitalization of forest using the Internet of Things (IoT). *Journal of King Saud University—Computer and Information Sciences*, S1319157821000483. https://doi.org/10.1016/j.jksuci.2021.02.009

Stopfer, L., Kaulen, A., & Purfürst, T. (2024). Potential of blockchain technology in wood supply chains. *Computers and Electronics in Agriculture, 216*, Article 108496. https://doi.org/10.1016/j.compag.2023.108496

Tironi, M., & Rivera Lisboa, D. I. (2023). Artificial intelligence in the new forms of environmental governance in the Chilean State: Towards an eco-algorithmic governance. *Technology in Society, 74*, Article 102264. https://doi.org/10.1016/j.techsoc.2023.102264

Van Tatenhove, J., Arts, B., & Leroy, P. (Eds.). (2000). *Political modernisation and the environment* (Vol. 24). Springer Netherlands. https://doi.org/10.1007/978-94-015-9524-7

Vidal, C., Alberdi, I., Redmond, J., Vestman, M., Lanz, A., & Schadauer, K. (2016). The role of European National Forest Inventories for international forestry reporting. *Annals of Forest Science, 73*(4), 793–806. https://doi.org/10.1007/S13595-016-0545-6/FIGURES/3

Westerlaken, M., Gabrys, J., Urzedo, D., & Ritts, M. (2023). Unsettling participation by foregrounding more-than-human relations in digital forests. *Environmental Humanities, 15*(1), 87–108. https://doi.org/10.1215/22011919-10216173

Westerlund, M., Isabelle, D. A., & Leminen, S. (2021). The acceptance of digital surveillance in an age of big data. *Technology Innovation Management Review, 11*(3).

Young, J. C. (2019). Rural digital geographies and new landscapes of social resilience. *Journal of Rural Studies, 70*, 66–74. https://doi.org/10.1016/J.JRURSTUD.2019.07.001

Young, O. R., Yang, J., & Guttman, D. (2020). Meeting cyber age needs for governance in a changing global order. *Sustainability, 12*(14), 5557. https://doi.org/10.3390/SU12145557

Zgłobicki, W., Karczmarczuk, K., & Baran-Zgłobicka, B. (2020). Intensity and driving forces of land abandonment in Eastern Poland. *Applied Sciences, 10*(10), 3500. https://doi.org/10.3390/APP10103500

Ziaja, S. (2021). How algorithm assisted decision making is influencing environmental law and climate adaptation. *SSRN Electronic Journal*. https://doi.org/10.2139/ssrn.3898812

Zuboff, S. (2019). The age of surveillance capitalism. The Fight for a Human Future at the New Frontier of Power. New York: Public Affairs

Zwitter, A., & Hazenberg, J. (2020). Decentralized network governance: Blockchain technology and the future of regulation. *Frontiers in Blockchain, 3*, Article 507920. https://doi.org/10.3389/FBLOC.2020.00012/BIBTEX

Chapter 9
Measuring What Matters: The Renewed Role of Criteria and Indicators (C&I) in Forest Policy and Governance

Stefanie Linser, Michael Köhl, Markus Lier, Kit Prins, and Andrzej Talarczyk

Abstract Over the past three decades, criteria and indicators (C&I) for sustainable forest management (SFM) have emerged as universal policy tools for monitoring, assessing, and reporting on forest sustainability. Used in over 170 countries, C&I frameworks have promoted a shared understanding of SFM, informed evidence-based policymaking, and helped to fulfill international reporting obligations. Lessons learned from international and regional C&I processes highlight the importance of adaptability, participatory approaches, and cross-sectoral integration when it comes to addressing emerging challenges, such as climate change and biodiversity loss. Recent developments, including the Global Core Set of Forest-related Indicators, the EU Taxonomy, and EuropaBON, emphasise the increasing importance of C&I

S. Linser (✉)
Department of Economics and Social Sciences, BOKU University, Institute of Forest, Environmental and Natural Resource Policy, Vienna, Austria
e-mail: stefanie.linser@boku.ac.at

Centre for Bioeconomy, BOKU University, Vienna, Austria

EFI Forest Policy Research Network, Vienna, Austria

M. Köhl
University of Hamburg, Institute for Wood Science—World Forestry, Hamburg, Germany
e-mail: michael.koehl@uni-hamburg.de

M. Lier
Natural Resources Institute Finland (Luke), Joensuu, Finland
e-mail: markus.lier@luke.fi

K. Prins
Independent Expert, Geneva, Switzerland
e-mail: kit.prins@gmail.com

A. Talarczyk
Forest and Natural Resources Research Centre, Warsaw, Poland
e-mail: andrzej@talarczyk.com

F. Aggestam (ed.), *Governing Forests Through Transformative Change*, Advances in Global Change Research 70, https://doi.org/10.1007/978-3-032-19072-7_9

in aligning forest-related monitoring with global sustainability issues and international objectives. Technological advances, such as remote sensing, artificial intelligence (AI), and big data, have further enhanced the capacity of C&I frameworks to provide real-time, high-resolution data. This enables more dynamic and adaptive forest management. However, challenges remain, including harmonising methodologies, addressing resource constraints, and ensuring equitable access to digital tools. Effective communication between data providers, stakeholders, and policymakers, as well as political and institutional commitments, are critical to overcoming these barriers. Furthermore, C&I frameworks are increasingly recognised as essential tools for enabling transformative change in forest governance. By addressing systemic barriers, fostering inclusive and participatory processes, and integrating new priorities such as climate adaptation and biodiversity restoration, C&I can drive fundamental shifts in governance systems. Even more, international collaboration, capacity building, and the sharing of lessons learned from sectors beyond forestry will be crucial in ensuring that C&I frameworks remain relevant and effective in an ever-changing world.

9.1 Introduction

This chapter explores the evolving role of criteria and indicators (C&I) for sustainable forest management (SFM), emphasising their dual functionality as both reporting mechanisms and universal policy tools. It demonstrates how C&I frameworks, which are traditionally considered to be instruments for monitoring and reporting, can also be used as transformative governance tools to bridge the gap between global sustainability goals and national and local implementation. It begins by tracing the historical development of SFM and the emergence of C&I frameworks, emphasising their origins in global policy initiatives such as Agenda 21 and the Forest Principles. It then delves into the conceptual underpinnings of C&I, examining their role in defining what is important to measure (criteria) and how to measure it (indicators). Subsequent sections explore the application of C&I in transformative forest governance and across scales—from global to local—while addressing their integration into international frameworks, such as the Sustainable Development Goals (SDGs), and their development within intergovernmental processes, including Forest Europe and the Montréal Process. The chapter also discusses recent advancements, including the use of digital technologies and the development of policy-relevant indicators. Finally, it identifies challenges and opportunities for improving the effectiveness of C&I frameworks in addressing emerging forest governance issues such as climate change, biodiversity loss, and the transition to a bio-based economy.

9.1.1 The Emergence of Indicators in Global Forest Policy

Agenda 21, the Programme of Action for the 21st Century, was adopted by 179 governments at the United Nations Conference on Environment and Development (UNCED) in Rio de Janeiro, Brazil, in June 1992. This document outlined an action plan for sustainable development at global, regional, national, and local levels, highlighting the challenges of attaining and measuring sustainability. To bridge the data gap, Chapter 40.4 of Agenda 21 proposed the development and use of indicators of sustainable development to provide a foundation for decision-making at all levels (United Nations, 1992a). Subsequent decades have witnessed the development of indicators as the most appropriate tools for measuring, monitoring, assessing, and reporting progress towards sustainability goals covering a spectrum of subjects, ranging from global to local scales (Linser, 2021). Agenda 21 also laid the foundation for many indicator-based sectoral monitoring activities, with subsequent revisions of sectoral indicator sets. Nonetheless, the focus meanwhile shifted towards striking a balance between sector-specific monitoring and the need for integrated, cross-sectoral approaches to address complex challenges.

The further explanations of indicators for sustainable forest management first require a brief overview of the genesis of sustainable forest management.

9.1.2 Addressing SFM in a Global Context

Although sustainable forest management (SFM) was discussed and applied in Central European forestry as early as the 18th Century (Carlowitz, 1713), its international anchoring occurred much later. It evolved as a critical concept following the publication of the Report of the World Commission on Environment and Development: Our Common Future, which introduced the idea of sustainable development as meeting the needs of the present without compromising the ability of future generations to meet their own needs (Brundtland, 1987). The report emphasised the importance of forests in maintaining ecological balance, supporting livelihoods, and mitigating climate change, while also warning against deforestation and unsustainable exploitation practices. This laid the groundwork for global discussions on forest conservation. The momentum continued at the above-mentioned UNCED in 1992, where forests were a key focus. Although no legally binding forest convention was adopted, there was agreement on the "Forest Principles" (formally, the Non-Legally Binding Authoritative Statement of Principles for a Global Consensus on the Management, Conservation, and Sustainable Development of All Types of Forests) which emphasised the need for SFM to balance environmental, social, and economic objectives (United Nations, 1992b). Additionally, Agenda 21 outlined specific actions to promote SFM, including reforestation, afforestation, forest conservation, and the sustainable use of forest resources (United Nations, 1992a).

Since UNCED, different regional, international, and intergovernmental forest-related processes and organisations have defined SFM in slightly different ways. However, the following definitions are broadly supported:

> A dynamic and evolving concept [that aims] to maintain and enhance the economic, social and environmental values of all types of forests, for the benefit of present and future generations (United Nations, 2007).

> The stewardship and use of forests and forest lands in such a way, and at a rate, that maintains their biodiversity, productivity, regeneration capacity, vitality and their potential to fulfil now, and in the future, relevant ecological, economic and social functions, at local, national, and global levels, and that does not cause damage to other ecosystems (MCPFE, 1993).

From the definitions, it is clear that, while SFM will change over time, its purpose should, at a minimum, be to maintain all forest values in perpetuity (Linser, 2021).

The development of C&I frameworks has been instrumental in operationalising these definitions of SFM. Beyond their role in monitoring and reporting, C&I frameworks have evolved into essential policy tools that enable governments and stakeholders to establish priorities, allocate resources, and assess the effectiveness of forest management strategies. This policy-oriented function of C&I is particularly significant in addressing the complex and evolving challenges of SFM and transformative change.

In the years following UNCED—but even before 2000—initiatives, such as the establishment of the United Nations Forum on Forests (UNFF), certification bodies like the Forest Stewardship Council (FSC), international organisations like the Food and Agriculture Organization of the United Nations (FAO) or the International Tropical Timber Organization (ITTO), and intergovernmental processes like Forest Europe or the Montréal Process translated the largely philosophical ideals of SFM into specific criteria and indicator (C&I) frameworks as tools for systematic and continuous data collection, monitoring, assessment, and reporting on progress towards SFM, despite its widespread adoption, the strategic role of C&I in shaping forest policy and governance has often been underutilised. The following sections revisit their function as not only monitoring tools but also instruments of forest policy and governance.

9.2 What Are C&I for SFM?

Over the past 30 years, C&I have been used as a foundation to set up or amend forest monitoring networks such as many national forest inventories (NFIs) (Linser & Wolfslehner, 2022; Otrakcier et al., 2023) by providing a structured framework that organises and harmonises forest data (FAO & CPF, 2022a; Forest Europe, 2020). While many NFIs primarily focus on systematic data collection based on agreed-upon metrics, C&I for SFM goes beyond mere data collection by linking data to specific sustainability goals and policy objectives. Through this integration, C&I for SFM have emerged in a dual role: as a powerful reporting tool and as a policy instrument that facilitates decision-making in and about forest management.

C&I for SFM can be applied at the global, regional, international, national, subnational, and forest management unit levels. However, only very specific sets are scalable for various levels. C&I for SFM can also be customised to accommodate differences within and between countries, regions, and specific locations (Linser & Wolfslehner, 2022; Linser et al., 2018b), providing a comprehensive approach that integrates the ecological, economic, and social dimensions of sustainability.

9.2.1 Criteria Outline: WHAT is Important to Measure

A criterion is a specific condition or a more general target that must be met to confirm that forests are managed sustainably. It often involves the concept against which sustainability is assessed. For instance, this could be maintenance, enhancement, protection or conservation of the essential elements of SFM.

Broadly used definitions include:

> Criteria define the essential elements against which sustainability is assessed, with due consideration paid to the productive, protective and social roles of forests and forest ecosystems. Each criterion relates to a key element of sustainability, and may be described by one or more indicators (FAO, 2015a).
>
> A Criterion is defined as an aspect of forest management that is considered important and by which SFM may be assessed. A criterion accompanied by a set of related indicators describes a state or situation that should be met to comply with SFM (ITTO, 2016).
>
> Criteria characterise or define the essential elements or set of conditions or processes by which sustainable forest management may be assessed (MCPFE, 1998).

These definitions highlight the foundational role of criteria in shaping the scope and focus of SFM policies. By defining what is important to measure, criteria provide a structured framework for aligning forest management practices with broader sustainability objectives.

In addition to their role in defining SFM, criteria also serve as a foundation for policy development. By outlining the essential elements of SFM, policymakers can identify areas requiring intervention, set priorities, and establish benchmarks for progress. This makes criteria a critical component of forest policy aimed at achieving SFM.

As can be seen from Table 9.1, the three processes, which cover completely different regions of the world, share quite similar criteria on identical topics. Forest Europe, the pan-European process, does not have a criterion on enabling conditions for SFM. However, the pan-European set does contain five qualitative indicators that are not assigned to a specific criterion but relate to forest policy and governance. Each criterion is associated with one qualitative indicator, related to the respective policies, institutions, and instruments (see Sect. 9.2.2.2).

Table 9.1 Examples of the criteria of some regional processes. Corresponding criteria are shown in the same colour

Criteria for the sustainable management of tropical forests (ITTO, 2016)	Pan-European criteria for SFM (MCPFE, 1998)	The Montréal process criteria for the conservation and sustainable management of temperate and boreal forests (Montréal Process, 2015)
C1: Enabling conditions for sustainable forest management	C1: Maintenance and appropriate enhancement of forest resources and their contribution to global carbon cycles	C1: Conservation of biological diversity
C2: Extent and condition of forests	C2: Maintenance of forest ecosystem health and vitality	C2: Maintenance of the productive capacity of forest ecosystems
C3: Forest ecosystem health and resilience	C3: Maintenance of productive functions of forests (wood and non-wood)	C3: Maintenance of forest ecosystem health and vitality
C4: Forest production	C4: Maintenance, conservation and appropriate enhancement of biological diversity in forest ecosystems	C4: Conservation and maintenance of soil and water resources
C5: Forest biological diversity	C5: Maintenance and appropriate enhancement of protective functions in forest management (notably soil and water)	C5: Maintenance of forest contribution to global carbon cycles
C6: Soil and water protection	C6: Maintenance of other socio-economic functions and conditions	C6: Maintenance and enhancement of long-term multiple socio-economic benefits to meet the needs of societies
C7: Economic, social and cultural aspects		C7: Legal, institutional and economic framework for forest conservation and sustainable management

9.2.2 *Indicators Pinpoint HOW to Measure SFM*

Indicators reduce large amounts of data to a simpler form, to make complex circumstances feasibly measurable and comprehensible to decision-makers and the public (Linser, 1999, 2002; McCool & Stankey, 2001; Shields, 2002). In short, the purpose of an indicator is to facilitate simplification (Ott, 1978). The result is a graded information system, see Fig. 9.1.

However, indicators are not just tools for simplification; they are also instruments for action. By providing actionable insights, indicators enable policymakers

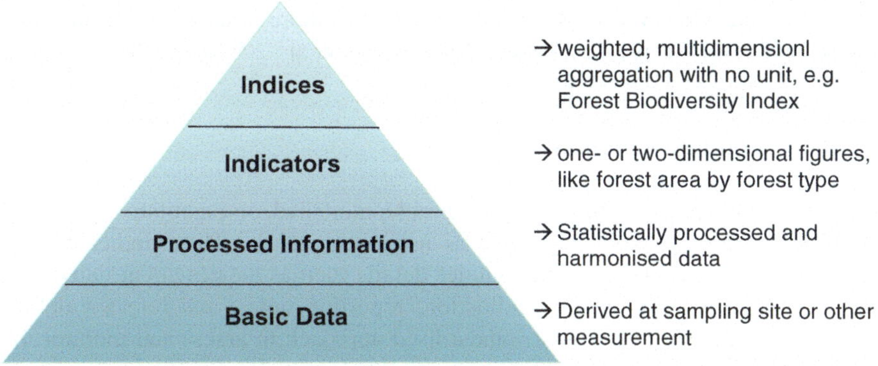

Fig. 9.1 Data pyramid. *Source* Adrianse (1995)

to identify gaps, set priorities, and implement targeted interventions. This makes them indispensable for translating the principles of SFM into tangible outcomes.

Indicators may be quantitative, with figures regularly monitored, or otherwise qualitative, with descriptive information obtained through surveys. Both types of indicators focus on the status and changes of SFM. Commonly used definitions include:

> Indicators are variables or parameters that enable the measurement of particular aspects of criteria. Indicators help monitor the status of, and trends in, forests in quantitative, qualitative or descriptive terms (FAO, 2015a).
>
> An indicator is a quantitative, qualitative or descriptive attribute that, when measured and monitored periodically, indicates the direction of change in a criterion. Indicators identify the information needed for assessing and monitoring change, both in the forest itself (outcome indicators) and as part of the environmental and forest management systems used (input and process indicators). A time series of the values of any measurable or clearly descriptive indicator can provide information on the direction of change, either towards or away from SFM. The indicators cannot by themselves establish the sustainability of management, however (ITTO, 2016).
>
> Indicators show changes over time for each criterion and demonstrate the progress made towards their specific objectives (MCPFE, 1998).

Indicators play a crucial role in bridging science, policy, and practice by providing measurable data that inform policy decisions and practical management strategies. In this way, indicators act as a bridge between monitoring and governance, ensuring that data collected through C&I frameworks is not only descriptive but also actionable. This dual role enhances their utility as tools for both assessing progress and driving policy reforms.

Interest in C&I for SFM has increased in recent years. In the context of the SDGs and emerging global challenges, such as climate change, biodiversity loss, and deforestation, the contribution of C&I to monitoring, assessing, and reporting on forest conditions and trends is becoming increasingly significant (Linser, 2024).

Whilst the criteria for SFM at local, national, regional, and international levels are increasingly aligned, the sets of indicators for SFM vary widely, ranging from the more context-specific at the local level to the more generic at national and international levels (Linser & O'Hara, 2019; Linser et al., 2018a, 2018b).

9.2.2.1 Quantitative Indicators

Quantitative indicators for SFM are measurable, modelled, or quantitatively derived variables that are typically expressed as numerical values. These indicators are accompanied by metadata, which includes details such as measurement units, data sources, the periodicity of data collection, reporting notes, and relevant definitions (Table 9.2). They provide a standardised approach to assess and monitor the sustainability of forest management practices over time.

Quantitative indicators are essential tools for monitoring and assessing SFM, but they face several significant challenges that limit their effectiveness as policy instruments. One of the primary issues is that indicators were usually developed to monitor relevant aspects of forests and forestry rather than to assess progress towards SFM, and lack any means of assessing the sustainability of the management, remaining purely descriptive. Secondly, there is still a lack of reliable, harmonised, and thus consistent data across regions, countries, and time periods. Many countries, not only in the Global South, struggle with resource and capacity constraints, which hinder the collection of high-quality, up-to-date data. This often results in gaps in datasets, making it difficult to conduct meaningful comparisons or track trends over time (MacDicken et al., 2015; McDonald & Lane, 2004). Additionally, methodological inconsistencies in how indicators are measured and reported further complicate their use. For instance, forest area change might be assessed using satellite imagery in one region and field surveys in another, leading to discrepancies that undermine the comparability and reliability of the data (FAO, 2020).

Another feature of quantitative indicators is their tendency to simplify and reduce complex ecological, social, and economic systems into a single figure or parameter, which is intended to convey a specific meaning. Although this simplification is often

Table 9.2 Examples for quantitative indicators under Criterion 1 of the Forest Europe process (Forest Europe, 2015)

Criteria	Indicator	Measurement unit (common examples)
Criterion 1: maintenance and appropriate enhancement of forest resources and their contribution to global carbon cycles	1.1 Forest area	ha, share of total land area
	1.2 Growing stock	m3/ha
	1.3 Age structure and/or diameter distribution	area (ha) per development phase volume (m^3) per development phase volume (m^3) by diameter classes
	1.4 Forest carbon	tonnes of C/ha

necessary to achieve focus and enhance communication, there is a risk that the true picture will be distorted. By reducing these dynamics to single numerical values, indicators often fail to provide a broader picture. For example, an indicator such as "forest area" does not account for the quality of the forest, its biodiversity, or the ecosystem services it provides. Thus, only a set of well-selected quantitative indicators can provide a more detailed picture.

Furthermore, the universal design of many indicator sets results in a lack of contextual relevance. For instance, in certain countries of a region with weak forest law enforcement, an indicator that measures illegal logging or in countries that suffer increasingly from climate change-induced droughts, an indicator on desertification might provide added value for political decision-making. Such a lack of contextual sensitivity can result in indicator sets that fail to adequately capture the unique challenges and priorities of specific countries or regions.

Political and institutional factors also play a role in limiting the effectiveness of quantitative indicators. In some cases, data may not be based on measurements, but estimations; they may be selectively reported or manipulated to align with policy goals, providing only data for those indicators that show positive trends while neglecting those that highlight challenges or failures. For instance, many countries do not report on threatened forest species (Lier et al., 2020). This undermines the credibility of indicators and their utility for evidence-based policymaking. Finally, quantitative indicators often face temporal and spatial limitations related to their reporting. Indicators based on certain reporting periods often fail to capture recent incidents. For example, the State of Europe's Forests 2020 report does not include the severe storm damage of autumn 2018, as the data collection period only spanned from 1990 to 2015 (Forest Europe, 2020). Additionally, the damages vary spatially, exhibiting, for instance, North–South gradients for drought and forest fire damages (Ferretti et al., 2020; Köhl et al., 2024).

These challenges underline the need for complementary approaches, such as qualitative indicators or regional subsets, to provide a more comprehensive and nuanced understanding of sustainable forest management.

9.2.2.2 Qualitative or Descriptive Indicators

"Qualitative indicators" (the—possibly misleading—term used by Forest Europe and the Montréal Process) consist of variables that are described rather than measured numerically. They primarily focus on the overall policies, institutions, and instruments in place for SFM (Table 9.3). These indicators are typically informed by questions expecting descriptive answers, offering insights into the frameworks and strategies that support SFM.

Despite their value in capturing the institutional and policy dimensions of forest management, several challenges limit the effectiveness and reliability of qualitative indicators.

Table 9.3 Examples for qualitative indicators under Criterion 1 of the Forest Europe process (Forest Europe, 2015)

Overarching framework	No	Indicator
Criterion 1: Maintenance and appropriate enhancement of forest resources and their contribution to global carbon cycles	C1	Policies, institutions and instruments to maintain and appropriately enhance forest resources and their contribution to global carbon cycles
Forest policy and governance	1	National Forest Programmes or equivalent
	2	Institutional frameworks
	3	Legal/regulatory framework: National (and/or sub-national) and International commitments
	4	Financial and economic instruments
	5	Information and communication

- Nature of qualitative indicators: Unlike quantitative indicators, which rely on measurable data, qualitative indicators are descriptive, usually non-quantifiable, and without standardisation. This can lead to inconsistencies in how indicators are interpreted and reported, for instance, when different stakeholders or institutions have varying perspectives on what constitutes "effective" policies or governance. Reporting authorities rarely consider their policy frameworks ineffective.
- Lack of standardisation: Another challenge is the lack of standardisation in the design and application of qualitative indicators. While qualitative indicators are valuable for capturing context-specific information, their flexibility can result in a lack of comparability across regions or periods. For instance, the evaluation of governance structures or stakeholder participation processes may vary significantly between countries, making it challenging to draw meaningful comparisons (Secco et al., 2014). This is, in fact, unsurprising, as policies and institutions adapt to the country's social, economic, and political conditions and traditions.
- Constrained data availability and reliability: Qualitative indicators often struggle with issues of data availability and reliability. Collecting qualitative data typically requires extensive engagement with experts or stakeholders, such as interviews or surveys, which can be time-consuming and resource-intensive. In many cases, particularly in countries with limited institutional capacity, this data may be incomplete, outdated, or unavailable altogether. It is also surprising to discover how little information is shared between government departments, so that national forest correspondents are often unable to provide information on developments in other departments or ministries.
- Limited actionability: Qualitative indicators are often criticised for their limited ability to provide actionable insights. While they are effective at highlighting governance issues, they may lack the precision needed to inform specific management decisions or interventions. For example, knowing that a country has a forest policy in place does not necessarily provide information about its implementation, enforcement, or outcomes. This gap between available policy frameworks

and on-the-ground realities can reduce the practical utility of qualitative indicators in advancing reporting and assessing SFM (Lier et al., 2021).

9.3 Application and Implementation of Indicators for SFM Across Scales

Having established the dual role of C&I as both reporting mechanisms and governance tools in Sect. 9.2, this section examines how these frameworks are applied and implemented across various scales, ranging from local to global. It is highlighted how C&I frameworks not only monitor progress but also actively shape forest management policies and practices, bridging the gap between global sustainability goals and local realities.

9.3.1 Rio Follow-Up

In the Rio follow-up process, all the resulting UN Conventions developed indicators to track progress in the implementation by various member states. The Convention on Biological Diversity (CBD) applies indicators to measure and assess the achievement of its 2020 Aichi Targets. The UN Framework Convention on Climate Change (UNFCCC) has indicators to report achievement of the climate targets. The indicators of the UN Convention to Combat Desertification (UNCCD) provide information on progress toward meeting its long-term objectives in areas affected by drought, desertification, and land degradation. All these sets include some forest-related indicators (see Table 9.4).

While these indicators primarily serve as reporting tools, they also influence policy by identifying areas requiring intervention and guiding resource allocation. For example, forest-related indicators under the CBD and UNFCCC have shaped national strategies for biodiversity conservation and climate change mitigation, demonstrating the governance potential of C&I frameworks.

9.3.2 Sustainable Development Goals (SDG)

In 2015, the UN General Assembly adopted the 2030 Agenda for Sustainable Development, which encompasses 17 SDGs and 169 targets to be achieved by all countries by 2030. There are 231 indicators to assess progress against set targets (United Nations, 2018). The 2030 Agenda acknowledges the important role that forests play in achieving the SDGs. Two SDGs mention forests explicitly. SDG15, "Life on land", assesses progress towards sustainable forest management in indicator 15.2.1 (FAO,

Table 9.4 International conventions and frameworks with forest-related indicators (selection)

Convention/framework	Forest-related indicators
Convention on Biological Diversity Indicators for the strategic plan for biodiversity 2011–2020 and the **Aichi Biodiversity Targets** (CBD, 2016)	• Trends in tree cover • Forest area as a proportion of total land area • Proportion of important sites for terrestrial and freshwater biodiversity that are covered by protected areas, by ecosystem type • Progress towards sustainable forest management • Trends in the fragmentation of forest and other natural habitats • Trends in the degradation of forest and other natural habitats • Trends in restoration of degraded ecosystems, including forests
Convention on Biological Diversity Indicators for the **Kunming-Montreal Global Biodiversity Framewor**k (CBD, 2022)	• Forest area as a proportion of total land area • Forest distribution • Tree cover loss • Continuous global mangrove forest cover • Trends in mangrove forest fragmentation • Trends in mangrove extent • Forest Fragmentation Index • Forestry production and trade (wood fuel) • Increase in secondary natural forest cover, annual tropical primary tree cover loss, forest landscape integrity index • Area of forest under sustainable management: total forest management certification by the Forest Stewardship Council and the Programme for the endorsement of forest certification
United Nations Framework Convention on Climate Change Global set of climate change statistics and indicators (United Nations, n.d.)	• Forest area • Area deforested • Timber removals • Forest area affected by fire • Crown defoliation • Carbon stock in soil and biomass • Impact on non-wood forest products and other plants
United Nations Convention to Combat Desertification Updated monitoring framework (UNCCD, 2023)	• Trends in land cover • Trends in land productivity

2015b). SDG6, "Clean water and sanitation", calls for restoration and protection of forests in one of its targets. Nevertheless, 14 of the 17 SDGs contain forest-related indicators or indicators with a forest component (Linser & Lier, 2020) (see Table 9.5). The FAO, with its data collections for the forest sector, serves as the custodian agency for the data and information for these indicators.

Table 9.5 Direct forest-related indicators of the Sustainable Development Goals

Framework	Forest-related indicators
United Nations Sustainable Development Goals (United Nations, 2018)	2.3.1 Volume of production per labour unit by classes of farming/pastoral/forestry enterprise size 7.2.1 Renewable energy share in the total final energy consumption 15.1.1 Forest area as a proportion of total land area 15.1.2 Proportion of important sites for terrestrial and freshwater biodiversity that are covered by protected areas, by ecosystem type 15.2.1 Progress towards sustainable forest management 1. The forest area net change rate 2. Aboveground biomass stock in forests 3. Proportion of forest area located in legally established protected areas 4. Proportion of forest area under long-term forest management plans 5. Forest area under independently verified forest management certification schemes 15.3.1 Proportion of land that is degraded over the total land area

The integration of forest-related indicators into the SDG framework highlights their role as governance tools that link forest management to broader sustainability objectives. By aligning national forest policies with SDG targets, countries can ensure that their forest management practices contribute to global development goals, such as poverty reduction, climate action, and biodiversity conservation.

9.3.3 International and Regional C&I for SFM Processes

Indicators have been a valuable tool in monitoring and guiding governance in many sectors and countries. However, the post-Rio move to monitor progress towards sustainable development using indicators is an ambitious approach for an area that has not yet been systematically monitored or quantified. In this domain, the forest sector has been a pioneer in developing and applying indicators to monitor progress. By the late 1990s, the first sets of indicators for sustainable forest management had already been developed by regional C&I processes for every region of the world, demonstrating how indicators can be adapted to address specific governance needs at different regional scales and providing tailored guidance for member countries. By the year 2000, 171 countries were actively participating in eleven regional processes to develop C&I for SFM. Today, however, only six C&I processes remain, proactively coordinating and supporting their member countries (Linser et al., 2018a).

The indicator sets of these processes serve several important purposes: First, they act as reporting tools for describing and diagnosing SFM for the public and decision-makers. Second, they function as communication tools to facilitate dialogue on SFM, clarify complex issues, and streamline forestry debates both within and among countries. Third, they serve as tools for collecting and processing priority information, linking forestry to other sectors and global initiatives. Additionally, they provide frameworks for policymaking, including monitoring, goal setting, decision-making, and identifying enabling conditions for SFM implementation. These frameworks also support the development of policies on forest conservation, management, and sustainable development. Furthermore, C&I processes provide assessment tools for analysing program effectiveness and describing national progress. Finally, they contribute to clarifying issues related to the environment and trade, including product certification (Baycheva et al., 2013; Caswell et al., 2014; Julve Larrubia et al., 2017; Linser, 2002; Linser & Wolfslehner, 2015a; MacDicken et al., 2015; Prabhu et al., 2001; Wolfslehner et al., 2016).

Through the development of C&I, the forest sector has been a leader in addressing the complex issues associated with sustainable forest management. In the following, the three most active and important regional C&I for SFM processes are introduced.

9.3.3.1 The Montréal Process

The Montréal Process Criteria and Indicators for the Conservation and Sustainable Management of Temperate and Boreal Forests cover 90% of the world's temperate and boreal forests. Founded in 1993, its first C&I set was developed in 1995. It provides a comprehensive framework for assessing and promoting sustainable forest management in these regions (Linser et al., 2018a). Following a revision in 2009, the framework now consists of seven criteria and 54 indicators, including 10 descriptive indicators. The 12 member countries that are part of the Montréal Process which include non-European countries in the temperate and boreal zones—Argentina, Australia, Canada, Chile, China, Japan, the Republic of Korea, Mexico, New Zealand, Russia, the United States, and Uruguay—use this set of criteria and indicators to prepare country reports on progress toward sustainable forest management. These reports are valuable tools for monitoring, assessing, and reporting on the state of forests as well as for reporting progress towards sustainable forest management at both national and international levels (Montréal Process Working Group, 2023).

9.3.3.2 The International Tropical Timber Organization

ITTO, the International Tropical Timber Organization, has played a crucial role in promoting the sustainable management of tropical forests through its criteria and indicators for sustainable forest management, first developed in 1992. The organisation represents 75% of the world's tropical forests. The set contains seven criteria and 58 indicators, about half of which are descriptive indicators. The indicators were last revised in 2016 to reflect advancements in forest management practices and to align with global sustainability goals (Linser et al., 2018a). ITTO's C&I have been widely adopted and adapted by member countries, fostering a shared understanding of sustainable forestry practices. They have also contributed to international efforts to mitigate climate change, combat deforestation, and support the livelihoods of communities dependent on tropical forests. By establishing a clear and practical framework, ITTO's pioneering work on C&I has set a global standard for sustainable forest management, ensuring that tropical forests can continue to provide vital ecological services and resources for future generations (ITTO, 2016).

9.3.3.3 Forest Europe

Forest Europe (previously known as the Ministerial Conference for the Protection of Forests in Europe, or MCPFE) is the corresponding forest policy process for the sustainable management of the pan-European forests. It was established in 1990 as a regional political initiative to promote sustainable forest management in its 44 member countries and the European Union. The initial pan-European C&I set was developed between 1993 and 1998. It was revised several times in response to new fields of interest and emerging issues, such as climate change. The six criteria for SFM remain unchanged. The pan-European indicator set, which consists of 34 quantitative and 11 qualitative indicators since its last revision in 2014–2015, has served as the structure for the State of Europe's Forests reports and formed also the basis for national indicator sets of 23 European countries (Linser & Wolfslehner, 2022; Linser et al., 2015).

Since the first Ministerial Conference on the Protection of Forests in Europe in 1990, significant changes in forests and society have influenced information requirements within the forest sector and for external communication. These requirements have expanded to include aspects such as social issues and qualitative aspects, addressing forest governance, transformative changes in forest governance, policy frameworks, and cross-sectoral policy areas to tackle complex challenges like climate change, biodiversity loss, and the bioeconomy. Despite this progress, a lack of awareness persists regarding the integration of new requirements and the need to communicate sustainable forest management to other sectors (Lier et al., 2021; Linser, 2024; Prins et al., 2023; Wolfslehner et al., 2016). When the first set of C&I was adopted in 1998, ministers stressed that the indicators were neither final nor comprehensive, given the multifunctionality of forests. They emphasised the need for ongoing analysis, amendment, and development (Linser et al., 2015).

Forest Europe recognised the necessity to address not only demands from the forest sector but also those from other sectors such as energy, biodiversity, and the bioeconomy. To assess international commitments, such as the European Goals for Forests and the European 2020 Targets for Forests (Linser & Wolfslehner, 2015b), the indicator set was revised in 2014–2015 to include previously unaddressed issues, including forest degradation. The revision process involved consultations with experts and stakeholders to reflect diverse national contexts and align with ongoing European initiatives. Political relevance, significance, data availability, cost-effectiveness, technical feasibility, validity, and reliability were the evaluation criteria for the indicators. The latest revision resulted in 34 quantitative and 11 qualitative indicators (Forest Europe, 2014, 2015).

The pan-European C&I set has served as the foundation for the State of Europe's Forests (SoEF) Reports (2003, 2007, 2011, 2015, 2020, and 2025), supporting regional and national policy formulation, analysis, and monitoring. Discussions on the use of a selection of key indicators were initiated in 2017 by the Forest Europe Expert Group on the Implementation of the Updated Pan-European Indicators for SFM (Forest Europe, 2017a; Linser & Wolfslehner, 2022). Thus far, this proposal has not been taken up.

9.3.3.4 Lessons Learned from the International and Regional C&I Processes

The development and implementation of C&I for SFM over the past three decades have provided valuable lessons for advancing sustainability in the forest sector and beyond. One key lesson is the importance of adaptability. C&I frameworks must evolve to address emerging challenges, such as climate change, biodiversity loss, and the growing demand for bioeconomy solutions, while maintaining their comprehensive and balanced approach. The iterative revisions of C&I sets, as seen in the Forest Europe process, demonstrate the need for continuous reconsideration, evaluation, and regular improvement to ensure relevance and alignment with global and regional priorities, even when the resulting changes to the text are relatively minor.

Another critical lesson is the value of participatory processes in addressing complex issues where stakeholder values sometimes differ significantly. Engaging diverse forest-related stakeholders, including forest managers, scientists, decision-makers, and other relevant stakeholders, has been essential for ensuring that C&I frameworks reflect diverse perspectives and are practical for implementation. This fosters ownership, enhances legitimacy, and strengthens the utility of C&I as tools for decision-making, reporting and communication.

The adaptation of forest-related C&I to align with broader policy frameworks has also highlighted the need for cross-sectoral collaboration. Forest-related indicators must align with targets in other sectors, such as energy, biodiversity, and climate policy, to ensure coherence and maximise impact, and elevate forest issues to the highest policy levels. However, challenges remain in harmonising terminology, data collection, and monitoring approaches across sectors and regions.

Ultimately, the C&I processes highlight the significance of political and institutional commitment. Sustaining these frameworks requires long-term investment in capacity building, data collection, and reporting systems. Despite resource constraints in some countries, the forest sector's leadership in developing C&I has demonstrated the potential for these tools to drive progress toward monitoring, assessment and reporting. Sharing these lessons with other sectors can further enhance global efforts to achieve the SDGs and address interconnected environmental, social, and economic challenges.

One of the most important lessons learned is the need to explicitly recognise and leverage the dual role of C&I as both reporting mechanisms and policy tools. While their monitoring and assessment functions are well-established, their potential to shape policy, guide resource allocation, and drive governance reforms remains underexplored. Addressing this gap could significantly enhance their impact on sustainable forest management.

9.3.4 Recent Indicator Developments

In the second decade of this Century, C&I for SFM regained significant attention due to emerging climate change and biodiversity-related challenges in forest management (Baycheva et al., 2013; Linser, 2024; Linser et al., 2015, 2018a).

9.3.4.1 The Global Core Set of Forest-Related Indicators

In addition to regional and international forest processes, there have been other initiatives to develop forest-related indicator sets. For example, the FAO led a global project on SFM indicators (FAO, 2015c), which resulted in the development of the Global Core Set of indicators. The 21 indicators show trends towards achieving the globally agreed-upon goals and targets of the 2030 Agenda for Sustainable Development (including monitoring SDG 15.2.1), the UN Strategic Plan for Forests, the UNFF Global Objectives for Forests, and the obligations from the Rio Conventions. The indicators are to be reported by all countries worldwide. Its main aim was to reduce the reporting burden on the countries and to obtain harmonised, comparable data to be presented in the FAO's State of the World's Forests reports (FAO, 2022; FAO & CPF, 2022b). According to a recent assessment of the global core set's uptake, it is a useful and flexible tool that could reduce the reporting burden on countries and increase understanding of SFM; however, more work is required to achieve its full potential. This could be achieved through increased capacity building, promotion, and cooperation in data collection, dissemination, and communication (FAO & CPF, 2022a).

9.3.4.2 Indicators for Sustainable Finance

Sustainable finance has emerged as a crucial component in the global effort to tackle challenging environmental, social, and governance (ESG) issues. It aims to direct financial flows toward activities that support sustainable development, mitigate climate change, and foster social well-being, all while ensuring financial stability and long-term economic growth. Despite its promising prospects, sustainable finance faces several challenges. Standardisation of ESG metrics remains inconsistent, creating difficulties in comparing and evaluating investments. This makes it difficult for investors to assess whether an investment is sustainable or whether the sustainability claims are misleading, potentially promoting "greenwashing".

The European Union (EU) established the EU Taxonomy as a classification system to define what types of economic activities can be considered environmentally sustainable. Its main objective is to create a common, transparent framework that guides investors, companies, and policymakers in identifying and supporting investments that contribute to environmental objectives, particularly in the context of the EU's climate and sustainability goals. The EU Taxonomy outlines clear technical criteria that economic activities must fulfil to be classified as environmentally sustainable. These criteria are based on scientific evidence and aim to ensure that investments genuinely contribute to sustainability. To date, the European Commission's efforts to publish a taxonomy for the forestry sector have been unsuccessful. This is due, in part, to the question of how to assess the maintenance and enhancement of biodiversity within the context of SFM. Therefore, no indicator system is currently available from the EU Taxonomy that can be used to make an objective assessment of sustainable forest management.

Although the financial sector is not fundamentally opposed to investing in forestry activities, a degree of uncertainty remains due to the anticipated complexity of managed forest ecosystems and the critical attitude of some NGOs towards forest utilisation (BirdLife International, 2022). There are also inherent risks such as drought, insect infestations, forest fires, and storms. As a result, there is an increasing demand for indicators to objectively assess sustainability in terms of ESG aspects.

Cadman and Köhl (2025) studied a range of initiatives related to financing and SFM, specifically sustainable forest financing and the principles and criteria underlying it. They present a preliminary, indicative framework with principles and criteria for sustainable forest financing, outlining the associated criteria for the areas of environment, economy, investment, society, governance, and cross-cutting issues. Kulionis et al. (2024) studied the impact of biodiversity on finance by analysing the MSCI All Country World Index. They showed the influence companies have with their financial engagement on biodiversity and the reciprocal effects. They concluded that companies neglecting those impacts risk financial setbacks.

Indicators for sustainable finance are essential tools for monitoring progress, identifying gaps, and guiding strategic decision-making towards a more sustainable economy. Although challenges remain, continued development and harmonisation of these indicators are crucial. They improve transparency and accountability and also catalyse the alignment of financial flows with global sustainability goals. As the field evolves, integrating comprehensive, reliable, and comparable indicators will be fundamental in ensuring that finance truly serves as a driver for sustainable development.

9.3.4.3 EuropaBON

The European Biodiversity Observation Network (EuropaBON) was a project funded by the European Union's Horizon 2020 research and innovation program from 2020 to 2024. The primary aim of EuropaBON was to design and implement a coordinated biodiversity monitoring system for Europe, addressing the fragmented and inconsistent nature of current biodiversity data collection. By harmonising monitoring efforts, EuropaBON sought to provide robust, policy-relevant data to support the European Green Deal, the EU Biodiversity Strategy for 2030, and international commitments such as the CBD. Forest-related variables were an essential component of EuropaBON's work, as forests are among the most biodiverse ecosystems in Europe. The project focused on improving the quality and comparability of forest biodiversity data, including indicators called in this context 'essential biodiversity variables' on forest cover, species diversity, and ecosystem health, to inform SFM and conservation policies (Junker et al., 2023).

9.3.4.4 Transformative Change Discourse and the Role of C&I in Forest Governance

Transformative change is defined as fundamental, strategic, and systemic shifts that address the root causes of unsustainability and inequality in forest governance. Literature highlights the growing importance of transformative change discourse in forest governance, emphasising the need for fundamental shifts in systems, institutions, and power relations to achieve sustainability and equity (Larson et al., 2021). C&I are increasingly recognised as essential tools for enabling, assessing, and operationalising such transformation (FAO & CPF, 2022a; Linser et al., 2018b; Prins et al., 2023). The transformative change discourse underscores that C&I are not merely technical tools but are embedded in broader governance debates about power, justice, and legitimacy. Their design and implementation can either enable or constrain transformation, depending on how well they reflect diverse stakeholder values, address systemic barriers, and support adaptive learning. Participatory and context-sensitive indicator systems are considered crucial for promoting inclusive and democratic forest governance. The adaptability of C&I frameworks positions them as critical tools for enabling and assessing transformative changes in forest governance (Prins et al., 2023; Visseren-Hamakers et al., 2021). By integrating new priorities, such as climate adaptation and biodiversity restoration, and fostering cross-sectoral collaboration, C&I frameworks can help drive systemic shifts in forest governance.

Recent broadly participatory C&I revision processes and development of new indicator sets propose hierarchical and holistic indicator frameworks that take into account the climate change and biodiversity crisis, resilience and trade-offs with economic aspects, allowing for more nuanced assessments of transformative change (Linser, 2024; Linser et al., 2025a, 2025b; Nikinmaa et al., 2023).

9.3.5 National Uptakes

Many countries have developed national sets of C&I for SFM, incorporating some or all indicators from regional or international processes alongside those of national importance. These national C&I sets reflect stakeholder values, objectives, priorities, and goals specific to each country. They are intended to serve as a basis for evidence-based policymaking, decision-making, and communication. A national C&I set helps define SFM within a national context, reveals the multifunctionality of forests, and supports the preparation of forest development and management plans. It facilitates monitoring, assessment, and periodic reporting on the state of forests and the forestry sector, while also aiding communication with government officials, stakeholders, and the general public through various channels such as press releases, indicator-based policy briefs, leaflets, and social media. Additionally, national indicator sets support international reporting obligations, such as those for various United Nations Economic Commission for Europe (ECE) and FAO reports, the SDG progress reports, the Montréal Process Overview Reports, or the SoEFs reports. In 2011, ITTO emphasised that C&I for SFM are the most decisive policy instruments for monitoring, assessing, and reporting on the implementation of SFM at the national level (ITTO, 2011). Through indicator-based reports, awareness is raised about forest-related issues in sectors such as biodiversity, bioeconomy, and rural development. National C&I sets also strengthen the development or revision of forest legislation, policies, and programmes, while monitoring their implementation. The indicators provide incentives for practical SFM, encourage dialogue within and beyond the forest sector, and demonstrate the societal benefits of forests. Core users of indicator-based national data and information include government organisations, forest owners, environmental groups, and the scientific community (Linser & O'Hara, 2019).

9.3.5.1 National Uptake of the Pan-European C&I

The definition of SFM and the development of related C&I were considered among the most significant achievements of the Forest Europe process by its signatories (Forest Europe, 2017b; Pülzl & Mayer, 2015). Linser and Wolfslehner (2022) investigated the implementation of the pan-European C&I for SFM across Forest Europe member countries. They found that more than half of the signatory countries had implemented C&I for SFM at the national level, covering 52% of the pan-European forest area. In six countries (13%), a national C&I set was under development or planned. However, eleven countries (24%), including those with large forest areas such as Ukraine, Sweden, and Germany, reported no intention to apply SFM indicators at the national level.

Overall, international C&I have provided a valuable structure and framework for deriving national indicators. The findings confirm their importance for monitoring and reporting on the status and trends of forests and forestry in Europe. However,

Forest Europe has yet to transition from descriptive reporting to target-based assessments. While this was not originally intended for the indicators, it is increasingly demanded by decision-makers and stakeholders. Future development of SFM indicators should prioritise their suitability for assessing objectives, goals, and targets, as the ability to monitor national efforts has become a critical tool for both international and national governance (Linser & Wolfslehner, 2022; Linser et al., 2018b).

9.3.6 Key Milestones in the Development of C&I for SFM

1992	**UNCED and Agenda 21** • *UNCED in Rio de Janeiro highlights the need for sustainable development and proposes the use of indicators in Agenda 21 (Chapter 40.4)* • *The Forest Principles emphasise the importance of SFM*
	ITTO begins developing C&I for tropical forests • *The International Tropical Timber Organization (ITTO) initiates the development of C&I for the sustainable management of tropical forests. It publishes the world's first C&I for SFM set consisting of seven criteria and 66 indicators*
1993	**Forest Europe C&I process established** • *Forest Europe (formerly MCPFE) began developing C&I for SFM for the pan-European region; the set was adopted in 1998*
1994	**Montréal Process Working Group launched** • *The Montréal Process began to develop C&I for temperate and boreal forests outside of Europe, as set out in 1995*
	UNCCD indicators adopted • *Forest-related indicators, such as land cover, land productivity, and carbon stocks, are included in the United Nations Convention to Combat Desertification (UNCCD) as part of a broader monitoring framework for land degradation*
1995	**First Montréal Process C&I Set** • *The Montréal Process publishes its first set of C&I, consisting of 7 criteria and 67 indicators*
	Intergovernmental Panel on Forests (IPF) • *The IPF is established under the UN Commission on Sustainable Development to address forest-related issues, including the development of indicators for SFM. Active until 2000*
1997	**Intergovernmental Forum on Forests (IFF)** • *The IFF succeeds the IPF and continues to promote the development and harmonisation of forest-related indicators, alongside other activities*
1998	**Pan-European C&I Set finalised** • *Forest Europe finalises its first pan-European C&I set, consisting of 6 criteria and 35 indicators endorsed by the Lisbon Ministerial Conference*
2000	**United Nations Forum on Forests (UNFF) established** • *The UNFF is created to promote the management, conservation, and sustainable development of forests, including the use of indicators to monitor progress toward global forest goals*

(continued)

(continued)

2001	**1st Pan-European C&I revision** • *Forest Europe revises its C&I set, resulting in 35 quantitative and 17 qualitative indicators*
2003	**First State of Europe's Forests Report** • *Forest Europe publishes the first State of Europe's Forests report, based on the pan-European C&I set*
2007	**UNFCCC and Forest-related Indicators** • *The United Nations Framework Convention on Climate Change (UNFCCC) incorporates forest-related indicators into the REDD+ mechanism, focusing on monitoring deforestation and forest degradation to reduce carbon emissions and enhance forest carbon stocks*
2009	**Montréal Process revision** • *The Montréal Process revises its C&I framework to reflect advancements in forest management practices*
2010	**CBD Aichi Biodiversity Targets adopted** • *The CBD adopts the Aichi Biodiversity Targets as part of its Strategic Plan for Biodiversity 2011–2020. Several forest-related indicators are included to monitor progress toward biodiversity conservation*
2014	**2nd Pan-European C&I revision** • *Forest Europe revises its C&I set, resulting in 34 quantitative and 11 qualitative indicators*
	First joint data collection under CFRQ • *The Collaborative Forest Resources Questionnaire (CFRQ) partnership, comprising the FAO, UNECE, Forest Europe, ITTO, the Montréal Process, and the Observatory of the Central African Forests Commission, conducts its first joint data collection, covering 88% of the global forest area. This joint data collection was repeated in 2018 and 2023*
2015	**Paris Agreement and SDGs** • *The Paris Agreement under the UNFCCC emphasises the role of forests in climate change mitigation, increasing the demand for robust forest-related indicators* • *The adoption of the Sustainable Development Goals (SDGs) includes forest-related indicators, such as SDG 15.2.1, which focuses on sustainable forest management*
	Global Core Set of Forest-related Indicators initiated • *The FAO begins developing the Global Core Set of Forest-related Indicators to reduce the reporting burden on the countries and to obtain harmonised, comparable data; the set consists of 21 indicators*
	EU Green Deal adopted • *The European Green Deal sets ambitious goals for climate neutrality and biodiversity conservation, driving the need for policy-relevant forest indicators*
2016	**SEMAFOR Project launched** • *The UNECE/FAO Team of Specialists on Monitoring SFM launches the System for the Evaluation of the Management of Forests (SEMAFOR) project* • *SEMAFOR utilises its innovative approach, leveraging the pan-European C&I, to assess progress toward SFM in 20 European countries. This involves setting thresholds or targets for areas of concern and engaging national experts to contextualise the data. A study published in 2017*
	ITTO C&I revision • *The ITTO revises its C&I framework to align with global sustainability goals*

(continued)

(continued)

2017	**UN Strategic Plan for Forests (2017–2030)** • *The UNFF adopts the Strategic Plan for Forests, which includes the Global Forest Goals and Targets, further emphasising the role of indicators in tracking progress*
2020	**EU Taxonomy and indicator needs** • *The EU Taxonomy is introduced as a classification system to define environmentally sustainable economic activities, including some relevant to forestry* • *The taxonomy underscores the necessity for robust, measurable indicators to evaluate sustainability, particularly in areas such as biodiversity, carbon sequestration, and ecosystem services*
	State of Europe's Forests 2020 report highlights gaps • *The State of Europe's Forests 2020 report identifies gaps in data collection and reporting, emphasising the need for more dynamic and adaptive indicators*
2022	**Kunming-Montreal Global Biodiversity Framework** • *The CBD adopts the Kunming-Montreal Global Biodiversity Framework, which includes forest-related indicators which aim to monitor progress toward biodiversity conservation, ecosystem restoration, and sustainable use of forest resources*
2023	**UNCCD updated monitoring framework** • *The UNCCD updates its monitoring framework to include improved indicators for land cover, land productivity, and carbon stocks*
2024	**9th Forest Europe Ministerial Conference initiates a review of C&I** • *The Ministers responsible for forests and forestry initiate a review of criteria and indicators for decision-making on SFM to strengthen monitoring, reporting, and assessment of forest resources and SFM in Europe*

9.3.7 Sustainability Assessments with Assigned Targets

Although SFM was first defined for pan-Europe at the Helsinki Ministerial Conference nearly thirty years ago, no widespread consensus has been reached yet on how to objectively monitor and assess progress towards SFM (rather than simply describing trends without assessing whether they are sustainable or not). Countries have emphasised the diverse range of national circumstances and the lack of consensus on what indicator values should be considered sustainable. Individual indicator targets or thresholds have been developed, on an experimental basis, for the SoEF Report 2011 (Forest Europe et al., 2011), for the Austrian set of C&I for SFM (Linser, 2020) and in the SEMAFOR project (System for the Evaluation of the Management of Forests) (UNECE/FAO, 2016).

Initiated by the ECE/FAO Team of Specialists on Monitoring SFM, the SEMAFOR project led to an innovative regional assessment method to assess progress made by 20 European countries toward SFM, aiming to answer what areas of concern about sustainability in each country are and how these areas of concern are being addressed now. The SEMAFOR concept was based on using the Pan-European indicators to describe progress towards SFM in European countries, setting common thresholds for "concern" about recorded trends, leading to a dialogue with national experts who, for example, put the recorded data in context or cited policy measures being put in place to address identified issues. The final text was agreed upon between secretariat analysts and national correspondents. SEMAFOR combines objective and transparent measurement with a modulated approach considering national circumstances to produce credible and meaningful results, moving beyond description to assessment (UNECE/FAO, 2016).

Currently, the assignment of thresholds or targets to various indicators has garnered significant interest and is increasingly being requested for comprehensive sustainability assessments (Linser et al., 2018b; Linser & O'Hara, 2019; Onida, 2021). While future revisions of indicators may address these demands, it is important to recognise that reaching participatory agreements on such thresholds or targets becomes increasingly challenging as the scale and visibility of the indicator assessments grow. At larger scales, achieving consensus may be nearly impossible due to the substantial regional variations in ecosystems, climatic conditions, and socio-economic contexts. These complexities underscore the need for flexible approaches that consider local and regional specificities while maintaining coherence with broader sustainability objectives. There should also be a strong belief among participating countries that the process, which can easily become embarrassing or conflictual, will produce desirable results, notably more evidence-based policies, international visibility and prestige, and a clear demonstration that international commitments regarding sustainability are being implemented.

9.3.8 Local Level Indicator Sets and Forest Certification

The widespread concern about sustainability and deforestation visible in Rio also led to several initiatives aimed at certifying that specific forests were being managed sustainably. These certifications were designed to reassure final consumers that they were not promoting deforestation when purchasing wood products. The first of these initiatives to be implemented at the international level was the FSC, as well as the Programme for the Endorsement of Forest Certification (PEFC), and national certification schemes outside the international systems, such as those in Brazil and Indonesia. All are grounded in a set of basic principles and a detailed country-specific standard, which is implemented by certifying agencies and chain-of-custody rules. Certification serves as a decision-making tool at the enterprise level, guiding forest managers in implementing sustainable practices. Additionally, it provides a means for continuous improvement and enables access to local and foreign markets. These schemes are considered voluntary market-based instruments (Maesano et al., 2018; Mikulková et al., 2015; Rametsteiner & Simula, 2003).

Various more recent Horizon Europe projects have funded the development of local-level forest-related indicator sets that focus only on certain aspects of SFM, like indicators for climate-smart forestry (Santopuoli et al., 2021), indicators for forest resilience (Lloret et al., 2024), essential forest mitigation indicators (Linser et al., 2025b) and indicators for integrative forest management (Linser et al., 2025a).

Some of those indicator sets are also scalable to the regional or national level (with separate national-level measurement units), making them important tools for intergovernmental comparisons.

9.4 Challenges and Opportunities for Criteria and Indicators

While Sect. 9.3 explored the application and implementation of C&I frameworks across scales, this section examines the challenges and opportunities that arise in their development and use. The dual role of C&I as both reporting mechanisms and governance tools, highlighted earlier, presents unique demands. On the one hand, their effectiveness as reporting tools depends on robust data collection, harmonisation, and standardisation of definitions. On the other hand, their role as policy instruments requires adaptability, stakeholder engagement, and alignment with broader governance frameworks. Balancing these functions is critical to ensuring that C&I frameworks remain relevant and effective and continue to contribute to monitoring and policy formation.

The development and implementation of C&I for SFM face numerous challenges but also present significant opportunities, particularly where C&I frameworks are more frequently applied, and data collection and harmonisation efforts are relatively advanced. A key issue remains the trade-offs between comprehensiveness and feasibility in indicator frameworks. While comprehensive systems provide a detailed picture of sustainability, they often require many resources, which makes them less feasible for continuous application. Even in regions with better-developed monitoring and data collection systems, challenges persist in harmonising methodologies across diverse political and ecological contexts. Existing monitoring systems still employ some incompatible definitions and approaches, which complicate data integration and limit the comparability of results.

Institutional and governance challenges also hinder progress. Although verbal support for using SFM is generally strong, insufficient prioritisation of monitoring, reporting and assessment, coupled with limited funding, can undermine the effectiveness of C&I frameworks. Specialised staff often face high turnover, and resources allocated to SFM monitoring are not always sufficient to meet the demands of comprehensive inventories. Financial and technical constraints include non-flexible data collection procedures in case of emerging issues, such as large-scale damaging events, and limited facilities for open-access information sharing. It is also unclear to what extent high-level policy decisions on forest-relevant matters rely on C&I-based evidence or whether more traditional tools (ad hoc studies, parliamentary enquiries, etc.) are used as a basis for policy making.

Despite these challenges, many countries and regional processes are well-positioned to lead the way in transforming C&I into transparent, continuous (with

5–10-year reporting cycles), and efficient systems for monitoring all aspects of sustainable forest management.

By leveraging their existing capacities, the reporting countries and international or regional processes can focus on improving data collection systems, enhancing knowledge transfer, and addressing remaining technological and methodological gaps. Strengthening collaboration among stakeholders and fostering regional harmonisation can further improve the effectiveness of C&I frameworks. By addressing these challenges and striking a balance between comprehensiveness and feasibility, C&I for SFM can become a powerful tool for advancing sustainable forest management and achieving broader sustainability goals.

The challenges associated with C&I frameworks are diverse, ranging from technical and methodological barriers to institutional and governance constraints. However, these challenges also present opportunities for innovation and improvement, particularly in the context of emerging digital technologies and increasing attention to sustainability issues. By addressing these barriers and leveraging new opportunities, C&I frameworks can be further strengthened to meet the evolving demands of SFM policy in a rapidly changing world.

9.4.1 Partnership in Data Collection

Indicator-based data on forests and forestry provide a comprehensive reference and information framework on the past and current state of forests and forestry, serving as an input for future modelling. The FAO Forest Resource Assessment (FRA) is the primary source of information on the state of the world's forests, with regular reporting dating back to 1948. In 2012, the FAO initiated a new Collaborative Forest Resources Questionnaire (CFRQ) partnership comprising six organisations or regional and international processes (FAO, 2025a): the FAO, ECE, Forest Europe, ITTO, the Montréal Process, and the Observatory of the Central African Forests Commission. The CFRQ is the successful result of the collaborative commitment of these organisations and indicator processes to streamline and harmonise forest-related data collection, thereby reducing the reporting burden on the national correspondents. However, the remaining C&I processes are not part of the CFRQ primarily because they are at different stages of development, particularly in terms of harmonised data availability. Joint data collection under the CFRQ was first conducted in 2014, and subsequently repeated in 2018 and 2023, covering 88% of the global forest area in 103 countries. The Global Core Set of 21 Forest-related Indicators was developed by the Collaborative Partnership on Forests (CPF) to streamline the collection of comparable information relevant to all countries worldwide (see Sect. 9.3.4.1).

The ECE/FAO Joint Pan-European Data Collection is a long-standing collaborative effort between the FAO, ECE, and Forest Europe (FAO, 2025b). Since the first SoEF report in 2003, these three secretariats have collaborated to gather comprehensive national data on forest resources and the environmental, social, and economic aspects of SFM across the pan-European region. This joint initiative serves as the primary source of information for the Forest Europe indicators, directly feeding into its regular SoEF reports. The primary goal is to minimise the reporting burden on individual countries by harmonising data collection, ensuring consistency through shared definitions and methodologies (often coordinated with the FAO's FRA), and providing robust and comparable data to support evidence-based forest policymaking and monitoring of SFM. This collaboration involves a well-established network of national correspondents and experts, and it has evolved to incorporate new digital tools for data submission and review, further optimising national and international resources.

Another example of collaboration in forest data collection is the Forest Data Partnership, which was initiated in October 2021. The five-year partnership is coordinated by the World Resources Institute (WRI) with support from USAID, the U.S. Department of State, Google, SERVIR (a joint NASA-USAID initiative), and numerous other leading organisations, governments, and private sector partners (WRI, n.d.). Its goal is to develop a consistent, open-source, publicly available, and validated geospatial data ecosystem related to forest-risk and restoration indicators. It focuses on aligning stakeholders, innovating data approaches, and deploying data delivery mechanisms. The initiative aims to facilitate credible and systematic monitoring, verification, and disclosure to drive progress in reducing deforestation and restoring degraded lands, particularly in light of global commitments, such as the COP26 Glasgow Leaders Declaration on Forests and Land Use, and regulations like the EU Deforestation Regulation (EUDR).

9.4.2 Indicator-Based Digital Transformation in Forest Monitoring

Digital transformation is revolutionising how indicators of SFM are developed, refined, and applied, moving from traditional, often laborious, and infrequent assessments to a more dynamic, precise, and integrated approach. This shift, often termed "Forest 4.0", leverages advanced technologies to overcome previous limitations, offering unprecedented opportunities for enhanced monitoring, reporting, and adaptive management (Damaševičius et al., 2024). Advances in remote sensing technologies, artificial intelligence (AI), and big data analytics provide unprecedented opportunities to improve the precision, efficiency, and scalability of forest monitoring systems. Remote sensing technologies, such as satellite imagery and LiDAR, provide high-resolution, real-time data on forest cover, biomass, forest damage, and many other indicators, enabling the development of more accurate and dynamic models (Lechner et al., 2020; White et al., 2016). For example, instead of just "forest cover change", it is now possible to develop indicators for "canopy gap dynamics," "forest structural complexity," or "specific tree species distribution changes", providing a richer picture of biodiversity and health. AI and machine learning algorithms can process vast amounts of data to identify patterns, predict trends, and automate the analysis of complex forest-related datasets (Holzinger et al., 2022; Lulli et al., 2019). These models can forecast forest growth, identify areas at high risk of disease, fire, or illegal logging, and predict the impact of management interventions. This enables indicators to transition from being purely retrospective (what happened?) to being predictive (what might happen?), facilitating proactive management strategies and the development of early warning indicators for potential sustainability breaches (Wang et al., 2025). Digital platforms facilitate the integration of diverse data types—from satellite imagery and drone data to ground-based sensor readings, traditional forest inventories, and even social media data for community engagement—leading to large-scale forest models and local to continental-scale decision support systems (Lexer & Vacik, 2016). This integrated data environment enables the creation of more holistic indicators that reflect the complex interactions between the ecological, social, and economic dimensions of SFM, moving beyond siloed information.

Automating data collection and analysis by digital tools significantly reduces the expenses and time associated with traditional monitoring. This allows for more frequent and extensive data collection, enabling the refinement of indicators across larger spatial scales and more diverse forest types, making them more broadly applicable and robust (Ehrlich-Sommer et al., 2024).

Digital platforms and standardised data structures (like those promoted by CFRQ) make it easier to collect, aggregate, and report SFM indicators to national and international bodies. Blockchain and other technologies are even being investigated to ensure the traceability and immutability of data, increasing trust and transparency in sustainability claims for certified forest products or carbon credits. This facilitates robust reporting for frameworks like the EU Taxonomy (Komdeur & Ingenbleek, 2021).

Real-time data and predictive analytics empower forest managers to make more informed, adaptive decisions. Digital dashboards and Geographic Information Systems (GIS) provide easily digestible visualisations of indicator performance, allowing managers to identify trends, pinpoint areas of concern, and adjust management practices proactively to maintain or enhance sustainability. This transition enables SFM to shift from static planning to dynamic adaptation. The accessibility of near-real-time forest data through user-friendly platforms, such as the FRA Platform (FAO, 2025c), INForest (UNECE, 2023), and Global Forest Watch (GFW, n.d.), democratises information. This increased transparency can foster greater engagement from local communities, indigenous groups, environmental organisations, and consumers, allowing them to monitor SFM performance, hold stakeholders accountable, and participate more effectively in decision-making processes.

However, the digital transformation also presents challenges. Ensuring equitable access to these advanced tools is critical, particularly for stakeholders in resource-constrained regions. Ethical considerations, such as data privacy, ownership, and the potential for misuse, must also be taken into account to create trust and ensure responsible use of digital technologies.

The unprecedented volume and granularity of data generated by digital tools (e.g., LiDAR, high-frequency satellite imagery, drone surveys) can create a "flood of data".

While offering more detail, this can make it difficult to discern meaningful signals from noise, leading to challenges in identifying the most relevant and efficient indicators for SFM. Forest managers and policymakers may struggle to filter, synthesise, and interpret this vast information effectively. Instead of clarifying, too much data without proper analytical frameworks can obscure trends, making the selection and prioritisation of truly impactful SFM indicators more complex. There is a risk of developing a multitude of new, highly specific indicators that overwhelm reporting capacities without necessarily adding value to policy or management decisions. There is a risk that SFM indicators, while becoming more precise in certain aspects, might lose their ability to capture the complex, synergistic relationships within a forest system, or fail to adequately integrate the human elements (social well-being, livelihoods) that are inherently less quantifiable. Indicators could become overly technical, losing relevance to broader policy goals or community understanding. Many new SFM indicators derived from digital transformation rely heavily on complex algorithms, AI, and machine learning to process raw data (e.g., automated deforestation alerts, biomass estimates from LiDAR, species classification from imagery). The "black box" nature of some of these algorithms can result in a lack of transparency regarding how an indicator's value was determined. This can erode trust, especially if the outputs contradict ground observations or local knowledge. If stakeholders (policymakers, local communities, certification bodies) do not fully understand or trust the methodology behind an indicator, its credibility and acceptance as a reliable measure of SFM are compromised. This is particularly critical for indicators used for reporting against international commitments or for market-based mechanisms, such as carbon credits, where high levels of transparency and verifiability are paramount (Damaševičius et al., 2024).

Examples of technology-enhanced monitoring systems include the Global Forest Watch platform, which uses satellite data to track deforestation in near real-time (GFW, n.d.), the Copernicus Forest Environmental Monitoring and Management System which is part of the Copernicus Earth Observation Programme, providing open-access Earth observation data to support forest management and policy (Copernicus, n.d.), EOSDA Forest Monitoring, a commercial platform that uses satellite imagery to offer services like forest health assessment via vegetation indices, tree species identification, illegal logging detection, and carbon stock estimation for various stakeholders (EOS Data Analytics, 2025), and Live EO, a commercial AI-powered system to detect and assess deforestation, and to gather supplier data for EUDR compliance (LIVE EO, 2024). These systems demonstrate the potential of digital tools to transform forest monitoring, making it more inclusive, transparent, and effective in addressing global forest challenges.

9.4.3 Policy Implications

Indicator sets define the scope of SFM and identify topics that require monitoring and assessment. The results of this monitoring and assessment could facilitate the revision

Box 9.1 Key takeaways on digital transformation in forest monitoring

Technological advances	Remote sensing (e.g., satellite imagery, LiDAR) enables real-time, high-resolution data collection
	AI and machine learning automate analysis and predict trends
	Digital platforms integrate diverse data types for holistic monitoring
Benefits of digital tools	Improved precision, scalability, and reduced costs
	Real-time data supports adaptive management and early interventions
	Increased transparency and accessibility foster community engagement
Challenges	Data overload complicates indicator selection
	Limited access for resource-constrained regions
	"Black box" algorithms may reduce trust and transparency in indicator data
	Overly technical indicators risk undermining uptake for decision-making due to a lack of comprehensibility
Technology-enhanced systems	Global Forest Watch, Copernicus, EOSDA Forest Monitoring, Live EO, among others

of national forest strategies and programs by providing evidence-based insights into progress toward goals and targets outlined in these policy documents. Additionally, C&I-based information can be used to express and assess progress towards quantitative targets set in forest-related strategies and to provide a foundation for setting new targets in regional strategies and action plans, such as the EU Forest Strategy or the ITTO Strategic Action Plan (Lier et al., 2021, 2022). At the global level, this approach is equally relevant, as one of the SDG indicators measures progress toward SFM, contributing to the achievement of the 2030 Agenda for Sustainable Development. The Global Core Set of Forest-related Indicators, in conjunction with the regular FAO FRA, could play a crucial role in tracking progress toward implementing the International Arrangement on Forests, particularly in achieving the UNFF Global Objectives on Forests (FAO & CPF, 2022a). C&I sets—and the Global Core Set in particular—act as "norm setters" and stimulants, informing governments of the data to be collected for each indicator to meet their reporting obligations. The regular collection and reporting of data for the Global Core Set are expected to establish a voluntary de facto worldwide minimum standard for forest and forestry information that all nations should aim to meet (cf. Section 9.3.4.1). This process enhances transparency and accountability in addition to increasing capacity building in data collection and reporting, particularly in countries with limited resources, thereby strengthening global efforts to achieve SFM.

9.4.4 Policy-Relevant Indicators—The EU View

Policy-relevant indicators should play a crucial role as governance tools, supporting informed decision-making, policy formulation, evaluation, and effective communication. They strengthen transparency, continuity, and a holistic approach to forest governance while fostering multi-sectoral collaboration through the sharing of data. Indicators are essential for monitoring progress, informing policy, and ensuring the accountability of all stakeholders, particularly in providing evidence-based insights to guide policy decisions.

Between 2019 and 2024, under the European Green Deal, several forest-related and forest-focused policy instruments were published. To mention just a few, the Nature Restoration Regulation, the Carbon Removal Certification Framework, the Taxonomy Regulation, the EUDR, the Renewable Energy Directive (RED III), the EU Biodiversity Strategy, the Farm to Fork Strategy, the EU Forest Strategy, and the EU Strategy on Adaptation to Climate Change. Each of the above policy instruments has objectives and goals, as well as actions and, in some cases, targets, either qualitative or quantitative (Lier et al., 2021, 2022). These EU policy instruments can be divided into two categories: regulations that explicitly mention indicators, and communications, directives, and regulations that do not include any reference to indicators per se in their texts, recitals, or articles. The latter group forms the bigger share. While the Nature Restoration Law includes seven forest-related indicators, the other policy instruments lack clear indicators, which might hinder their implementation. Several possible methods exist for measuring the progress of policy instruments, as demonstrated by Lier et al. (2021, 2022). The results of the publications demonstrate that, for instance, the pan-European C&I for SFM (Forest Europe, 2015) can inform progress toward many of the targets in forest-related and forest-focused policymaking. However, adjustments to several indicators would be necessary to meet the information needs of EU policymakers fully. This highlights the potential of these indicators to avoid additional data collection while contributing to the monitoring and revision of EU forest-related policies, particularly under the umbrella of the European Green Deal. However, inconsistencies in terminology, conceptual language, and monitoring approaches across policy instruments remain a challenge, underscoring the need for greater synergy and alignment in forest-related target setting and reporting (Aggestam & Pülzl, 2018, 2020; Alberdi et al., 2020; EC JRC & EEA, 2021; Hetemäki et al., 2017; Lier et al., 2021).

9.4.5 Emerging Trends in Indicator Development

The development of forest-related indicators is evolving to address emerging global challenges and priorities, integrating new dimensions such as climate change adaptation, biodiversity decline, and the provision of all ecosystem services. Indicators are increasingly being designed to capture the complex interactions between forests and these critical issues, providing a more holistic understanding of forest dynamics and their contributions to global SDGs. For example, recent indicator development processes now aim to assess the roles of forests in carbon sequestration, climate change adaptation, and the provision of ecosystem services, while also monitoring biodiversity conservation efforts (Linser et al., 2025b; Nabuurs et al., 2024).

Another emerging trend is the growing emphasis on participatory processes in the co-development of indicators. Engaging stakeholders, including forest managers, policymakers, scientists, and local communities, ensures that indicators are relevant, practical, and reflective of diverse perspectives. This participatory approach not only enhances the legitimacy and acceptance of indicators but also fosters collaboration across sectors and scales, strengthening their utility in decision-making.

Additionally, there is a shift toward continuous or event-related reporting to address the dynamic challenges that forests face, as they are increasingly pressured by climate change and biodiversity emergencies. Indicator-based monitoring, assessment, and reporting must be flexible and capable of evolving in response to these changes to provide timely, informed support for sustainable forest management and serve as timely and effective tools for management and policy-level decision-making.

9.5 Conclusions

Over the past three decades, C&I for SFM has been developed and implemented worldwide, establishing a global framework for monitoring, reporting and evaluating SFM. These efforts have fostered a common understanding of SFM and supported the creation of extensive databases. They have also continuously expanded the scope of C&I in response to changing realities and challenges. Forest-related indicators have been successfully incorporated into the wider SDG framework (FAO, 2015b). Furthermore, C&I frameworks are increasingly being adapted to address emerging forest-related sustainability challenges such as biodiversity conservation, climate change mitigation and adaptation, and bioeconomy development. This demonstrates their relevance to related sectors.

In recent years, C&I have been recognised as dynamic governance instruments that go beyond their traditional role as reporting mechanisms. They now have the potential to influence policy, guide resource allocation and encourage cross-sectoral collaboration. This dual role as both monitoring tools and policy instruments positions C&I as transformative tools for addressing the complex challenges of SFM in the 21st Century. However, to realise their full potential, several essential changes are needed:

- Institutional commitment: Stronger political ownership and accountability are essential to ensure that C&I frameworks are prioritised and adequately resourced.
- Integration with digital platforms: Leveraging technologies such as remote sensing, AI, and big data can enhance the precision, scalability, and timeliness of C&I, enabling real-time monitoring and predictive insights.

- Clearer target setting: Establishing harmonised terms, definitions, and measurable targets will improve the comparability and policy relevance of C&I.
- Stakeholder engagement: Participatory processes must remain central to ensure that C&I reflect diverse perspectives and foster collaboration across sectors.

There are opportunities in the integration of digital tools, which can reduce costs, improve data accuracy and enable more frequent reporting cycles. It is also crucial to share lessons learned across sectors and foster international collaboration in order to address global challenges such as biodiversity conservation, climate change adaptation and the development of sustainable, bio-based economies. This will ensure that C&I frameworks remain relevant and effective in a rapidly changing world.

References

Adriaanse, A. (1995). In search of balance. A conceptual framework for sustainable development indicators. In *Accounting for change. Papers from an International Seminar* (pp. 3–10). London, UK: Toynbee Hall, The New Economics Foundation.

Aggestam, F., & Pülzl, H. (2020). Downloading Europe: A regional comparison in the uptake of the EU forest action plan. *Sustainability, 12*, 3999. https://doi.org/10.3390/su12103999

Aggestam, F., & Pülzl, H. (2018). Coordinating the uncoordinated: The EU forest strategy. *Forests, 9*, 125. https://doi.org/10.3390/f9030125

Alberdi, I., Bender, S., Riedel, T., Avitable, V., Boriaud, O., Bosela, M., Camia, A., Cañellas, I., Castro Rego, F., Fischer, C., Freudenschuß, A., Fridman, J., Gasparini, P., Gschwantner, T., Guerrero, S., Kjartansson, B. T., Kucera, M., Lanz, A., Marin, G., ... Hernández, L. (2020). Assessing forest availability for wood supply in Europe. *Forest Policy and Economics, 111*, Article 102032. https://doi.org/10.1016/j.forpol.2019.102032

Baycheva, T., Inhaizer, H., Lier, M., Prins, K., & Wolfslehner, B. (Eds.). (2013). *Implementing criteria and indicators for sustainable forest management in Europe*. European Forest Institute.

BirdLife International. (2022). *EU green taxonomy—Burning biodiversity.*

Brundtland, G. H. (1987). *Report of the world commission on environment and development: Our common future (No. A/42/427).* Geneva, Switzerland: United Nations Publications.

Cadman, T., & Köhl, M. (2025). Getting sustainable investments in the forest sector straight: Towards an integrating framework. *Journal of Sustainable Finance and Accounting, 5*, Article 100017. https://doi.org/10.1016/j.josfa.2025.100017

Carlowitz, H. C. V. (1713). Sylvicultura Oeconomica oder Haußwirthliche Nachricht und naturgemäße Anweisung zur Wilden Baum-Zucht. J.F. Braun, Leipzig, Germany.

Caswell, S., Tomaselli, I., & Hirakuri, S. (2014). Indicating progress: Uses and impacts of criteria and indicators for sustainable forest management (No. 42), Technical Series. ITTO, Yokohama, Japan.

CBD. (2022). Monitoring framework for the Kunming-Montreal global biodiversity framework. CBD/COP/DEC/15/5.

CBD. (2016). Indicators for the strategic plan for biodiversity 2011–2020 and the aichi biodiversity targets. CBD/COP/DEC/XIII/28.

Copernicus. (n.d.) *FOREMMS—Forest environmental monitoring and management system.*

Damaševičius, R., Mozgeris, G., Kurti, A., & Maskeliūnas, R. (2024). Digital transformation of the future of forestry: an exploration of key concepts in the principles behind Forest 4.0. *Frontiers in Forests and Global Change, 7*, 1424327. https://doi.org/10.3389/ffgc.2024.1424327

EC JRC, EEA. (2021). *EU ecosystem assessment: Summary for policymakers.* Luxembourg: Publications Office.

Ehrlich-Sommer, F., Hoenigsberger, F., Gollob, C., Nothdurft, A., Stampfer, K., & Holzinger, A. (2024). Sensors for digital transformation in smart forestry. *Sensors, 24*, 798. https://doi.org/10.3390/s24030798

EOS Data Analytics. (2025). *Forest monitoring using remote sensing: Solution by EOSDA.*

FAO. (2025a). The collaborative forest resources questionnaire (CFRQ) [WWW Document]. *Collaborative Forest Resources Questionnaire CFRQ*, https://www.fao.org/forest-resources-assessment/partnerships/cfrq-partners/en/

FAO. (2025b). Pan-European partnership [WWW Document]. *Robust forest data ensured by joint collection FAO UNECE and forest Europe pan-Europe partnership*, https://www.fao.org/forest-resources-assessment/highlights/robust-forest-data-ensured-by-joint-collection-with-the-fao--unece-and-forest-europe-pan-european-partnership/en

FAO. (2025c). *FRA platform*, https://fra-data.fao.org/

FAO. (2022). *The state of the world's forests 2022 forest pathways for green recovery and building inclusive, resilient and sustainable economies.* Rome: FAO. https://doi.org/10.4060/cb9360en

FAO. (2020). *Global forest resources assessment 2020.* Rome: FAO. https://doi.org/10.4060/ca9825en

FAO. (2015a). *C&I: A tool for enhancing SFM from policy to practice (C&I Toolbox).* Rome, Italy: FAO.

FAO. (2015b). *Keeping an eye on SDG 15.* Rome, Italy: FAO

FAO. (2015c). *Strengthening criteria and indicators for sustainable forest management in policy and practice (Final report submitted to the FAO, unpublished).* Rome, Italy.

FAO, CPF. (2022a). *An assessment of uptake of the global core set of forest-related indicators.* In S. Linser, & K. Prins (Eds.). Rome, Italy: FAO. https://doi.org/10.4060/cc2223en

FAO, CPF. (2022b). *Status of, and trends in, the Global Core Set of Forest-related Indicators.* Rome, Italy: FAO. https://doi.org/10.4060/cb9963en

Ferretti, M., Waldner, P., Verstraeten, A., Schmitz, A., Michel, A., Marchetto, A., Hansen, K., Pitar, D., Gottardini, E., Calatayud, V., Nikinmaa, L., Köhl, M., Marchetti, M., & Linser, S. (2020). Criterion 2: Maintenance of forest ecosystem health and vitality. In *State of Europe's Forests 2020* (pp. 51–87). Bratislava, Slovakia: Forest Europe Liaison Unit Bratislava.

Forest Europe. (2020). State of Europe's forests 2020. In M. Köhl, S. Linser & K. Prins (Eds.), Bratislava, Slovakia: Forest Europe Liaison Unit Bratislava.

Forest Europe. (2017a). *Minutes of the kick-off meeting of the expert group on implementation of updated pan-European indicators for sustainable forest management.* Bratislava, Slovakia: Forest Europe Liaison Unit Bratislava.

Forest Europe. (2017b). *Future direction of forest Europe; Final Report: Survey results and recommendations for action.* Bratislava, Slovakia: Forest Europe Liaison Unit Bratislava.

Forest Europe. (2015). *Final report of the UPI Advisory Group, FOREST EUROPE Advisory Group on the updating of the pan-European indicators for sustainable forest management.* Madrid, Spain: Forest Europe Liaison Unit Madrid.

Forest Europe. (2014). *Minutes of the Forest Europe expert group to propose improvements in tools for sustainable forest management, 4–5 June 2014 in Helsinki.* Madrid, Spain: Forest Europe Liaison Unit Madrid.

Forest Europe, UNECE, FAO. (2011). *State of Europe's Forests 2011. Status and trends in sustainable forest management in Europe.* Oslo, Norway: Forest Europe Liaison Unit Oslo.

GFW. (n.d.). *Forest monitoring designed for action*, https://www.globalforestwatch.org/

Hetemäki, L., Hanewinkel, M., Muys, B., Ollikainen, M., Palahí, M., & Trasobares, A. (2017). *Leading the way to a European circular bioeconomy strategy (From Science to Policy).* Joensuu, Finland: European Forest Institute. https://doi.org/10.36333/fs05

Holzinger, A., Saranti, A., Angerschmid, A., Retzlaff, C. O., Gronauer, A., Pejakovic, V., Medel-Jimenez, F., Krexner, T., Gollob, C., & Stampfer, K. (2022). Digital transformation in smart farm and forest operations needs human-centered AI: Challenges and future directions. *Sensors, 22*, 3043. https://doi.org/10.3390/s22083043

ITTO. (2016). *Criteria and indicators for the sustainable management of tropical forests (No. 21), ITTO policy development series*. Yokohama, Japan: International Tropical Timber Organization.

ITTO. (2011). *Twenty-five success stories: Illustrating ITTO's 25-year quest to sustain tropical forests*. Yokohama, Japan: ITTO.

Julve Larrubia, C., Kane, K.R., Wolfslehner, B., Guldin, R., & Rametsteiner, E. (2017). *A way to strengthen results-based management of national forest programmes* (FAO Forestry Policy and Institutions Working Paper No. 37). Rome, Italy: FAO.

Junker, J., Beja, P., Brotons, L., Fernandez, M., Fernández, N., Kissling, W. D., Lumbierres, M., Lyche Solheim, A., Maes, J., Morán-Ordóñez, A., Moreira, F., Musche, M., Santana, J., Valdez, J., & Pereira, H. (2023). D4.1. List and specifications of EBVs and EESVs for a European wide biodiversity observation network. https://doi.org/10.3897/arphapreprints.e102530

Köhl, M., Koch, F., Linser, S., Melin, M., Robertson, G., & Talarczyk, A. (2024). *Reporting on forest damages and disturbances in the UNECE region*. United Nations Publications.

Komdeur, E. F. M., & Ingenbleek, P. T. M. (2021). The potential of blockchain technology in the procurement of sustainable timber products. *International Wood Products Journal, 12*, 249–257. https://doi.org/10.1080/20426445.2021.1967624

Kulionis, V., Pfister, S., & Fernandez, J. (2024). Biodiversity impact assessment for finance. *Journal of Industrial Ecology, 28*, 1321–1335. https://doi.org/10.1111/jiec.13515

Larson, A. M., Mausch, K., Bourne, M., Luttrell, C., Schoneveld, G., Cronkleton, P., Locatelli, B., Catacutan, D., Cerutti, P., Chomba, S., Djoudi, H., Ihalainen, M., Lawry, S., Minang, P., Monterroso, I., Myers, R., Naito, D., Pham, T. T., Reed, J., ... Stoian, D. (2021). Hot topics in governance for forests and trees: Towards a (just) transformative research agenda. *Forest Policy and Economics, 131*, Article 102567. https://doi.org/10.1016/j.forpol.2021.102567

Lechner, A. M., Foody, G. M., & Boyd, D. S. (2020). Applications in remote sensing to forest ecology and management. *One Earth, 2*, 405–412. https://doi.org/10.1016/j.oneear.2020.05.001

Lexer, M. J., Vacik, H. (2016). The AFM-ToolBox to support adaptive forest management under climate change. In J. Papathanasiou, N. Ploskas, & I. Linden (Eds.), *Real-world decision support systems, integrated series in information systems* (pp. 223–241). Cham: Springer International Publishing. https://doi.org/10.1007/978-3-319-43916-7_10

Lier, M., Köhl, M., Korhonen, K. T., Linser, S., & Prins, K. (2021). Forest relevant targets in EU policy instruments—Can progress be measured by the pan-European criteria and indicators for sustainable forest management? *Forest Policy and Economics, 128*, Article 102481. https://doi.org/10.1016/j.forpol.2021.102481

Lier, M., Köhl, M., Korhonen, K. T., Linser, S., Prins, K., & Talarczyk, A. (2022). The new EU forest strategy for 2030: A new understanding of sustainable forest management? *Forests, 13*, 245. https://doi.org/10.3390/f13020245

Lier, M., Schuck, A., Van Brusselen, J., Derks, J., Parviainen, J., Bastrup-Birk, A., Gasparini, P., Bozzano, M., Oggioni, S.D., Raši, R., Vogt, P., Biala, K., Köhl, M., Konczal, A., de Konig, J., Alberdi, I., Vořísek, P., Schwarz, M., & Linser, S. (2020). Criterion 4: Maintenance, conservation and appropriate enhancement of biological diversity in forest ecosystems. In *State of Europe's Forests 2020*. Bratislava: Forest Europe Liaison Unit Bratislava.

Linser, S. (2024). Potential for revision of forest Europe indicators under criterion 4 maintenance, conservation and appropriate enhancement of biological diversity in forest ecosystems. *Forest Europe Background Paper*. Germany: Forest Europe Liaison Unit Bonn. https://doi.org/10.13140/RG.2.2.14518.72008/1

Linser, S. (2021). *Criteria and indicators—An universal policy tool for sustainable forest management, policy briefs of the UNECE forestry and timber section*. Geneva, Switzerland: UNECE.

Linser, S. (2020). *Indikatoren für nachhaltige Waldbewirtschaftung 2020*. Vienna, Austria: BMLRT

Linser, S. (2002). Critical analysis of the basics for the assessment of sustainable development by indicators. Schriftenreihe Freiburger Forstliche Forschung, Germany: Freiburg.

Linser, S. (1999). Theoretical background of indicators and indicator systems for the assessment of sustainable development. In A. Niskanen & J. Väyrynen (Eds.), *Regional Forest Programmes: A*

Participatory Approach to Support Forest-Based Rural Development, EFI Proceedings. Joensuu, Finland: European Forest Institute.
Linser, S., Dabic, I., Mazziotta, A., Parry, J., et al. (2025a). Participatory agreed set of key indicators for monitoring and reporting IFM (No. D5.2), TRANSFORMIT Deliverable 5.2.
Linser, S., Hemming, D., Barrio Giullo, R., Anand, J., Amihaesei, V. A., Cheval, S., et al. (2025b). Report on a novel set of Essential Forest Mitigation Indicators (EFMIs) (OptFor-EU D1.2).
Linser, S., & Lier, M. (2020). The contribution of sustainable development goals and forest-related indicators to national bioeconomy progress monitoring. *Sustainability, 12*, 2898. https://doi.org/10.3390/su12072898
Linser, S., & O'Hara, P. (2019). *Guidelines for the development of a criteria and indicator set for sustainable forest management*. Geneva: United Nations Publications.
Linser, S., & Wolfslehner, B. (2022). National implementation of the forest Europe indicators for sustainable forest management. *Forests, 13*, 191. https://doi.org/10.3390/f13020191
Linser, S., & Wolfslehner, B. (2015a). *Meeting the goals for European forests and the European 2020 targets for forests. Report on the mid-term evaluation of the goals for European forests and the European 2020 targets for forests*. Madrid, Spain: Forest Europe Liaison Unit Madrid.
Linser, S., & Wolfslehner, B. (2015b). *Report on the mid-term Evaluation: Meeting the goals for European forests and the European 2020 targets for forests*. Madrid, Madrid, Spain: Forest Europe Liaison Unit.
Linser, S., Wolfslehner, B., Asmar, F., Bridge, S. R. J., Gritten, D., Guadalupe, V., Jafari, M., Johnson, S., Laclau, P., & Robertson, G. (2018a). 25 years of criteria and indicators for sustainable forest management: Why some intergovernmental C&I processes flourished while others faded. *Forests, 9*, 515. https://doi.org/10.3390/f9090515
Linser, S., Wolfslehner, B., Bridge, S. R. J., Gritten, D., Johnson, S., Payn, T., Prins, K., Raši, R., & Robertson, G. (2018b). 25 Years of criteria and indicators for sustainable forest management: How intergovernmental C&I processes have made a difference. *Forests, 9*, 578. https://doi.org/10.3390/f9090578
Linser, S., Wolfslehner, B., & Pülzl, H. (2015). The genesis of the pan-European criteria and indicators and their further development towards emerging policy needs. In *Proceedings. Presented at the XIV World Forestry Congress*. Durban, South Africa: FAO.
LIVE EO. (2024). *Unlock the potential of earth observation*, https://www.live-eo.com/
Lloret, F., Hurtado, P., Espelta, J. M., Jaime, L., Nikinmaa, L., Lindner, M., & Martínez-Vilalta, J. (2024). ORF, an operational framework to measure resilience in social–ecological systems: The forest case study. *Sustainability Science, 19*, 1579–1593. https://doi.org/10.1007/s11625-024-01518-1
Lulli, A., Oneto, L., & Anguita, D. (2019). Mining big data with random forests. *Cognitive Computation, 11*, 294–316. https://doi.org/10.1007/s12559-018-9615-4
MacDicken, K. G., Sola, P., Hall, J. E., Sabogal, C., Tadoum, M., & de Wasseige, C. (2015). Global progress toward sustainable forest management. *Forest Ecology Management, Changes in Global Forest Resources from 1990 to 2015, 352*, 47–56. https://doi.org/10.1016/j.foreco.2015.02.005
Maesano, M., Ottaviano, M., Lidestav, G., Lasserre, B., Matteucci, G., Scarascia Mugnozza, G., & Marchetti, M. (2018). Forest certification map of Europe. *iForest—Biogeosciences and Forestry, 11*, 526–533. https://doi.org/10.3832/ifor2668-011
McCool, S., & Stankey, G. (2001). Representing the future: A framework for evaluating the utility of indicators in the search for sustainable forest management. In R. Raison, A. Brown, & D. Flinn (Eds.), *Criteria and indicators for sustainable forest management* (pp. 93–109). CAB International, Wallingford, UK.
McDonald, G. T., & Lane, M. B. (2004). Converging global indicators for sustainable forest management. *Forest Policy and Economics, 6*, 63–70. https://doi.org/10.1016/S1389-9341(02)00101-6
MCPFE. (1998). Resolution L2. In *Ministerial Conference on the Protection of Forests in Europe*. Lisbon, Portugal: Liaison Unit Lisbon.

MCPFE. (1993). Resolution H1 general guidelines for the sustainable management of forests in Europe. In *Ministerial Conference on the Protection of Forests in Europe.* Helsinki, Finland: Liaison Unit Helsinki.

Mikulková, A., Hájek, M., Štěpánková, M., & Ševčík, M. (2015). Forest certification as a tool to support sustainable development in forest management. *Journal of Forest Science, 61*, 359–368. https://doi.org/10.17221/16/2015-JFS

Montréal Process. (2015). *The Montréal process criteria and indicators for the conservation and sustainable management of temperate and boreal forests* (5th ed.). Montréal Process Liaison Office.

Montréal Process Working Group. (2023). *Montreal process: Synthesis of indicator trends 1990 to 2020 and future outlooks.* New Zealand: Scion.

Nabuurs, G.-J., Begemann, A., Linser, S., Paillet, Y., Pettenella, D., Zu Ermgassen, S., & European Forest Institute (2024). *Sustainable finance and forest biodiversity criteria (From Science to Policy).* European Forest Institute. https://doi.org/10.36333/fs16

Nikinmaa, L., Lindner, M., Cantarello, E., Gardiner, B., Jacobsen, J. B., Jump, A. S., Parra, C., Plieninger, T., Schuck, A., Seidl, R., Timberlake, T., Waring, K., Winkel, G., & Muys, B. (2023). A balancing act: Principles, criteria and indicator framework to operationalize social-ecological resilience of forests. *Journal of Environmental Management, 331*, Article 117039. https://doi.org/10.1016/j.jenvman.2022.117039

Onida, M. (2021). Forest and forest policy between the EU and its member states. *ELNI Reviewed Environmental Law Network International,* 22–30.

Otrakcier, T., Linser, S., Bastrup-Birk, A., Gaworska, M., Köhl, M., Korhonen, K. T., & Talarczyk, A. (2023). *Reporting on forests and sustainable forest management in the Caucasus and central Asia: Focus on criteria and indicators.* United Nations Geneva Timber and Forest Study. https://doi.org/10.18356/9789210023825

Ott, W. R. (1978). *Environmental Indices: Theory and Practice.* Ann Arbor, Michigan, USA: Ann Arbour Science.

Prabhu, R., Ruitenbeek, H. J., Boyle, T. J. B., & Colfer, C. J. P. (2001). Between voodoo science and adaptive management: the role and research need of indicators for SFM. In R. Raison, A. Brown & D. Flinn (Eds.), *Criteria and Indicators for Sustainable Forest Management.* Wallingford, United Kingdom: CAB International.

Prins, K., Köhl, M., & Linser, S. (2023). Is the concept of sustainable forest management still fit for purpose? *Forest Policy Economics, 157*, Article 103072. https://doi.org/10.1016/j.forpol.2023.103072

Pülzl, H., & Mayer, P. (2015). Assessment of the achievements and added value of the Forest Europe process, (Unpublished Report). In *Proceedings of the Forest Europe Expert Level Meeting.* Spain: Santiago de Compostela.

Rametsteiner, E., & Simula, M. (2003). Forest certification—An instrument to promote sustainable forest management? *Journal of Environmental Management, 67*, 87–98. https://doi.org/10.1016/S0301-4797(02)00191-3

Santopuoli, G., Temperli, C., Alberdi, I., Barbeito, I., Bosela, M., Bottero, A., Klopčič, M., Lesinski, J., Panzacchi, P., & Tognetti, R. (2021). Pan-European sustainable forest management indicators for assessing climate-smart forestry in Europe. *Canadian Journal of Forest Research, 51*, 1741–1750. https://doi.org/10.1139/cjfr-2020-0166

Secco, L., Da Re, R., Pettenella, D. M., & Gatto, P. (2014). Why and how to measure forest governance at local level: A set of indicators. *Forest Policy Economics, 49*, 57–71. https://doi.org/10.1016/j.forpol.2013.07.006

Shields, D. (2002). The role of values and objectives in communicating indicators of sustainability. *Ecological Indicators, 2*, 149–160. https://doi.org/10.1016/S1470-160X(02)00042-0

UNCCD. (2023). Improving the procedures for communication as well as the quality and formats of reports to be submitted to the conference of the parties, ICCD/CRIC(21)/7.

UNECE. (2023). *INForest—Your data and knowledge platform for forests in the UNECE region,* https://forest-data.unece.org/

UNECE/FAO. (2016). Pilot project on the system for the evaluation of the management of forests (SEMAFOR) (No. 66). In *Geneva Timber and Forest Discussion Paper*. Geneva, Switzerland: United Nations Publications.

United Nations. (2018). Global Indicator framework for the sustainable development goals and targets for the 2030 agenda for sustainable development. A/RES/71/313 E/CN.3/2018/2. New York, USA.

United Nations. (2007). Non-legally binding instrument on all types of forests (later renamed as the United Nations Forest Instrument). A/RES/62/98 (Resolution adopted by the General Assembly 62nd sess.). New York.

United Nations. (1992a). *AGENDA 21, United Nations conference on environment & development Rio de Janerio, Brazil, 3 to 14 June 1992.* New York: United Nations Publications.

United Nations. (1992b). Non-legally binding authoritative statement of principles for a global consensus on the management, conservation, and sustainable development of all types of forests.

United Nations. (n.d.). *Global set of climate change statistics and indicators (CISAT)*, https://unstats.un.org/unsd/envstats/climate%20change/CISAT_Part_2.xlsx

Visseren-Hamakers, I. J., Razzaque, J., McElwee, P., Turnhout, E., Kelemen, E., Rusch, G. M., Fernández-Llamazares, Á., Chan, I., Lim, M., Islar, M., Gautam, A. P., Williams, M., Mungatana, E., Karim, M. S., Muradian, R., Gerber, L. R., Lui, G., Liu, J., Spangenberg, J. H., & Zaleski, D. (2021). Transformative governance of biodiversity: Insights for sustainable development. *Current Opinion in Environment Sustainability, 53*, 20–28. https://doi.org/10.1016/j.cosust.2021.06.002

Wang, T., Zuo, Y., Manda, T., Hwarari, D., & Yang, L. (2025). Harnessing artificial intelligence, machine learning and deep learning for sustainable forestry management and conservation: Transformative potential and future perspectives. *Plants, 14*, 998. https://doi.org/10.3390/plants14070998

White, J. C., Coops, N. C., Wulder, M. A., Vastaranta, M., Hilker, T., & Tompalski, P. (2016). Remote sensing technologies for enhancing forest inventories: A review. *Canadian Journal of Remote Sensing, 42*, 619–641. https://doi.org/10.1080/07038992.2016.1207484

Wolfslehner, B., Linser, S., Pülzl, H., Bastrup-Birk, A., Camia, A., & Marchetti, M. (2016). *Forest bioeconomy—A new scope for sustainability indicators (From Science to Policy).* Joensuu, Finland: European Forest Institute. https://doi.org/10.36333/fs04

WRI. (n.d). *The forest data partnership* [WWW Document], https://begonia-robin-tem5.squarespace.com/s/forest-data-partnership-brochure.pdf

Part IV
Insights, Implications, and Pathways Forward

Chapter 10
Emerging Trends and the Future of Forest Policy Research

Filip Aggestam

Abstract This chapter reviews the history of forest policy research, outlining how it has expanded beyond a focus on timber production to cover biodiversity, climate, ecosystem services, and justice. It then assesses the implications of this expansion for transformative change in European forest governance. The chapter presents a dual challenge in terms of advancing the understanding of long-standing issues, such as deforestation, degradation, illegal logging, tenure, and fragmentation, while also analysing emerging dynamics, including digitalisation, shifting ownership, evolving narratives, and participation. The analysis is structured around six emerging trends, namely, integrative and cross-sectoral governance, discursive and normative shifts, communication and knowledge systems, digitalisation and datafication, tools for reflexivity and learning, and institutions, ownership, and power relations. Based on these, the chapter outlines some theoretical futures, demonstrating how institutional, discursive, network/coalition, rational, and reflexive approaches, when used in combination, can help to explain transformative change more effectively. It then argues for a reorientation of transformative research, emphasising methodological pluralism, reflexivity, normative engagement and long-term comparative perspectives. Forest policy research has developed enabling capacities (e.g., monitoring, indicators and data systems) as well as steering mechanisms (e.g., policy mixes and coordination). The chapter concludes by presenting forest policy research as analytical and transformative with ethical responsibilities.

10.1 Introduction

The scope of what "forest policy research" means in practice has changed significantly over the past century, shifting from a focus on timber production and forestry regulations to addressing and covering a much wider set of societal concerns, such

---

F. Aggestam (✉)
ORBIS Green Policy and Governance, Forest Policy Research Network—European Forest Institute (FPRN-EFI), Vienna, Austria
e-mail: aggestam@orbis-gpg.com

F. Aggestam (ed.), *Governing Forests Through Transformative Change*, Advances in Global Change Research 70, https://doi.org/10.1007/978-3-032-19072-7_10

as biodiversity loss, climate change, ecosystem services and social justice (Acosta-Muñoz et al., 2024; Raum & Potter, 2015). For example, over the past two decades, reviews and state-of-the-art assessments have demonstrated a significant increase in research on forest ecosystem services, changes in forest governance, including multi-level and cross-sectoral dynamics, and social aspects as they relate to forest management (Art & Visseren-Hamakers, 2012; Kleinschmit et al., 2016; Nummelin et al., 2021). At the same time, there has also been an increase in criticism of what is often labelled as technocratic forestry, often coupled with advocacy for more reflexive, interdisciplinary and participatory approaches as well as for learning in policymaking processes (Krott & Giessen, 2014). These developments in forest policy research mirror a wider transformation in governance and natural resource management, where problem framing increasingly integrates (and accounts for) varied socio-economic and environmental considerations.

This widening scope is driven, in part, by an improved understanding of our natural environment and our place within it. It is also, in part, driven by increasing material and political pressures being put on forests (e.g., wood for paper and pulp, furniture, energy production, and carbon sequestration). For instance, global assessments have emphasised the central importance of forests for planetary health. They contain more than 80% of all terrestrial species, regulate the climate by storing huge amounts of carbon, maintain hydrological cycles, and sustain and support the livelihoods of people (FAO, 2020; IPBES, 2019; IPCC, 2022). However, these critical functions are under increasing threat due to accelerating climate change, biodiversity decline, land-use competition and ecosystem degradation (FAO, 2020; IPCC, 2022). The European Green Deal and other relevant policy initiatives, such as the EU Forest Strategy for 2030 and the Nature Restoration Regulation (EC, 2019, 2021; EP & CoE, 2024), demonstrate how forests have become more important to EU policy (see Chap. 3 and 5), particularly when considering mitigation, adaptation, resilience, and the circular bioeconomy. These developments have expanded the scope of EU forest-related policy and have raised the stakes for policy coherence and legitimacy.

Taken together, the chapters in this book reflect the expanded policy agenda and changing priorities noted above. They cover the fundamentals of EU forest governance, including definitions of key terms such as "forest" and the distinctions between policy and governance (see Chap. 1). They also explore the institutional and discursive dynamics of EU forest governance (see Chap. 3), as well as cross-cutting levers such as communication, digitalisation, monitoring frameworks, and ownership and tenure (see Chap. 6–9). However, despite the introduction of new instruments and narratives, the preceding chapters demonstrate that many governance challenges persist, such as fragmentation, conflicting policy objectives, power asymmetries, and varied human and financial capacities.

Forest policy researchers face a dual challenge:

1. Continue to improve our understanding of long-standing concerns and problems, including deforestation, degradation, illegal logging, land tenure disputes, and fragmented governance, while assessing how these issues interact with other factors, such as climate change and biodiversity loss.

2. Analyse how emerging issues, such as the adoption of digital technologies and data infrastructures, shifting ownership forms and actor constellations, evolving public narratives and frames, and new expectations regarding accountability and participation, may impact forest governance and practices.

Meeting this dual challenge will require methodological pluralism (quantitative, qualitative, and participatory) and theoretical innovations that can bridge institutional, interpretive, and transitional frameworks, as well as relevant value-based commitments.

10.2 Emerging Trends

This chapter discusses six emerging trends that exemplify this dual challenge: (i) integrative and cross-sectoral governance; (ii) Discursive and normative shifts; (iii) Communication and knowledge systems; (iv) Digitalisation and datafication; (v) Tools for reflexivity and learning; and (vi) Institutions, ownership, and power relations. These trends have been developed from contributions to this book and are complemented by the theoretical considerations presented in Chap. 3.

(i) **Integrative and Cross-Sectoral Governance**

The EU cases in Chaps. 4 and 5 demonstrate that policy fragmentation remains a common phenomenon. Objectives relevant to forests are spread across policy domains, such as climate, agriculture, trade, energy, and biodiversity. The overlap between these domains commonly leads to trade-offs, conflicts, inefficiencies, and incoherence (Pülzl et al., 2014, 2018). While sectoral silos once represented distinct policy pathways, they now tend to generate conflicting targets that affect forests. For instance, one common example is the conflict between bioenergy and conservation targets.

Building on existing frameworks relating to policy integration, reflexive governance and adaptive policy mixes (Howlett & Rayner, 2013; Jordan & Lenschow, 2010; Ostrom, 1990; Voß & Bornemann, 2011), future forest-specific applications and comparative analyses could (i) examine how integration is (or is not) achieved across different regions (e.g., comparing the EU with federal states such as Canada and the US, and decentralised regions such as Switzerland and Indonesia), (ii) investigate longitudinal trajectories in forest-related policy, such as shifting EU climate, biodiversity and trade regulations, (iii) study which interests are being promoted or sidelined in integrative efforts, and (iv) examine hybrid arrangements (e.g., voluntary partnerships, polycentric approaches and cross-sectoral efforts) that could bridge institutional divides while remaining adaptive (Howlett & Rayner, 2013; Ostrom, 2010). These suggestions reframe the research agenda by extending integration theory into the forest governance domain.

(ii) **Discursive and Value-Based Shifts**

Policy changes often follow a change in narratives. For example, the way forests are framed, in terms of issues such as timber production, biodiversity hotspots, climate sinks and cultural landscapes, has significant implications for the legitimacy of forest-relevant policies. In short, frames can determine what is seen as urgent, who is held responsible, and what is considered acceptable. The frame analysis in Chap. 5 demonstrate how shifts in the discourse around forests can lead to changes in governance. It also illustrates how frames can persist and stabilise existing paradigms or value systems over time.

Although considerable research has been conducted in this area, future studies could focus on how discourses emerge, spread, and become institutionalised within the forest-based sector (Dryzek, 2005; Dryzek & Pickering, 2018). This could include (i) analysing how epistemic communities, NGOs, and social movements express new frames around forests and embed them in relevant policies, (ii) investigating how crises, such as wildfires, pest outbreaks, or climate tipping points, can open or close policy windows (Rose et al., 2017; Zahariadis, 2018), (iii) further exploring the use of AI-based methods to expand studies on frames and narratives embedded in policies, particularly from a longitudinal perspective (Hajer, 1993, 1997), and (iv) linking these types of studies with institutional change research to clarify the conditions under which shifts in narratives can lead to transformative change. Furthermore, this type of research could further the application of discursive theory to forest-specific domains as well as providing the basis for new methodological approaches (e.g. AI-based methods and longitudinal scaling).

(iii) **Communication and Knowledge Systems**

As discussed in Chap. 6, communication plays a central role in forest governance rather than only supporting it. More specifically, the structure of communication can, for example, influence the legitimacy and power of an actor or instrument. To illustrate, an individual or organisation that possess strong communication capabilities, such as access to the media, effective communication skills, and digital literacy, can affect the governance of forests. This highlights the importance of examining communication not merely as a background variable but as a fundamental aspect of power in forest governance.

While previous research has provided a valuable conceptual foundation for areas such as agenda-setting, framing, misinformation and deliberation (Entman, 2008; Habermas, 1998; Lewandowsky et al., 2017), less is known about how these processes operate in forest-specific contexts and across governance scales. Future research could focus on (i) investigating how forest issues are represented and contested across traditional and digital media in different cultural and political settings, (ii) assess the impact of misinformation and polarisation on forest-related policymaking, such as restoration, bioenergy, and protected areas, (iii) explore how deliberative and participatory communication practices can be institutionalised in forest governance; and (iv) analyse the role of emerging digital platforms to assess who has a voice and whose knowledge is legitimised. This type of research could provide new insights into how forest-relevant knowledge is produced, contested, and institutionalised. Comparative and cross-regional studies could be particularly valuable in demonstrating variations in communication dynamics.

(iv) **Digitalisation and Datafication**

Digital technologies are transforming forest governance, and their adoption is becoming increasingly prevalent in the forest-based sector. With this in mind, Chap. 7 illustrates how AI, blockchain technology, big data, and the Internet of Things are being used to enhance monitoring, traceability, and stakeholder engagement (Nitoslawski et al., 2019, 2021). For instance, blockchain technology has been proposed as a way to increase transparency in the timber supply chain. At the same time, AI-based tools are increasingly being used to predict forest fire risks, and digital platforms are facilitating participatory mapping and citizen science initiatives. However, digitalisation has been criticised for amplifying epistemic power and embedding inequalities (Mol, 2008). For example, data infrastructures tend to benefit those with resources, marginalising smallholders, indigenous peoples, and communities that lack access to resources.

These factors highlight the need for forest policy research to consider not only the technological aspects of digitalisation, but also its socio-economic effects. Future research could focus on: (i) analysing the political economy of forest data infrastructures, considering who controls the data, which knowledge is included or excluded, and how authority is redistributed (Yeung, 2018); (ii) investigating the inclusivity of digital participation tools, for example by examining whether citizen science platforms genuinely empower marginalised communities, (iii) developing mixed-methods approaches that combine technical evaluations of digital tools with participatory elements to capture technical performance and social legitimacy, and (iv) comparing the levels of forest digitalisation in different regions in order to evaluate the conditions for adoption, governance, and equity. Expanding research into the digitalisation of the forest-based sector could enable policy researchers to progress from evaluating "what technology can do" to assessing "for whom and what consequences".

(v) **Tools for Reflexivity and Learning**

Monitoring frameworks, such as criteria and indicators (C&I), continue to play a significant role in forest governance and management. Chapter 8 reconsiders these frameworks as policy instruments that can facilitate learning and adaptation through feedback loops, rather than mainly acting as a reporting tool. The main argument is that indicators can serve as a common language that transcends sectoral boundaries and levels, linking global goals with local practices and facilitating comparisons across policy environments. However, despite the extensive development of indicator frameworks, uptake in policy learning remains limited.

Future research could focus on how C&I are used (or ignored) in practice as well as the conditions under which they facilitate reflexivity. This could include (i) work on participatory indicator design to ensure indicators reflect varied stakeholder values rather than simply technocratic priorities (Cash et al., 2003; Pülzl et al., 2011),

(ii) regular crosswalks to determine how forest-related indicators align with SDG indicators and the KM-GBF (Art & Visseren-Hamakers, 2012; Visseren-Hamakers & Glasbergen, 2007), (iii) explore how new technologies can be integrated to improve forest monitoring while addressing concerns about interoperability, data ownership, and legitimacy (Fassnacht et al., 2024; Linser et al., 2018), and (iv) assesse whether and how indicator systems evolve in response to global change, and whether this facilitates learning across regions, e.g., through processes such as Forest Europe or other regional initiatives (Wijewardana, 2008; Wolfslehner et al., 2020). In summary, there is much that can be done to further assess how C&I can facilitate adaptive learning and move beyond being about static compliance mechanism.

(vi) **Institutions, Ownership, and Power Relations**

Ownership (and user/property rights) remains a fundamental consideration when discussing forest governance. Chapter 5 explains different types of ownership, including state, municipal, corporate, communal, smallholder, and lifestyle, which influence values, capacities, and priorities related to forests (Weiss et al., 2017, 2019). Ownership structures also shape forest management practices and the discourse surrounding forest use, including associated costs and benefits.

Although research has documented significant differences between ownership categories, there are still gaps in our knowledge. Future research could (i) continue to expand our understanding of the practical implications of varied forest ownership structures, including gender-based aspects (Sikor & Lund, 2009), intergenerational change, and urban–rural ownership dynamics (Matilainen et al., 2019), (ii) analyse emerging ownership forms, such as carbon trusts, conservation easements, and investor-driven land acquisitions, and consider the implications these changes may have for forest governance (Fairhead et al., 2012; McDermott et al., 2010), and (iii) consider collective and hybrid ownership arrangements, such as co-management, joint forest management and new commons institutions, which blur the boundaries between the public and private sectors (Lawrence et al., 2020). Forest ownership is a fundamental aspect of the transformative agenda, whether its about climate-smart forestry, ecosystem restoration or the transition to the bioeconomy. For this reason, it remains as a highly relevant topic for future research.

10.3 Theoretical Futures

Different theoretical approaches have played an important role in the development of forest policy research, providing conceptual frameworks that help us understand the forest-based sector, as well as the interactions between actors, institutions, and perspectives within complex forest governance systems (Aggestam, 2025; Aggestam et al., 2025). However, as socio-economic and environmental crises increase (in number and impacts) and governance systems become more polycentric and digitally driven, forest policy research may require theoretical innovation. In other words, future-oriented forest policy research could build on established frameworks while adapting them to explain and understand transformative change in forest governance.[1] The theoretical approaches outlined in Chap. 2 form a foundation for looking at and theorising transformation (Art & Visseren-Hamakers, 2012; Leach et al., 2007; Ostrom, 1990, 2010; Sabatier, 2007).

Institutional and actor-centred theories remain crucial for studying the stability and change of structures within European forest governance. For example, institutional policy analysis can provide a conceptual framework for assessing how formal and informal rules, norms, and decision-making processes affect policy reforms that may be relevant to transformative change (Hajer, 1993; Schmidt, 2008). Alternatively, the Advocacy Coalition Framework (ACF) provides an approach to examining how belief systems and coalitions influence policy learning and adaptation over time (Sabatier, 1988; Weible, 2006; Hysing & Olsson, 2008). These different perspectives provide some of the tools and approaches needed to better understand multi-level forest governance, such as explaining how different power and policymaking centres interact over time and across scales (Ostrom, 2010). For instance, future research could focus on how hybrid institutional arrangements, such as certification schemes, voluntary partnerships, or cross-sectoral coalitions, have evolved in response to pressures relating to climate change, biodiversity loss and changing markets (Cashore et al., 2004; Jordan & Lenschow, 2010). In addition, comparative and longitudinal analyses could help explain the institutional conditions necessary for forest governance systems to become resilient or engage in transformative change. This would contribute to the increasing body of work underlying the theory of governance for transformation (Voß & Bornemann, 2011).

Discursive and constructivist theories provide other complementary perspectives, explaining how meaning-making and narrative changes over time can shape policy change (including transformative change). For example, these approaches conceptualise policymaking as a competition between ideas, identities, and legitimacy through discourse (Arts & Buizer, 2009; Dryzek, 2005; Dryzek & Pickering, 2018; Hajer, 1997). In terms of forest management, these theoretical perspectives provide an approach to understanding how concepts such as sustainable forest management (SFM), the bioeconomy, and nature restoration are socially constructed and promoted by different stakeholder groups (Mather-Gratton et al., 2021; Pülzl et al., 2014, 2018). Future work could extend these ideas by using longitudinal and computational methods to map changes in discourse across forest-focused and forest-related policies (or other types of relevant media). For instance, recent advances in AI-based frame analysis have enabled researchers to analyse much larger datasets and explore changing narratives and institutional uptake in new ways (Aggestam et al., 2025; Curry et al., 2024; Ryazanov et al., 2024). These methods may help bridge the gap between interpretive and quantitative approaches. This may in turn improve our understanding of how ideas and legitimacy evolve in relation to transformative change.

[1] Defined in Chap. 1 as the reconfiguration of capacities, steering mechanisms, values, and power relations.

Similarly, network and coalition-based approaches can provide valuable tools for examining transformative change as a process of interactions between heterogeneous actors. For example, a policy network analysis can help explain the impact of different relationships between actors, such as the exchange of resources, trust, and influence in policymaking (Knoke & Yang, 2008; Rhodes, 1997; Thacher, 2005). These approaches, when combined with ACF, can also help explain how coalitions form, persist, and evolve in relation to specific forest-based issues. This can be demonstrated by studies of certification systems and the integration of forest-water policy (Baulenas et al., 2021; Elliott & Schlaepfer, 2001). As digitalisation expands the forest governance space (e.g., through online platforms and data infrastructures), network theory could be used to analyse how digital transformations reshape participation and accountability. For instance, future work could consider how (or whether) digitally mediated forest governance promotes inclusion and transparency, or if it creates new power imbalances. This is an important question regarding the governance transformations highlighted in this book.

Along the same lines of reasoning, rational and reflexive theories could also play an important role in theorising transformative change. Although rational policy analysis is often criticised for its technocratic perspective (Hoefer, 2021), its focus on systematic evaluation and goal-oriented design may be relevant when considered together with reflexive governance (Krott & Giessen, 2014; Voß & Bornemann, 2011). For example, as outlined earlier in this book, criteria and indicator systems can be viewed not only as compliance tools but also as a policy instrument that promotes reflexivity, monitoring, and policy learning at different levels. Understanding how these types of mechanisms evolve and change over time (e.g., in response to societal challenges) and how they affect the relationship between data, values, and policymaking could improve our theoretical and empirical understanding of transformative change and associated processes.

In a nutshell, the future of research on European forest governance and policy, as it relates to transformative change, may depend on its ability to evolve theoretically. For instance, the different theoretical perspectives highlighted in this chapter demonstrate the importance of theoretical pluralism. In other words, there is no single framework that can fully explain all the dimensions of transformative change (e.g., institutional, discursive, and material) in forest governance. Integrating different theories and perspectives enhance our understanding. To illustrate, institutional and discursive theories can reveal how power is exercised through formal rules and symbolic representation. By combining rational and critical approaches, they can demonstrate the interplay between efficiency, legitimacy, and justice. Integrative frameworks offer a way forward for work on sustainability transitions and transformative change. Importantly, these frameworks also emphasise the co-production of knowledge and governance.

10.4 Towards Transformative Research

The trends outlined in this chapter suggest that the future of forest policy research will be integrative. However, recognising the need for integration is only the first step. When considering transformative change, there is a need for deliberate and sustained reorientation in how research is conducted. More specifically, researchers contribute not only to explaining but also to enabling transformative change in forest governance:

1. Methodological pluralism should become an organising principle for forest policy researchers. The complexity of forest governance requires methods that are combined to capture multidimensional factors, such as linking quantitative, qualitative, computational, and participatory approaches. Integrating mixed-methods approaches could lead to a more comprehensive analysis of how forest governance systems evolve and interact with environmental and socio-economic transformations.
2. Reflexivity is vital. Researchers are (at the end of the day) not detached observers, they are actors that are placed within the forest governance systems they study. By framing research questions, selecting indicators, and disseminating results, researchers influence how problems and solutions are perceived. Forest policy researchers must, therefore, recognise their position and critically engage with the processes of knowledge co-production that link science, policy, and practice. While genuine co-production may not always be possible, some degree of reflexivity can certainly help to strengthen the relevance and added value of research on transformative change.
3. Normative engagement should be at the forefront of future work on transformative change. Questions of justice, equity, legitimacy, and inclusion are key factors to consider when thinking about transformation. This implies that forest policy researchers should explore how governance systems allocate benefits and costs, which knowledge is valued in policymaking processes, and how different stakeholder groups can participate in shaping the future of the forest-based sector. Regardless of the focus, incorporating these considerations into analytical frameworks could ensure that the work is more socially relevant and impactful.
4. Long-term and comparative approaches are fundamentally important. Societal transformations evolve over decades, shaped by path-dependent trajectories, policy legacies, and varied cultural backgrounds. Longitudinal studies (and the range of new tools being made available) can reveal how forest governance capacities and discourses change over time, while comparative research across regions, countries, and governance levels can pinpoint different trajectories of change. This temporal and spatial expansion is essential for improving our understanding of transformative processes.

Taken together, forest policy researchers should recognise that their work has analytical and transformative elements. Forests are central to many of today's most pressing societal challenges, where environmental, social, and economic issues are intertwined. In other words, forests are not only ecological systems, but also areas of work in which competing perspectives of sustainability, justice, and development are negotiated. By combining methodological pluralism, theoretical innovation, reflexivity, and normative engagement, forest policy researchers could play a crucial role in developing new and adaptive governance systems.

However, this also implies that researchers are not neutral or value-free observers. Researchers have ethical and practical responsibilities, whether or not they are acknowledged. As scholars, such as Habermas (1998), have highlighted, power, perspectives and values shape knowledge creation, which in turn influences policymaking. Recognising this interplay between knowledge and governance suggests the need for experimentation with new epistemologies and an openness to transdisciplinary collaboration.

References

Acosta-Muñoz, C., Navarro-Cerrillo, R., Bonet-García, F., Ruiz-Gómez, F., & González-Moreno, P. (2024). Evolution and paradigm shift in forest health research: A review of global trends and knowledge gaps. *Forests, 15*(8), 1279. https://doi.org/10.3390/f15081279

Aggestam, F. (2025). *European forest policy and governance: An integrative analytical framework* (1st ed.). Routledge. https://doi.org/10.4324/9781003198987

Aggestam, F., Trummer, J., & Wilkes-Allemann, J. (2025). Sixty years of EU forest policy discourse: An AI-based frame analysis of policy documents. *Forest Policy and Economics, Under Review*.

Art, B., & Visseren-Hamakers, I. (2012). Forest governance: A state of the art review. In B. Arts, S. Van Bommel, M. Ros-Tonen, & G. Verschoor (Eds.), *Forest-people interfaces* (pp. 241–257). Wageningen Academic Publishers. https://doi.org/10.3920/978-90-8686-749-3_15

Arts, B., & Buizer, M. (2009). Forests, discourses, institutions. *Forest Policy and Economics, 11*(5–6), 340–347. https://doi.org/10.1016/j.forpol.2008.10.004

Baulenas, E., Kruse, S., & Sotirov, M. (2021). Forest and water policy integration: A process and output-oriented policy network analysis. *Environmental Policy and Governance, 31*(5), 432–450. https://doi.org/10.1002/eet.1951

Cash, D. W., Clark, W. C., Alcock, F., Dickson, N. M., Eckley, N., Guston, D. H., Jäger, J., & Mitchell, R. B. (2003). Knowledge systems for sustainable development. *Proceedings of the National Academy of Sciences, 100*(14), 8086–8091. https://doi.org/10.1073/pnas.1231332100

Cashore, B., Auld, G., & Newsom, D. (2004). *Governing through markets: Forest certification and the emergence of non-state authority*. Yale University Press, JSTOR. www.jstor.org/stable/j.ctt1npqtr

Curry, N., Baker, P., & Brookes, G. (2024). Generative AI for corpus approaches to discourse studies: A critical evaluation of ChatGPT. *Applied Corpus Linguistics, 4*(1), Article 100082. https://doi.org/10.1016/j.acorp.2023.100082

Dryzek, J. S. (2005). *The politics of the earth: Environmental discourses* (2nd ed.). Oxford University Press.

Dryzek, J. S., & Pickering, J. (2018). *The politics of the anthropocene* (1st ed.). Oxford: Oxford University Press. https://doi.org/10.1093/oso/9780198809616.001.0001

EC. (2019). *COM (2019) 640 final: The European green deal*. European Commission. Available: https://eur-lex.europa.eu/legal-content/EN/TXT/?uri=uriserv:OJ.C_.2020.104.01.0003.01.ENG. Accessed 8 April 2025.

EC. (2021). *COM (2021) 572 final: New EU forest strategy for 2030*. European Commission.

Elliott, C., & Schlaepfer, R. (2001). Understanding forest certification using the advocacy coalition framework. *Forest Policy and Economics, 2*(3–4), 257–266. https://doi.org/10.1016/S1389-9341(01)00043-0

Entman, R. M. (2008). Media framing biases and political power: Explaining slant in news of Campaign 2008. *Journalism, 11*(4), 389–408. https://doi.org/10.1177/1464884910367587

EP & CoE. (2024). Regulation (EU) 2024/1991: Nature restoration and amending Regulation (EU) 2022/869. *Official Journal of the European Union*, https://eur-lex.europa.eu/eli/reg/2024/1991/oj

Fairhead, J., Leach, M., & Scoones, I. (2012). Green grabbing: A new appropriation of nature? *Journal of Peasant Studies, 39*(2), 237–261. https://doi.org/10.1080/03066150.2012.671770

FAO. (2020). Global forest resources assessment 2020: Main report. *Food and Agriculture Organization of the United Nations.*

Fassnacht, F. E., White, J. C., Wulder, M. A., & Næsset, E. (2024). Remote sensing in forestry: Current challenges, considerations and directions. *Forestry: An International Journal of Forest Research, 97*(1), 11–37. https://doi.org/10.1093/forestry/cpad024

Habermas, J. (1998). *Between facts and norms: Contributions to a discourse theory of law and democracy.* MIT Press.

Hajer, M. (1993). Discourse coalitions and the institutionalisation of practice: The case of acid rain in Great Britain. In F. Fischer & J. Forester (Eds.), *The argumentative turn in policy analysis and planning.* Duke University Press.

Hajer, M. (1997). *The politics of environmental discourse: Ecological modernisation and the policy process.* Oxford University Press.

Hoefer, R. (2021). Theory in social policy research: Rationality and its discontents. *Journal of Policy Practice and Research, 2*(4), 233–237. https://doi.org/10.1007/s42972-021-00045-y

Howlett, M., & Rayner, J. (2013). Patching vs packaging in policy formulation: Assessing policy portfolio design. *Politics and Governance, 1*(2), 170–182.

Hysing, E., & Olsson, J. (2008). Contextualising the advocacy coalition framework: Theorising change in Swedish forest policy. *Environmental Politics, 17*(5), 730–748. https://doi.org/10.1080/09644010802421471

IPBES. (2019). *Global assessment report on biodiversity and ecosystem services of the Intergovernmental Science—Policy Platform on Biodiversity and Ecosystem Services.* IPBES Secretariat.

IPCC. (2022). *Climate change and land: IPCC special report on climate change, desertification, land degradation, sustainable land management, food security, and greenhouse gas fluxes in terrestrial ecosystems* (1st ed.). Cambridge University Press. https://doi.org/10.1017/9781009157988

Jordan, A., & Lenschow, A. (2010). Environmental policy integration: A state of the art review. *Environmental Policy and Governance, 20*(3), 147–158. https://doi.org/10.1002/eet.539

Kleinschmit, D., Böcher, M., & Giessen, L. (2016). Forest policy analysis: Advancing the analytical approach. *Forest Policy and Economics, 68*, 1–6. https://doi.org/10.1016/j.forpol.2016.05.001

Knoke, D., & Yang, S. (2008). *Social network analysis.* SAGE Publications, Inc. https://doi.org/10.4135/9781412985864

Krott, M., & Giessen, L. (2014). Learning from practices—Implications of the "practice based approach" for forest and environmental policy research. *Forest Policy and Economics, 49*, 12–16. https://doi.org/10.1016/j.forpol.2013.04.013

Lawrence, A., Deuffic, P., Hujala, T., Nichiforel, L., Feliciano, D., Jodlowski, K., Lind, T., Marchal, D., Talkkari, A., Teder, M., Vilkriste, L., & Wilhelmsson, E. (2020). Extension, advice and knowledge systems for private forestry: Understanding diversity and change across Europe. *Land Use Policy, 94*, Article 104522. https://doi.org/10.1016/j.landusepol.2020.104522

Leach, M., Scoones, I., & Stirling, A. (2007). *Pathways to sustainability: An overview of the steps centre approach,* http://sro.sussex.ac.uk/26103/. Accessed 25 June 2020 (STEPS Approach Paper). STEPS Centre.

Lewandowsky, S., Ecker, U. K. H., & Cook, J. (2017). Beyond misinformation: Understanding and coping with the "post-truth" era. *Journal of Applied Research in Memory and Cognition, 6*(4), 353–369. https://doi.org/10.1016/j.jarmac.2017.07.008

Linser, S., Wolfslehner, B., Bridge, S. R. J., Gritten, D., Johnson, S., Payn, T., Prins, K., Raši, R., & Robertson, G. (2018). 25 Years of criteria and indicators for sustainable forest management: How intergovernmental C&I processes have made a difference. *Forests, 9*(9), 578.

Mather-Gratton, Z. J., Larsen, S., & Bentsen, N. S. (2021). Understanding the sustainability debate on forest biomass for energy in Europe: A discourse analysis. *PLoS ONE, 16*(2), Article e0246873. https://doi.org/10.1371/journal.pone.0246873

Matilainen, A., Koch, M., Zivojinovic, I., Lähdesmäki, M., Lidestav, G., Karppinen, H., Didolot, F., Jarsky, V., Põllumäe, P., Colson, V., Hricova, Z., Glavonjic, P., & Scriban, R. E. (2019). Perceptions of ownership among new forest owners—A qualitative study in European context *Forest Policy and Economics, 99*, 43–51. https://doi.org/10.1016/j.forpol.2018.06.002

McDermott, C., Cashore, B., & Kanowski, P. (2010). *Global environmental forest policies* (0 ed.). Routledge. https://doi.org/10.4324/9781849774925

Mol, A. P. J. (2008). *Environmental reform in the information age: The contours of informational governance* (1st ed.). Cambridge University Press. https://doi.org/10.1017/CBO9780511491030

Nitoslawski, S. A., Galle, N. J., Van Den Bosch, C. K., & Steenberg, J. W. N. (2019). Smarter ecosystems for smarter cities? A review of trends, technologies, and turning points for smart urban forestry. *Sustainable Cities and Society, 51*, Article 101770. https://doi.org/10.1016/j.scs.2019.101770

Nitoslawski, S. A., Wong-Stevens, K., Steenberg, J. W. N., Witherspoon, K., Nesbitt, L., & Konijnendijk Van Den Bosch, C. C. (2021). The digital forest: mapping a decade of knowledge on technological applications for forest ecosystems. *Earth's Future, 9*(8), e2021EF002123. https://doi.org/10.1029/2021EF002123

Nummelin, T., Hänninen, R., & Kniivilä, M. (2021). Exploring forest sector research subjects and trends from 2000 to 2019 using topic modeling. *Current Forestry Reports, 7*(4), 267–281. https://doi.org/10.1007/s40725-021-00152-9

Ostrom, E. (1990). *Governing the commons: The evolution of institutions for collective action.* Cambridge University Press.

Ostrom, E. (2010). Beyond markets and states: Polycentric governance of complex economic systems. *American Economic Review, 100*(3), 641–672. https://doi.org/10.1257/aer.100.3.641

Pülzl, H., Kleinschmit, D., & Arts, B. (2014). Bioeconomy—An emerging meta-discourse affecting forest discourses? *Scandinavian Journal of Forest Research, 29*(4), 386–393. https://doi.org/10.1080/02827581.2014.920044

Pülzl, H., Prokofieva, I., Berg, S., Rametsteiner, E., Aggestam, F., & Wolfslehner, B. (2011). Indicator development in sustainability impact assessment: Balancing theory and practice. *European Journal of Forest Research, 131*(1), 35–46. https://doi.org/10.1007/s10342-011-0547-8

Pülzl, H., Wydra, D., & Hogl, K. (2018). Piecemeal integration: explaining and understanding 60 years of European union forest policymaking. *Forests, 9*, 719.

Raum, S., & Potter, C. (2015). Forestry paradigms and policy change: The evolution of forestry policy in Britain in relation to the ecosystem approach. *Land Use Policy, 49*, 462–470. https://doi.org/10.1016/j.landusepol.2015.08.021

Rhodes, R. A. W. (1997). *Understanding governance: Policy networks, governance, reflexivity and accountability.*

Rose, D. C., Mukherjee, N., Simmons, B. I., Tew, E. R., Robertson, R. J., Vadrot, A. B. M., Doubleday, R., & Sutherland, W. J. (2017). Policy windows for the environment: Tips for improving the uptake of scientific knowledge. *Environmental Science & Policy.* https://doi.org/10.1016/j.envsci.2017.07.013

Ryazanov, I., Öhman, C., & Björklund, J. (2024). How ChatGPT changed the media's narratives on AI: A semi-automated narrative analysis through frame semantics. *Minds and Machines, 35*(1). https://doi.org/10.1007/s11023-024-09705-w

Sabatier, P. A. (1988). An advocacy coalition framework of policy change and the role of policy-oriented learning therein. *Policy Sciences, 21*(2), 129–168. https://doi.org/10.1007/BF00136406

Sabatier, P. A. (Ed.). (2007). *Theories of the policy process* (2nd ed.). Westview Press.

Schmidt, V. A. (2008). Discursive institutionalism: The explanatory power of ideas and discourse. *Annual Review of Political Science, 11*(1), 303–326. https://doi.org/10.1146/annurev.polisci.11.060606.135342

Sikor, T., & Lund, C. (2009). Access and property: A question of power and authority. *Development and Change, 40*(1), 1–22. https://doi.org/10.1111/j.1467-7660.2009.01503.x

Thacher, D. (2005). Deliberative policy analysis: Understanding governance in the network society. *Journal of Planning Education and Research, 24*(4), 455–457. https://doi.org/10.1177/0739456X05276134

Visseren-Hamakers, I. J., & Glasbergen, P. (2007). Partnerships in forest governance. *Global Environmental Change, 17*(3–4), 408–419. https://doi.org/10.1016/j.gloenvcha.2006.11.003

Voß, J.-P., & Bornemann, B. (2011). The politics of reflexive governance: Challenges for designing adaptive management and transition management. *Ecology and Society, 16*(2), 9.

Weible, C. M. (2006). An advocacy coalition framework approach to stakeholder analysis: Understanding the political context of california marine protected area policy. *Journal of Public Administration Research and Theory, 17*(1), 95–117. https://doi.org/10.1093/jopart/muj015

Weiss, G., Lawrence, A., Hujala, T., Lidestav, G., Nichiforel, L., Nybakk, E., Quiroga, S., Sarvašová, Z., Suarez, C., & Živojinović, I. (2019). Forest ownership changes in Europe: State of knowledge and conceptual foundations. *Forest Policy and Economics, 99*, 9–20. https://doi.org/10.1016/j.forpol.2018.03.003

Weiss, G., Lawrence, A., & Nichiforel, L. (2017). How does forest ownership in Europe affect the forest-based bioeconomy? In G. Winkel (Ed.), *Towards a sustainable European forest based bioeconomy—Assessment and the way forward*. European Forest Institute.

Wijewardana, D. (2008). Criteria and indicators for sustainable forest management: The road travelled and the way ahead. *Ecological Indicators, 8*(2), 115–122. https://doi.org/10.1016/j.ecolind.2006.11.003

Wolfslehner, B., Pülzl, H., Kleinschmit, D., Aggestam, F., Winkel, G., Candel, J., Eckerberg, K., Feindt, P., McDermott, C., Secco, L., Sotirov, M., Lackner, M., & Lazya Roux, J. (2020). *European forest governance post-2020* (From Science to Policy 10). European Forest Institute. https://doi.org/10.36333/fs10.

Yeung, K. (2018). Algorithmic regulation: A critical interrogation. *Regulation & Governance, 12*(4), 505–523. https://doi.org/10.1111/rego.12158

Zahariadis, N. (2018). The multiple streams framework: Structure, limitations, prospects. In M. C. Weible & P. A. Sabatier (Eds.), *Theories of the policy process* (4th ed.). Routledge.

Chapter 11
Forests at the Crossroads: Navigating Pathways to Transformative Change

Filip Aggestam

Abstract This concluding chapter summarises the conceptual, empirical and cross-cutting insights presented throughout the book. The aim is to evaluate the conditions under which transformative change in forest governance may occur. Drawing on studies of European Union (EU) forest governance, forest ownership structures, discursive dynamics, digitalisation, communication, and monitoring frameworks, the chapter summarises recurring patterns that could influence governance outcomes. These patterns include persistent fragmentation, uneven implementation capacities, slow realignment of values, and the growing importance of knowledge, legitimacy, and coordination. The chapter then contextualises these findings within the broader debates on transformative change by introducing an integrative analytical framework, the "transformative feedback triangle", which analyses the interactions between governance capacities, alignment mechanisms, and institutional reconfiguration over time. Rather than viewing transformation as a discrete event, the ongoing, context-dependent nature of transformation is emphasised. The chapter concludes by outlining opportunities for future research.

11.1 Introduction

Forests are spaces where environmental processes, economic interests, and societal demands intersect and interact in increasingly complex ways. As demonstrated in the preceding chapters, persistent institutional structures, established policy pathways, evolving narratives, and intensifying pressures such as climate change, biodiversity loss, and land use change all influence forest policy and governance. These factors have broadened the scope of "forest-relevant" policies and revealed significant governance challenges, including sectoral and multi-level fragmentation, uneven implementation capacities, and conflicting views on the functions and societal roles of

F. Aggestam (✉)
ORBIS Green Policy and Governance, Forest Policy Research Network—European Forest Institute (FPRN-EFI), Vienna, Austria
e-mail: aggestam@orbis-gpg.com

F. Aggestam (ed.), *Governing Forests Through Transformative Change*, Advances in Global Change Research 70, https://doi.org/10.1007/978-3-032-19072-7_11

forests. At the same time however, forest policy research has also facilitated innovation, methodological development, and collective learning. This concluding chapter synthesises these insights to assess their implications for transformative change in forest governance. Furthermore, it explores how the book's contributions can inform a more integrated understanding of transformative change.

Ideas around transformative change have become increasingly important in global narratives and debates. The appear in the work of international bodies such as the Intergovernmental Science-Policy Platform on Biodiversity and Ecosystem Services (IPBES) and the Intergovernmental Panel on Climate Change (IPCC). The also appear in literature on sustainability transitions (Calvin et al., 2023; Korhonen-Kurki et al., 2025; Loorbach, 2010; O'Brien et al., 2023; Turnheim et al., 2015). However, their uptake and implementation remains uneven in forest policy and governance. Further more, transformative change is increasingly being used as a broad term, with limited conceptual or operational clarity. There is a risk that the concept will become a rhetorical placeholder (or buzzword) rather than an actual governance agenda (Apetrei et al., 2021; Lidskog & Sundqvist, 2022; Vogel & O'Brien, 2022). This is arguably, in part, because transformation covers far more than technological or incremental policy developments, it signifies substantial shifts across governance, including institutional structures, stakeholder configurations, knowledge systems, power relations, and underlying values (Patterson et al., 2017; Scoones & Stirling, 2020). In other words, genuine transformation requires radical changes to existing governance systems.

This book highlights different aspects of the challenge in each chapter. Chapters 1 and 2 establish a conceptual foundation by defining key terms, such as "forests", "forest policy", and "governance", also outlining the evolution of the theoretical approaches used in forest policy research. They chapters clarify how forest governance is shaped by varied environmental, socio-economic, and institutional dynamics, and how transformative change can be analysed through the interaction of structure, agency, discourse, and learning. Chapters 3 to 5 focus on forest governance in the European Union (EU). They chapters consider policy fragmentation, regulatory developments, implementation challenges, and long-term narrative and institutional changes. The work reveals that, despite consistent pressure for change have been increasing consistently, governance responses have remained partial and uneven. Chapters 6 to 9 examine cross-cutting issues, focusing on forest ownership and tenure, communication and legitimacy, digitalisation and data infrastructures, as well as the evolving role of criteria and indicators (C&I) in forest governance. They demonstrate how institutional design, communicative capacity, technological structures and monitoring tools influence systemic change. Finally, chapter 10 identifies emerging trends and future research needs relating to the study of transformative change.

This concluding chapter aims to summarise some of the many insights presented throughout the book and contextualise them within the broader literature on transformative change. There are four key objectives: (1) summarise the main findings to bring the chapters together, (2) identify cross-cutting themes and patterns that could influence the prospects for transformative change, (3) present an integrative model for understanding transformative forest governance that, builds on the frameworks introduced earlier in the book, and (4) identify key gaps and opportunities for future research and policy action.

11.2 What Have We Learned?

11.2.1 Conceptual Framing

As mentioned earlier, the introductory chapters provide the book's conceptual and theoretical background. Chapter 1 begins by defining key terms and establishing that forests are multidimensional socio-ecological systems characterised by interactions across levels, sectors, and knowledge domains, rather than being defined solely as ecosystems or policy domains. It highlights the absence of a universally accepted definition of "forests" in Europe and internationally, noting that legal, institutional, environmental, socio-economic, and cultural perspectives yield different definitions of what a forest is and how it should be used. These differences matter because they inform governance decisions, influence policy design, and determine which knowledge is considered relevant in policy debates.

Furthermore, Chapter 1 distinguishes between "forest-focused" and "forest-related" policies. This distinction underpins the subsequent analysis of the EU's governance arrangements around forests. The chapter also shows that EU forest policy remains fragmented, mainly because different sectors, such as agriculture, climate, environment, and energy, influence forest use. This fragmentation is partly rooted in historical path dependencies and the EU principle of subsidiarity. However, it is also related to the evolving concept of multifunctional use of forests (Sutherland & Huttunen, 2018). Against this backdrop, the chapter introduces the concept of transformative change as a systemic process.

Chapter 2 develops the conceptual basis of transformative change by reviewing how different theoretical approaches have been developed and used in forest policy research. Initially, forest policy research was characterised by technocratic rationality, emphasising state-centred regulation and sustainable timber production (Carlowitz, 2013). However, over time, this research area has evolved into an interdisciplinary field, incorporating institutionalism, discourse theory, political ecology, sustainability transitions, adaptive governance, and reflexive governance, among others (Art & Visseren-Hamakers, 2012; Folke, 2006; Voß et al., 2006), among others. This chapter aadvocates for theoretical pluralism, proposing that transformative change is best understood through frameworks capable of integrating structural constraints, actor strategies, knowledge dynamics, and discursive shifts.

The chapter's key contribution is the proposed adaptive and integrative framework for researching transformative forest governance. This framework is organised around four drivers of systemic change: structure, agency, discourse, and learning.

The chapter also emphasises the importance of reflexivity and methodological diversity. Together, the introductory chapters establish the conceptual basis for the book and, most notably by clarifying how the concept of transformation will be used.

11.2.2 Governance in Practice: Fragmentation, Implementation, and Discursive Shifts

Chapters 3 to 5 zoom in on the EU as a key region for forest governance research. The EU is an interesting case study, characterised by a unique multi-level governance structure, which involves involving subsidiarity and cross-sectoral policy interactions. While the pressures and demands for sustainable forest use have increased, these chapters demonstrate that relevant governance responses affecting forests vary significantly in terms of depth, coherence, and potential for transformative change.

Chapter 3 introduces the innovative Pressure-Response-Transformation (PRT) framework to EU forest governance. This chapter aims to explain how forest governance operates at the EU level. For instance, it explains how pressures such as climate change, biodiversity loss and the bioeconomics have prompted policy responses in the form of strategic instruments (e.g., the EU Forest Strategy), regulations (e.g., Land Use, Land-Use Change and Forestry (LULUCF) and the Regulation on Deforestation-free Products (EUDR)) and coordination mechanisms. The analysis concludes that EU forest governance is characterised by fragmented adaptation and policy layering. This means that the number of policy instruments affecting forests increases incrementally, rather than undergoing structural reconfiguration (Aggestam, 2025; Gordeeva et al., 2025; Pülzl et al., 2018). With regard to transformative change, the chapter highlights that recent policy developments, such as legally binding restoration targets and enhanced monitoring systems, suggest an underlying potential for transformation. However, the overall governance pattern remains unchanged.

Chapter 4 shifts the focus to implementation, which is arguably one of the most critical links between policy design and governance outcomes. Drawing on the implementation literature (Falkner et al., 2007; Hajer, 1993; Hupe & Hill, 2016), the chapter illustrates how forest-related policy implementation can vary significantly between EU Member States (MS) due to differences in capacity, political will, administrative cultures, and local contexts. While top-down compliance monitoring only captures of the picture, bottom-up and hybrid approaches reveal how local actors reinterpret, resist, or adapt to EU rules. The chapter emphasises that effective forest governance requires implementation capacities that are context-sensitive and involve multiple stakeholders, aas well as being able to respond to evolving challenges and demands.

Chapter 5 uses a novel AI-based frame analysis to explore discursive dynamics in new methodological territory. Through a longitudinal analysis spanning 60 years, the chapter identifies seven dominant frames, demonstrating the evolution of framing patterns during punctuated phases. The analysis introduces the concept of "transformative hotspots", which are periods during which discursive changes align with institutional change, enabling transformation. However, it also reveals that justice-, crisis-, and systems-oriented frames are primarily present in soft policy instruments, whereas legally binding legislation predominantly uses compliance-oriented frames. This "soft-hard" policy divide suggests that discursive innovation alone is insufficient for transformative change, unless it is clearly institutionalised through regulatory mechanisms.

Together, Chapters 3 to 5 provide a cross-cutting analysis of EU forest policy, complementing the existing literature on forest governance and introducing innovative analytical approaches. Although pressures are intensifying and capacities and monitoring infrastructures are improving, the chapters demonstrate that EU forest governance remains significantly fragmented. Institutional inaction persists, and value-based changes have yet to be incorporated into legally binding policy instruments and frameworks.

11.2.3 Cross-Cutting Dimensions of Forest Governance and Transformative Change

Chapters 6 to 9 shift the focus from the EU's overarching forest policy architecture to the underlying conditions that shape forest governance. They consider forest ownership and tenure arrangements, communication and legitimacy dynamics, the growing importance of digital technologies, and the use of indicators and monitoring frameworks as cross-cutting factors affecting forests management and use.

Chapter 6 focuses on ownership and land tenure, emphasising that the structure of forest is ownership a fundamental aspect of forest governance. The chapter reveals that the traditional distinction between public and private ownership conceals a complex mix of communal, municipal, corporate, and smallholder forms of ownership (ECE, 2020). It notes that the types of forest owners has increased due to social change, land restitution, demographic shifts, and evolving motivations for ownership. Nevertheless, ownership-related policies continue to assume a "normal" forest owner who is rational and commercially oriented, resulting in a misalignment between policy design and management objectives (Weiss et al., 2019). Nevertheless, the chapter highlights that ownership diversity is not an obstacle but rather a potential driver for change, including for increased experimentation and socially embedded forest management. In other words, governance systems that recognise and support the various types of forest owners could enhance the prospects of transformation.

Chapter 7 considers communication to be a structural component of forest governance (see, for example, Craig, 1999; Chadwick, 2013; McNair, 2017). It argues that communication is both constitutive of and instrumental in forest governance. Policies succeed (or fail) not only on a technical basis but also due to communicative infrastructures that set the conditions for participation, establishing legitimacy and trust in the process. The emergence of digital and hybrid media ecosystems, for example, is transforming the way forest-related issues are debated, contested, and legitimised. The chapter also identifies communication failures, such as polarised discourse, misinformation, inadequate engagement and technocratic communication, as obstacles to transformative change. These findings emphasise the potential of deliberative processes and co-production to enhance legitimacy and adaptive capacity.

Chapter 8 explores the relationship between digitalisation and the future of forest governance. More specifically, it considers digitalisation in the context of the Fourth Industrial Revolution, analysing how technologies such as remote sensing, the Internet of Things (IoT) devices, big data analytics, artificial intelligence (AI) and blockchain could transform governance capabilities (Gabrys et al., 2022; Mol, 2008). While digitalisation may enhance monitoring and transparency, concerns are raised about surveillance, inequity, data ownership and technocratic dominance. The chapter also presents the "Forest Governance 4.0" framework, which considers the technological, institutional, and political aspects required to support reflexive, participatory, and equitable governance systems through digital infrastructures. In this context, digitalisation is presented as a means of enabling transformative change, provided that normative, social, and political considerations guide the design and implementation of the technologies.

Chapter 9 analyses the evolving role of criteria and indicators (C&I). Initially used as reporting tools, C&I frameworks have evolved into instruments that enable reflexive governance, by helping to define policy priorities, track progress, and align objectives across levels (Baycheva-Merger & Wolfslehner, 2016; Lier et al., 2021; Linser et al., 2018). The chapter considers innovations such as the Global Core Set of Forest Indicators, the EU Taxonomy, and emerging biodiversity and restoration indicators. It emphasises that indicators influence forest governance by shaping what is considered important, what is monitored, and what constitutes success. Consequently, when aligned with justice-oriented, adaptive, and cross-sectoral approaches, C&I systems can facilitate transformation, however, if they are designed too narrowly, they can also reinforce compliance-oriented governance.

Beyond formal policy and institutional design, Chapters 6 to 9 demonstrate how the everyday governance infrastructures can impact the feasibility of transformation. Significant transformative potential is also situated in forest ownership and tenure arrangements, communicative structures and legitimacy (including media dynamics), technological systems (including digitalisation) and epistemic tools, such as indicators and monitoring frameworks.

11.3 Underlying Patterns and Governance Implications

Integrating insights from across the book makes it possible to identify the broader patterns that shape forest governance and the prospects for transformation. The different contributions reveal four cross-cutting themes:

1. **Persistent fragmentation and incremental change, alongside the emergence of integrative mechanisms.**

In line with a substantial body of research on EU forest governance, the policy fragmentation outlined in this book remains a defining feature of the overarching governance system (Pülzl et al., 2018; Kleinschmit et al., 2018; Wolfslehner et al., 2020; Aggestam, 2025; Gordeeva et al., 2025). Chapters 3 to 5 demonstrate that this reflects the diverse institutional arrangements and the multifunctional nature of forest use in Europe. Multiple policy domains, including agriculture, climate, biodiversity, energy and trade, claim some degree of authority over forests. This results in overlapping mandates, competing priorities and conflicting policy objectives. These dynamics are further reinforced by multi-level governance arrangements, particularly within the EU, where the principle of subsidiarity and varied national forestry traditions prevent (or limit) formal integration.

However, the book also highlights the gradual emergence of integrative mechanisms that aim to manage, rather than eliminate, fragmentation. Examples include increasingly sophisticated monitoring and reporting systems, cross-sectoral policy frameworks, multi-actor implementation networks and more harmonised criteria and indicators. These mechanisms contribute to improved coordination and alignment across governance levels and sectors. While fragmentation itself does not prevent transformative change, the analysis suggests that transformation is more likely to occur when governance structures mediate between different policy objectives and competing perspectives, enabling coordination across institutional and sectoral boundaries.

2. **Governance capacities are expanding faster than shifts in value orientations.**

The synthesis reveals a widening gap between the expansion of governance capacities and the limited progress made towards substantive, value-based, change. Several chapters demonstrate, the strengthening of forest governance's material and institutional capacities is demonstrated, including the development of more advanced data infrastructures and monitoring systems, more sophisticated enforcement and compliance mechanisms, and improved coordination instruments and strategic frameworks at the EU level. Taken together, these developments suggest an enhanced capacity to measure, track, and manage forest-related objectives across governance levels and spatial scales.

These capacity gains have, however, not been matched by equivalent changes in underlying values, priorities and governance logics. The value-based dimensions of forest governance, particularly those relating to justice, equity, participation, and the societal valuation of forest functions, remain weakly institutionalised. For example, Chapter 5 shows that narratives focusing on justice and crisis are more common in soft-law instruments and strategic documents, whereas legally binding policy instruments continue to prioritise technocratic and compliance-oriented approaches. Similarly, Chapter 6 suggests that prevailing forest ownership and tenure arrangements are still based on assumptions that only partially reflect current social expectations and power relations.

Overall, the findings suggest that, although expanding governance capacities is necessary for transformative change, it is not sufficient in itself. Transformation requires a more fundamental realignment of the values, power relations, and institutional logics that shape how forests are understood, managed, and used, as well as, improved tools, data, and coordination mechanisms.

3. **Forest governance is characterised by polycentricity and unevenness.**

A third insight is that forest governance is becoming increasingly polycentric and uneven. More specifically, forest governance varies significantly depending on the scale, institutional context, and the actors involved. For example, the implementation of forest-related policies can differ greatly between national administrative systems, reflecting variations in institutional capacity, legal traditions, and political priorities. Significant differences also exist in forest ownership and tenure arrangements across Europe, influencing governance opportunities and constraints in context-specific ways. Meanwhile, discursive analysis reveals that changes in dominant policy frames do not develop gradually, but rather occur in episodic bursts, often in response to external shocks or shifts in political focus.

These findings reinforce the view that transformation in forest governance is non linear, multi level, and context dependent. Structurally, governance arrangements are also becoming increasingly polycentric, with authority (or power) being spread across multiple levels and groups of actors. This setup may enables experimentation, learning and varied responses to forest-related challenges. However, this structural polycentricity operates within governance systems that remain strongly path-dependent and are shaped by institutional arrangements and technocratic policy-making processes.

Consequently, while polycentric configurations may facilitate innovation in certain conditions or contexts, they also present challenges in terms of coherence and coordination. For instance, dominant epistemic communities, regulatory standards, and administrative pathways continue to limit value-based diversity and distribute power unevenly across governance levels and structures. Therefore, it is argued that polycentricity both enables and limits transformative change.

4. **The central role of knowledge, communication and legitimacy in shaping transformative change.**

It is evident that knowledge, communication, and legitimacy play a fundamental role in determining the prospects for transformative change. Throughout the book, it is demonstrated that governance outcomes depend not only on institutional design and policy instruments, but also on how knowledge is produced, communicated, and contested. For example, Chapter 7 reveals that policy interventions lacking communicative legitimacy find it hard to gain social acceptance and political traction, particularly in situations involving value pluralism and conflicting forest use. Chapter 8 illustrates how emerging data infrastructures and digital governance tools can expand monitoring and coordination capacities, as well as reconfiguring power relations by favouring certain forms of knowledge and expertise. Chapter 9 demonstrates that indicators act as epistemic tools that determine priorities and define what is measured and valued, in turn, these factors then shape governance pathways over time.

These findings emphasise that transformation is as much an epistemic and communicative process as it is an institutional and technical one. This is consistent with the research on sustainability transformations, which highlights the significance of social learning, epistemic diversity, and deliberative engagement in enabling systemic change (Korhonen-Kurki et al., 2025; O'Brien et al., 2025; Scoones & Stirling, 2020). From this perspective, it is evident that transformative forest governance requires knowledge systems that acknowledge plurality, communication processes that facilitate inclusive dialogue, and governance structures that can promote legitimacy through deliberation, reflexivity and co-production.

11.4 Synthesising Pathways to Transformative Forest Governance

Building on the cross-cutting themes identified above and the conceptual foundation introduced earlier, the following sub sections provide a structured approach to analysing how transformative change can be enabled, managed, and embedded within forest governance processes.

11.4.1 Three Mechanisms for Transformative Governance

Earlier chapters conceptualised transformative change in terms of three dimensions: governance *of*, governance *for*, and governance *in* transformation. These dimensions are reframed here as three interactive mechanisms, rather than sequential ones. This aligns with Patterson et al.'s (2017) understanding that transformation is shaped by an interplay between enabling capacities, steering processes, and structural and value-based changes. This is also consistent with the IPBES perspective of transformative change as a systemic process involving concurrent changes in institutions, power relations, knowledge systems, and values (O'Brien et al., 2025).

1 Governance *for* Transformation: Enabling conditions
The first mechanism involves creating the **conditions** that enable for transformation. Based on IPBES (2025) and Meadows' (1999) "leverage points",[1] these conditions cover material capacities (e.g., finance, skills and technologies), epistemic capacities (e.g., knowledge systems, indicators, and communication), institutional capacities (e.g., coordination mechanisms and legal frameworks) and social legitimacy based on participation. While these conditions are becoming more prevalent in forest governance, particularly due to improvements in monitoring, data infrastructures, and coordination, Chapters 3, 7, 8 and 9 demonstrate that they are still unevenly developed and distributed. This restricts the potential for transformative change.

(continued)

[1] Meadows distinguishes between different places to intervene in complex systems, each with its own potential to produce change ("leverage points"). These are ordered from relatively low-leverage

(continued)

2 Governance *of* transformation: Steering and alignment
The second mechanism relates to **steering and alignment**. Steering refers to the ability of forest governance systems to promote change through policies, regulations and coordination across sectors and levels. Research on sustainability transitions shows that effective steering requires interventions to be aligned across policy domains, experimentation to be encouraged and the feedback loops to be established that link monitoring and learning to policy adjustments (Geels, 2002, 2021; Loorbach, 2010). Chapters 3 and 4 demonstrate the incremental progress made in this regard, as well as the ongoing challenges relating to fragmentation, conflicting policy objectives, and limited cross-sectoral integration.
3 Governance *in* Transformation: Reconfiguration and embedding
The third mechanism relates to **reconfiguration and embedding**. It covers the substantial institutional and value-based changes required for long-term transformation. This includes changes in governance logic, realignment of values, redistribution of power, and the establishment of new rights, responsibilities and accountability measures. As demonstrated in Chapters 5 and 6, this mechanism is currently the least developed aspect of forest governance. Although discursive shifts and strategic commitments have occurred, these are rarely accompanied by substantive changes to forest ownership structures, power relations, or institutional design.

The three mechanisms interact dynamically and represent different aspects of the same process. When the enabling conditions are right for effective management, transformative change may arise, creating opportunities for institutional and value-based change. However, without progress in reconfiguration and embedding, improvements in capacity and coordination may simply reinforce technocratic or incremental approaches rather than facilitating systemic change. This perspective highlights the multidimensional and co-evolving nature of the politics and governance of transformative change (O'Brien et al., 2025; Patterson et al., 2017). It emphasises that understanding transformative forest governance requires attention to all three mechanisms in different contexts, at different scales, and over different time periods (Box 1).

Box 1 Forests as a microcosm of global sustainability transformations.

Forests provide an interesting setting in which to study sustainability transformations, mainly, as they bring together ecological limits, economic strategies, and contested social values within the same governance system. Ecologically, forests play a vital role in climate regulation and biodiversity support, connecting governance to planetary boundary dynamics (IPBES, 2019; IPCC, 2022). Economically, forests lie at the heart of the bioeconomy and energy transition agendas, connecting land use, industrial policy, and decarbonisation pathways. Socially and culturally, forest use is shaped by varied tenure

---

interventions to higher-leverage interventions that change how the system functions, including underlying paradigms and perspectives.

arrangements, forest ownership structures, and livelihoods, making forests a space where rights, identities, and values converge.

When viewed through the above-outlined mechanisms, forest governance demonstrates how transformation depends on the interplay between enabling conditions, steering processes, and reconfiguration dynamics. In the EU, enabling capacities have expanded through monitoring and reporting, as well as increasing data infrastructures. Meanwhile, steering efforts are increasingly relying on policy mixes and cross-sectoral strategies to manage competing policy objectives. However, reconfiguration and embedding (e.g., shifts in governance logics, value priorities and power relations) remain uneven. This helps to explain why change often takes the form of layering and incremental adjustment rather than systemic reorientation. In this context, EU forest governance operates as both a policy domain and as a testing ground for the co-evolution of dynamics between ecological, institutional, technological and cultural systems (Olsson et al., 2004, 2014; Scoones & Stirling, 2020).

11.4.2 The Transformative Governance Triangle: An Integrative Conceptual Model

This model— triangle of transformative governance—is proposed as a way to synthesise patterns observed across governance systems, policy domains, and analytical perspectives, and to bring together the analytical insights from the previous chapters. Rather than introducing new categories, the conceptual model draws directly on the three mechanisms outlined in Sect. 4.1 above, demonstrating how different factors can interact to shape the scope and direction of transformative change in forest governance. Transformation is conceptualised as emerging from the feedback effects (both reinforcing and limiting) between the models three domains (Fig. 11.1).

- The **capacities and infrastructure** domain has emerged as one of the areas of forest governance that has developed most visibly. Syntheses demonstrate that investments in monitoring systems, data infrastructure, administrative capacity, and communication tools have increased significantly. This has strengthened the ability of governance systems' to monitor, coordinate, and manage forest-related policy objectives. However, the analysis also shows that these capacities often operate in isolation from more significant institutional and value-based changes. When these capacities are unevenly distributed or primarily oriented towards compliance and reporting, they tend to stabilise existing governance systems rather than facilitating transformation. Therefore, while capacities function as enabling conditions, they do not determine outcomes in the absence of interaction with other domains.

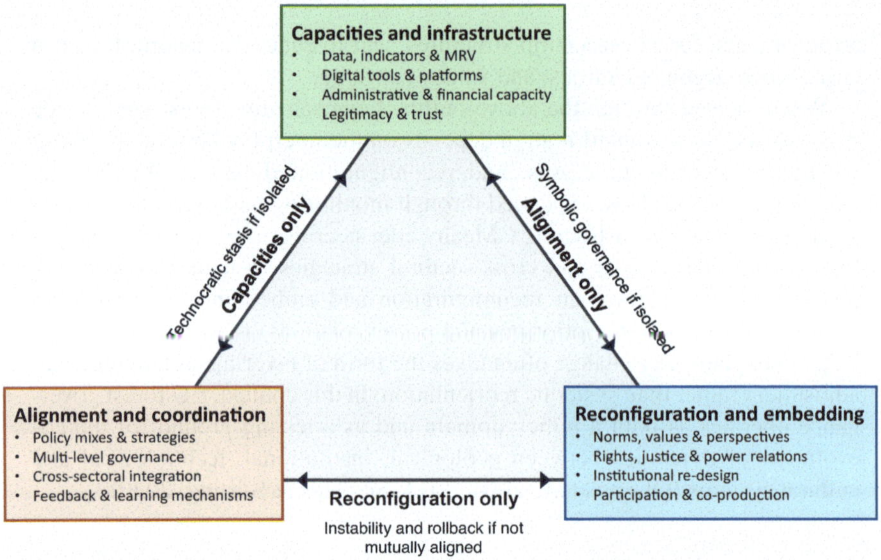

Fig. 11.1 An integrative model of governance dynamics shaping transformative change. *Source* own figure

- The **alignment and coordination** domain demonstrates how available capacities can be transformed into shared visions or perspectives. Throughout the book, the important mediating role of coordination mechanisms has been demonstrated, particularly in fragmented governance contexts involving interactions between multiple sectors, levels of governance, and stakeholder groups. Alignment processes determine whether expanded capacities support coherent transformative pathways or reinforce parallel (and sometimes contradictory) policy pathways. The synthesis suggests that although coordination has improved, it is still inconsistent and often insufficient to overcome established sectoral boundaries and competing mandates.
- The **reconfiguration and embedding** domain is the least developed area of the governance triangle. Significant institutional and value-based changes, such as shifts in governance logics, power redistribution, and redefinitions of rights and responsibilities, remain limited. Although commitments to sustainability, justice, and restoration have become more prevalent in discourse (including in soft policy instruments and strategies), these commitments are only partially reflected in institutional design and policymaking processes. Consequently, governance change remains vulnerable to reversal, particularly when value-based changes are not anchored in formal institutions and policy instruments, or when they are not aligned with capacity-building and coordination efforts.

The analytical contribution of the governance triangle is to make these interaction effects explicit. For instance, increasing capacities without institutional and value-based reconfiguration could lead to technocratic stasis, whereby improved tools and data merely reinforce existing governance pathways. Similarly, insufficient capacity in coordination efforts may result in symbolic governance, where strategic objectives lack operational power. Likewise, reconfiguration efforts lacking alignment mechanisms may generate instability, as discursive or value-based changes remain weakly embedded. Together, these dynamics emphasise that transformative change requires positive feedback loops across all three domains.

11.5 Future Directions for Research

This section identifies priority areas for future research, based on the discussion of emerging trends and the future of forest policy research in Chapter 10, and informed by analyses throughout the book. Rather than presenting these as isolated areas of work, Table 11.1 organises the topics as interconnected areas where conceptual development, methodological innovation, and governance practice converge.

Taken together, the contributions in this book are reflected in Table 11.1. These contributions highlight areas where future research appears important for advancing transformative forest governance. However, these contributions should not be considered as an exhaustive agenda. Many additional topics, shaped by evolving environmental conditions, societal priorities, and governance contexts, will continue to emerge as equally significant.

11.6 Concluding Remarks

Transformation should be understood as an ongoing and uneven process rather than a discrete event. It emerges from the interplay between governance capacities, coordination mechanisms, institutional and value-based structures (Fig. 11.1). Forests are an interesting case study for examining transformation because they sit at the intersection of multiple sectors and policy objectives, highlighting the tension between incremental adaptation and more fundamental change. The analysis also emphasises the important role of reflexivity in transformative governance. Reflexivity is defined as the capacity to question dominant perspectives and values, review institutional arrangements, engage with varied knowledge systems and address underlying power relations. However, transformative processes remain inherently uncertain and context-dependent, and no single governance pathway can be assumed to deliver transformation in across all situations.

Across the chapters, it has become clear that transformative change depends not only on the substantive content of governance interventions (such as policies, technologies, and indicators), but also on the processes through which forest governance

Table 11.1 Frontiers for research on transformative forest governance

Topic	Focus	Research direction	Policy relevance
Longitudinal and comparative governance analysis	Temporal and cross-context dynamics of forest governance	Long-term analysis of discursive change, comparative studies of ownership transformation, implementation trajectories across MSs, and interactions within cross-sectoral policy mixes	Supports evidence-based policy learning, anticipatory governance, and the evaluation of reform pathways
Mixed-methods and methodological innovation	Methodological development for complex governance systems	Integration of computational text analysis, qualitative interpretation, geospatial data, and participatory methods	Improves the robustness, scalability, and contextual sensitivity of policy analysis
Participatory monitoring and co-production	Democratisation of knowledge systems	Citizen science in forest monitoring, participatory restoration tracking, and co-developed indicators incorporating justice and equity	Enhances the legitimacy, accountability, and responsiveness of governance systems
Digital justice and governance	Political and ethical dimensions of digitalisation	Data ownership and sovereignty, algorithmic bias, equity implications of digital tools, and environmental footprint of digital infrastructures	Aligns digital governance with principles of fairness, transparency, and environmental responsibility
Ownership pluralism and polycentric restoration	Governance implications of diverse tenure regimes	Effects of ownership diversity on restoration outcomes, polycentric coordination mechanisms, and rights-based governance approaches	Supports socially robust restoration strategies and landscape-scale resilience
Transformation literacy and reflexive governance	Capacities for navigating systemic change	Development of systems thinking, communication literacy, conflict mediation, and reflexive decision-making	Strengthens institutional and societal capacity to engage with transformative governance processes

is implemented and on the stakeholders involved. Inclusive policy-making, communicative legitimacy, and epistemic diversity are central to building trust and sustaining effective governance over time. With this in mind, forest governance systems are under increasing pressure to transition from technocratic and compliance-oriented models towards adaptive, collaborative, and justice-based approaches.

The integrative framework developed in this chapter makes three key contributions to the literature on sustainability transformations. First, it links sector-specific analyses of forest governance to broader theories of systemic change by drawing on empirical insights from areas such as ownership, discourse, digitalisation, indicators, and implementation. Second, it conceptualises forests as co-evolving social-ecological governance systems, rather than as separate policy domains. This emphasises the importance of forests as sites for studying sustainability transformations. Third, it provides analytical tools to support research and practice, helping to assess whether governance systems primarily enable, steer, or limit transformative change, and to identify where targeted interventions may be needed to address bottlenecks.

Taken together, the findings suggest that achieving transformative forest governance requires sustained attention to learning, coordination and institutional change. Although the insights and analytical framework presented here primarily focus on forest governance, they can also be applied to other policy areas facing comparable challenges of fragmentation, complexity, and conflicting societal demands. While the pathways and outcomes of transformation will clearly remain context-specific, the analysis emphasises the importance of aligning governance processes more closely with the socio-economic and ecological foundations on which forest systems ultimately depend.

References

Aggestam, F. (2025). *European Forest Policy and Governance: An Integrative Analytical Framework* (1st ed.). Routledge. https://doi.org/10.4324/9781003198987

Apetrei, C. I., Caniglia, G., Von Wehrden, H., & Lang, D. J. (2021). Just another buzzword? A systematic literature review of knowledge-related concepts in sustainability science. *Global Environmental Change, 68*, Article 102222. https://doi.org/10.1016/j.gloenvcha.2021.102222

Art, B., & Visseren-Hamakers, I. (2012). Forest governance: A state of the art review. In B. Arts, S. Van Bommel, M. Ros-Tonen, & G. Verschoor (Eds.), *Forest-people interfaces* (pp. 241–257). Wageningen Academic Publishers. https://doi.org/10.3920/978-90-8686-749-3_15

Baycheva-Merger, T., & Wolfslehner, B. (2016). Evaluating the implementation of the Pan-European Criteria and indicators for sustainable forest management—A SWOT analysis. *Ecological Indicators, 60*, 1192–1199. https://doi.org/10.1016/j.ecolind.2015.09.009

Calvin, K., Dasgupta, D., Krinner, G., Mukherji, A., Thorne, P. W., Trisos, C., Romero, J., Aldunce, P., Barrett, K., Blanco, G., Cheung, W. W. L., Connors, S., Denton, F., Diongue-Niang, A., Dodman, D., Garschagen, M., Geden, O., Hayward, B., Jones, C., … Péan, C. (2023). *Climate Change 2023: Synthesis Report. Contribution of Working Groups I, II and III to the Sixth Assessment Report of the Intergovernmental Panel on Climate Change [Core Writing Team, H. Lee and J. Romero (eds.)]. IPCC, Geneva, Switzerland.* (First). Intergovernmental Panel on Climate Change (IPCC). https://doi.org/10.59327/IPCC/AR6-9789291691647

Carlowitz, H. C. (2013). Sylvicultura oeconomica. In L. Robin, S. Sörlin, & P. Warde, (Eds.), *The Future of Nature: Documents of Global Change* (pp. 67–77). Yale University Press. https://doi.org/10.12987/9780300188479-009

ECE. (2020). *Who owns our forests? Forest Ownership in the ECE Region*. United Nations Economic Commission for Europe.

Falkner, G., Hartlapp, M., & Treib, O. (2007). Worlds of Compliance: Why leading approaches to the implementation of EU legislation are only "sometimes-true theories." *European Journal of Political Research, 46*, 395–416.

Folke, C. (2006). Resilience: The emergence of a perspective for social–ecological systems analyses. *Global Environmental Change, 16*(3), 253–267. https://doi.org/10.1016/j.gloenvcha.2006.04.002

Gabrys, J., Westerlaken, M., Urzedo, D., Ritts, M., & Simlai, T. (2022). Reworking the political in digital forests. The cosmopolitics of socio-technical worlds. *Progress in Environmental Geography, 1*(1–4), 58–83. https://doi.org/10.1177/27539687221117836

Geels, F. W. (2002). Technological transitions as evolutionary reconfiguration processes: A multi-level perspective and a case-study. *Research Policy, 31*(8–9), 1257–1274. https://doi.org/10.1016/S0048-7333(02)00062-8

Geels, F. W. (2021). From leadership to followership: A suggestion for interdisciplinary theorising of mainstream actor reorientation in sustainability transitions. *Environmental Innovation and Societal Transitions, 41*, 45–48. https://doi.org/10.1016/j.eist.2021.10.021

Gordeeva, E., Wolfslehner, B., & Weber, N. (2025). 25 Years of EU Forest Policy—An analysis. *Forests, 16*(2), 256. https://doi.org/10.3390/f16020256

Hajer, M. (1993). Discourse Coalitions and the institutionalisation of Practice: The case of acid rain in Great Britain. In F. Fischer & J. Forester (Eds.), *The Argumentative Turn in Policy Analysis and Planning*. Duke University Press.

Hupe, P. L., & Hill, M. J. (2016). 'And the rest is implementation.' Comparing approaches to what happens in policy processes beyond Great Expectations. *Public Policy and Administration, 31*(2), 103–121. https://doi.org/10.1177/0952076715598828

IPBES. (2019). *Global assessment report on biodiversity and ecosystem services of the Intergovernmental Science—Policy Platform on Biodiversity and Ecosystem Services*. IPBES Secretariat.

IPCC. (2022). *Climate Change and Land: IPCC Special Report on Climate Change, Desertification, Land Degradation, Sustainable Land Management, Food Security, and Greenhouse Gas Fluxes in Terrestrial Ecosystems* (1st ed.). Cambridge University Press. https://doi.org/10.1017/9781009157988

Kleinschmit, D., Pülzl, H., Secco, L., Sergent, A., & Wallin, I. (2018). Orchestration in political processes: Involvement of experts, citizens, and participatory professionals in forest policy making. *Forest Policy Economy, 89*, 4–15. https://doi.org/10.1016/j.forpol.2017.12.011

Korhonen-Kurki, K., D'Amato, D., Belinskij, A., Lazarevic, D., Leskinen, P., Nylén, E.-J., Pappila, M., Penttilä, O., Pitzén, S., Pykäläinen, N., Turunen, T., & Vikström, S. (2025). Transformative governance: Exploring theory of change and the role of the law. *Earth System Governance, 23*, Article 100230. https://doi.org/10.1016/j.esg.2024.100230

Lidskog, R., & Sundqvist, G. (2022). Lost in transformation: The Paris Agreement, the IPCC and the quest for national transformative change. *Frontiers in Climate, 4*, Article 906054. https://doi.org/10.3389/fclim.2022.906054

Lier, M., Köhl, M., Korhonen, K. T., Linser, S., & Prins, K. (2021). Forest relevant targets in EU policy instruments—Can progress be measured by the pan-European criteria and indicators for sustainable forest management? *Forest Policy and Economics, 128*, Article 102481. https://doi.org/10.1016/j.forpol.2021.102481

Linser, S., Wolfslehner, B., Bridge, S. R. J., Gritten, D., Johnson, S., Payn, T., Prins, K., Raši, R., & Robertson, G. (2018). 25 Years of Criteria and indicators for sustainable forest management: How intergovernmental C&I processes have made a difference. *Forests, 9*(9), 578.

Loorbach, D. (2010). Transition management for sustainable development: A prescriptive complexity-based governance framework. *Governance, 23*(1), 161–183. https://doi.org/10.1111/j.1468-0491.2009.01471.x
Meadows, D. (1999). *Leverage Points: Places to Intervene in a System.* The Sustainability Institute. https://donellameadows.org/archives/leverage-points-places-to-intervene-in-a-system/
Mol, A. P. J. (2008). *Environmental Reform in the Information Age: The Contours of Informational Governance* (1st ed.). Cambridge University Press. https://doi.org/10.1017/CBO9780511491030
O'Brien, K., Garibaldi, L., Agrawal, A., Bennett, E., Biggs, R., Calderón Contreras, R., Carr, E. R., Frantzeskaki, N., Gosnell, H., Gurung, J., Lambertucci, S. A., Leventon, J., Chuan, L., Reyes García, V., Shannon, L., Villasante, S., Wickson, F., Zinngrebe, Y., Périanin, L., & Zaccagnini, M. E. (2025). *IPBES Transformative Change Assessment: Summary for Policymakers* (Version v.10.1.0). Zenodo. https://doi.org/10.5281/ZENODO.11382230
Olsson, P., Folke, C., & Berkes, F. (2004). Adaptive comanagement for building resilience in social-ecological systems. *Environmental Management, 34*(1), 75–90.
Olsson, P., Galaz, V., & Boonstra, W. J. (2014). Sustainability transformations: A resilience perspective. *Ecology and Society, 19*(4), art1. https://doi.org/10.5751/ES-06799-190401
Patterson, J., Schulz, K., Vervoort, J., Van Der Hel, S., Widerberg, O., Adler, C., Hurlbert, M., Anderton, K., Sethi, M., & Barau, A. (2017). Exploring the governance and politics of transformations towards sustainability. *Environmental Innovation and Societal Transitions, 24*, 1–16. https://doi.org/10.1016/j.eist.2016.09.001
Pülzl, H., Wydra, D., & Hogl, K. (2018). Piecemeal Integration: Explaining and Understanding 60 Years of European Union Forest Policy-Making. *Forests, 9*, 719.
Scoones, I., & Stirling, A. (2020). *The Politics of Uncertainty: Challenges of Transformation.* Routledge.
Sutherland, L.-A., & Huttunen, S. (2018). Linking practices of multifunctional forestry to policy objectives: Case studies in Finland and the UK. *Forest Policy Economy, 86*, 35–44. https://doi.org/10.1016/j.forpol.2017.10.019
Turnheim, B., Berkhout, F., Geels, F., Hof, A., McMeekin, A., Nykvist, B., & Van Vuuren, D. (2015). Evaluating sustainability transitions pathways: Bridging analytical approaches to address governance challenges. *Global Environmental Change, 35*, 239–253. https://doi.org/10.1016/j.gloenvcha.2015.08.010
Vogel, C., & O'Brien, K. (2022). Getting to the heart of transformation. *Sustainability Science, 17*(2), 653–659. https://doi.org/10.1007/s11625-021-01016-8
Voß, J.-P., Bauknecht, D., & Kemp, R. (2006). Reflexive governance for sustainable development. *Reflexive Governance for Sustainable Development.* https://doi.org/10.4337/9781847200266
Weiss, G., Lawrence, A., Hujala, T., Lidestav, G., Nichiforel, L., Nybakk, E., Quiroga, S., Sarvašová, Z., Suarez, C., & Živojinović, I. (2019). Forest ownership changes in Europe: State of knowledge and conceptual foundations. *Forest Policy and Economics, 99*, 9–20. https://doi.org/10.1016/j.forpol.2018.03.003
Wolfslehner, B., Pülzl, H., Kleinschmit, D., Aggestam, F., Winkel, G., Candel, J., Eckerberg, K., Feindt, P., McDermott, C., Secco, L., Sotirov, M., Lackner, M., & Lazya Roux, J. (2020). *European forest governance post-2020* (From Science to Policy 10). European Forest Institute. https://doi.org/10.36333/fs10

The manufacturer's authorised representative in the EU is Springer
Nature Customer Service Centre GmbH, Europaplatz 3, 69115 Heidelberg,
Germany. If you have any concerns regarding our products, please
contact ProductSafety@springernature.com

Printed and bound by CPI Group (UK) Ltd, Croydon, CR0 4YY
17/09/2026
02218456-0011